Changing the Atmosphere

Politics, Science, and the Environment
Peter M. Haas, Sheila Jasanoff, and Gene Rochlin, editors

Changing the Atmosphere

Expert Knowledge and Environmental Governance

edited by Clark A. Miller and Paul N. Edwards

The MIT Press
Cambridge, Massachusetts
London, England

This book was set in Sabon by Best-set Typesetter Ltd., Hong Kong, and was printed and bound in the United States of America.

Library of Congress Cataloging-in-Publication Data

Changing the atmosphere : expert knowledge and environmental governance / edited by Clark A. Miller and Paul N. Edwards.
    p. cm.—(Politics, science, and the environment)
  Includes bibliographical references and index.
  ISBN 0-262-13387-3 (alk. paper)—ISBN 0-262-63219-5 (pbk. : alk. paper)
  1. Environmental management. 2. Global environmental change.
3. Climatic changes—Environmental aspects. 4. Nature—Effect of human beings on. 5. Globalization. I. Miller, Clark A. II. Edwards, Paul N.
III. Series.

GE300 .C48 2001
551.6—dc21   00-069548

Contents

Series Foreword

As our understanding of environmental threats deepens and broadens, it is increasingly clear that many environmental issues cannot be simply understood, analyzed, or acted upon. The multifaceted relationships between human beings, social and political institutions, and the physical environment in which they are situated extend across disciplinary as well as geopolitical confines, and cannot be analyzed or resolved in isolation.

The purpose of this series is to address the increasingly complex questions of how societies come to understand, confront, and cope with both the sources and the manifestations of present and potential environmental threats. Works in the series may focus on matters political, scientific, technical, social, or economic. What they share is attention to the intertwined roles of politics, science, and technology in the recognition, framing, analysis, and management of environmentally related contemporary issues, and a manifest relevance to the increasingly difficult problems of identifying and forging environmentally sound public policy.

Peter M. Haas
Sheila Jasanoff
Gene Rochlin

Acknowledgments

The editors would like to express their gratitude to the many people and institutions who helped create *Changing the Atmosphere: Expert Knowledge and Environmental Governance.*

Above all, we thank our contributors, who patiently endured several delays and revisions of the book's format.

We are also grateful for editorial help from many people. Sheila Jasanoff offered especially thoughtful insights, as did her coeditors for the MIT Press series on Politics, Science, and the Environment, Peter Haas and Gene Rochlin. MIT Press editors Clay Morgan and Larry Cohen steered the book's progress and helped us work out a number of problems. Before he became personally involved with the book project as coauthor of chapter 7, Steve Schneider served as a referee, providing detailed and very helpful comments and criticisms. Several anonymous referees also offered excellent advice.

Many institutions provided helpful research assistance. We would like to acknowledge especially the World Meteorological Organization, the Intergovernmental Panel on Climate Change, the U.S. National Center for Atmospheric Research, the Hadley Centre for Climate Prediction and Research, the NASA Goddard Institute for Space Studies, the Center for the History of Physics, the NOAA Geophysical Fluid Dynamics Laboratory, and the Science, Technology & Society Program at Stanford University.

We also wish to acknowledge the generous financial support without which this work would not have been possible. The National Science Foundation supported some of Paul Edwards's work on this book under grants SBE-9310892 and SBR-9710616. NSF's Program in Ethics and

Values Studies also supported Clark Miller's research for this book, under grants SBR-9423373 and SBR-9601987. Additional support for Edwards's research came from the Alfred P. Sloan Foundation under a consulting arrangement with the Center for the History of Physics at the American Institute of Physics (on a different but related project). Part of Miller's research for this book was carried out as a fellow in the Global Environmental Assessment project at Harvard University (supported by NSF grant SBR-9521910) and as a visiting scholar at the International Institute for Applied Systems Analysis in Vienna.

Finally, we would like to thank the following individuals for contributions in many shapes and sizes: Anna Banchieri, Amy Cooper, Arthur Daemmrich, Michael Dennis, Karin Ellison, Joel Genuth, Margaret Harris, Gabrielle Hecht, Myanna Lahsen, Marybeth Long, Stacy Van-Deveer, Warren Washington, and Spencer Weart.

Contributors

Paul N. Edwards is Associate Professor of Information at the University of Michigan. He is the author of *The Closed World: Computers and the Politics of Discourse in Cold War America* and of numerous articles on the history, politics, and culture of information technology. He is currently working on a book, tentatively titled *The World in a Machine: Computer Models, Data Networks, and Global Atmospheric Politics*, about climate science and politics since 1950.

Dale Jamieson is Henry R. Luce Professor in Human Dimensions of Global Change at Carleton College. He has published widely on environmental philosophy, and is completing a book on the science and ethics of global environmental change.

Sheila Jasanoff is Professor of Science and Public Policy at the John F. Kennedy School of Government at Harvard University. She is the author of several important books on science, politics, and law, including *Controlling Chemicals: The Politics of Regulation in the United States and Europe*, *Risk Management and Political Culture*, *The Fifth Branch: Science Advisers as Policymakers*, and *Science at the Bar: Law, Science, and Technology in America*.

Chunglin Kwa is Lecturer in Science Dynamics at the University of Amsterdam. His current work involves the impact of global change on the development of various environmental sciences. He is also interested in the philosophical underpinnings and historical development of "complexity" in the sciences.

Clark A. Miller is Assistant Professor of Public Affairs and Science Studies in the LaFollette School of Public Affairs at the University of Wisconsin–Madison. His research focuses on expert institutions and their role in evolving structures of global environmental governance.

Stephen Norton is a Ph.D. candidate at the University of Maryland's program in the History and Philosophy of Science. He is presently completing a dissertation on the epistemology of modeling in the atmospheric sciences, which includes work on global warming and discovery of the Antarctic ozone hole.

Stephen H. Schneider is Professor of Biological Sciences, Senior Fellow at the Institute for International Studies, and Professor by Courtesy in the Department of Civil Engineering at Stanford University. From 1973 to 1996 he was a member of the scientific staff at the U.S. National Center for Atmospheric Research. Schneider has been a consultant to federal agencies and/or White House staff in the Nixon, Carter, Reagan, Bush, and Clinton administrations, and presently serves on the Intergovernmental Panel on Climate Change. In 1992 he received a MacArthur Fellowship. Schneider is the author of many scientific and popular articles and books on climate change, including *Global Warming: Are We Entering the Greenhouse Century?* and *Laboratory Earth: The Planetary Gamble We Can't Afford To Lose*. He is also founder and editor of the interdisciplinary journal *Climatic Change*.

Simon Shackley   is Research Programme Manager at the Tyndall Center for Climate Change Research, University of Manchester Institute of Science and Technology (UMIST). His research interests include the relationship between institutions and knowledge, and the implications that thereby arise for public policy. He has also conducted research on the impacts of, and responses to, climate change on a regional scale. He is the author of *Changing by Degrees: The Impacts of Climate Change in North West England* (Aldershot, England: Avebury, 2000).

Frederick Suppe   is Professor and Chair of Philosophy at Texas Tech University. He is also Emeritus Professor and Distinguished Scholar-Teacher at the University of Maryland. Presently he is writing *Venus Alive!: Modeling Scientific Knowledge*, a history of Venus planetary exploration from the perspective of the modeling on which virtually all our knowledge rests.

1

Introduction: The Globalization of Climate Science and Climate Politics

Clark A. Miller and Paul N. Edwards

... now I am become Death, the Shatterer of Worlds
—J. Robert Oppenheimer, 1945, quoting the *Bhagavad Gita*

The explosion at 5:30 A.M. on July 16, 1945, in Alamogordo, New Mexico, first hinted at the power of science to reshape world order, challenging us all to assume global responsibilities and become citizens of planet Earth. Harnessed by $2 billion and 200,000 American workers, Einstein's equation, $E = mc^2$, shattered worlds. Not the earth itself (at least not yet). Rather, worlds: imagined worlds in which the most horrific of human atrocities could never threaten the foundations of civilization. The subsequent forty years of cold war became, in a very real sense, an extended encounter with the idea that human actions could destroy the planet—and with how we order global politics in the face of such danger.

In the aftermath of Hiroshima, Nagasaki, and the nuclear tests that followed, it hardly seems surprising that nuclear fallout emerged in public discourses of the 1950s as the first global environmental threat, or that the first global environmental treaty—banning atmospheric tests of nuclear weapons—came disguised as arms control. Nuclear explosions were shots heard around the world (literally, if one had a seismic listening device). Overshadowed by Sputnik, missile gaps, and the Berlin Wall, humanity's second distant early warning began as a whisper, published in an obscure scientific journal read by at most a few thousand people:

Human beings are now carrying out a large scale geophysical experiment of a kind that could not have happened in the past nor be reproduced in the future. Within a few centuries we are returning to the atmosphere and oceans the concentrated organic carbon stored in sedimentary rocks over hundreds of millions of years. (Revelle and Suess 1957, 19)

Since 1957, however, the idea that humans are experimenting with critical, global-scale environmental processes has achieved a political salience at least as great as nuclear weapons once commanded. Climate change epitomizes, perhaps as much as any other policy issue, people's growing perception that the world itself is finite and indivisible, raising fundamental questions about how we govern ourselves as a global community. Our goal in this book is to probe these connections between environmental science and politics. We explore how scientific ideas about the climate have acquired so much political clout; how advances in scientific understandings of Earth systems are contributing to processes of political change in global society; how climatological expertise is being institutionalized in international politics; and what impacts all of this carries for the distribution of power and authority in global governance.

Climate Science and Environmental Governance

Scientific debates about climate change have often treated global warming as analogous to other environmental issues. Scientists have disputed complex issues of climate change detection, attribution (natural vs. human causes), and consequences, from how to validate climate models to whether rapid global warming might cause melting of the West Antarctic ice sheet or a shift in the North Atlantic current. They have sought to detect climatic changes, first in a warming "signal" in long-term, globally averaged temperatures and now, as well, in the statistical "fingerprint" of subtler shifts in climatic patterns (Santer et al. 1996b; Schneider 1994). Growing demands for policy action have prompted increasingly complex and sophisticated attempts to quantify the potential damage from climatic changes and the potential costs of policies to prevent them. Throughout, however, the basic tenor of the questions has remained much the same as if the issue were clean air or clean water: Are human actions changing the environment in dangerous ways? Can we afford to take the necessary measures to prevent such changes? Can we afford not to?

This book takes a different tack. Climate change, we argue, can no longer be viewed as simply another in a laundry list of environmental issues; rather, it has become a key site in the global transformation of world order (see Rayner and Malone 1998; Thompson, Rayner, and Ney 1998a, 1998b; O'Riordan and Jaeger 1996). In the three decades since the 1972 UN Conference on the Human Environment at Stockholm, states have constructed a suite of new international environmental regimes, with climate change taking center stage since the mid-1980s (Young 1998, 1994; Young, Demko, and Ramakrishna 1996; Lipschutz and Mayer 1996; Keohane and Levy 1996; Haas and Haas 1995; Litfin 1994; Lipschutz and Conca 1993; Haas, Keohane, and Levy 1993; Haas 1992, 1990b). These regimes—ensembles of political and scientific institutions and networks, often centered on a formal treaty and developed to address particular issues such as global warming, ozone depletion, acid rain, or desertification—have emerged as key focal points of contestation over the development of new norms and practices for making decisions of potentially worldwide reach. (For further clarification of this use of the term *regime*, borrowed from international relations scholarship, see Krasner 1982; Chayes and Chayes 1995.) In hotly disputed negotiations, participants have begun to redistribute power and authority in global society, not only among governments but also among an array of other actors: firms, experts, nongovernmental organizations, international institutions, and billions of individual citizens (Litfin 1998; see also Miller, forthcoming).

The atmospheric sciences have contributed in crucial ways to the creation and evolution of these new regimes, becoming deeply enmeshed in the constitutional structure of global environmental governance. Clearly, for instance, expert knowledge was a sine qua non of the Montreal Protocol on ozone-depleting substances and its successors (Benedick 1991). Yet for the emerging climate regime—this book's principal focus—continuing scientific controversy, and a dearth of simple solutions, render relations between expert knowledge and environmental governance even more important, and far more contested.

This book argues that contemporary debates about climate science are no longer simply about building accurate pictures of nature with which to ground good policy. They are also, and in the long run just as importantly, helping to set basic rules of standing and legislation for global environmental decisionmaking. Because of their immense scope, evolving climate governance processes may feed back into other, existing global environmental regimes as well. Put simply, debates about how to represent the earth's climate today serve also as key sites where people are "busily constructing ideas about what constitutes legitimate knowledge, who is entitled to speak for nature, and how much deference science should command in relation to other modes of knowing" (Jasanoff 1996b, xv).

Previous scholarship, even where it has examined science's relationship to international governance, has tended to treat the production of scientific knowledge as external to politics. Studies of "epistemic communities," for example, identify the authoritative knowledge claims of experts as a significant "power resource" in influencing the construction of environmental regimes (Haas 1990b, 55–56 and throughout; Haas 1992; Rowlands 1995). But the climate regime is more complex than most cases so far treated by the epistemic communities approach. It remains unclear whether the extent of scientific agreement, even on core issues such as global warming's extent, is sufficient to be called consensual knowledge. Multiple communities claim expertise, with different conclusions and different background commitments. Yet the epistemic communities approach typically takes consensus as the necessary condition of political influence, presuming that without it an expert group cannot speak with the united voice required for political authority. The model is linear. First, scientists agree among themselves; then they present their views to political actors; finally, politics responds.

We argue that issues sidelined by the epistemic communities approach are in fact central to a complete understanding of contemporary debates about global environmental governance—for example, how epistemic communities form, create consensual knowledge, and acquire political authority (see Jasanoff 1996c) as well as how knowledge and order become intertwined even in the absence of epistemic consensus. Scientific investigations of the global environment have rarely occurred in an influence-free vacuum, prior to or independent of politics. Instead, processes of knowledge creation, community formation, and expert institutionalization are themselves deeply political exercises, with substantial implications for broader debates concerning how people of vastly unequal technological capacity and means are going to live together on the planet (see, e.g., Jasanoff and Wynne 1998; Jasanoff 1990; Ezrahi 1990). Science, as it emerges from the chapters in this book, thus appears less an independent input to global governance than an integral part of it: a human institution deeply engaged in the practice of ordering social and political worlds.

As we intend the term, *governance* has a broad sweep, referring not simply to institutions conventionally viewed as political (i.e., "government") but to the full range of knowledge, technique, power, and practice constituting "the manner in which something is governed or regulated" (OED). This seems to us the most appropriate way to think about the vast, only partially coherent and integrated ensemble of local, regional, national, international, and global actors and activities seeking to understand, represent, and respond to planetary environmental change, where science in all its manifestations grows ever more important. As Oran Young has pointed out, we have no global government, yet we already have global governance: a tangled web of processes for muddling through on problems affecting the planet as a whole (Young, Demko, and Ramakrishna 1996; see also Young 1999).

In part, this perspective requires us to examine anew aspects of international relations often given short shrift by studies of international regimes, such as expert advisory meetings and technical standard setting. More often, it also means shifting our attention beyond the confines of diplomatic negotiations and intergovernmental institutions. Questions of what counts as credible knowledge, who speaks for nature, and how science relates to other policy-relevant ways of knowing are as frequently at stake in computer-modeling centers and laboratory field sites as in legislatures, regulatory agencies, and international organizations. To study constitutional change in global environmental governance, we must thus attend to myriad institutional settings in which people construct knowledge about the earth's climate and use it to inform their activities.

Not every chapter in this book treats governance in the same way, and some do not even employ the term. Yet the authors writing here do share a common sensibility and conviction. We believe that climate change matters, in part, because as communities around the world explore the problem and its remedies, they are building both new ideas about nature and society and new institutions for managing their collective lives on global scales. A deeper understanding of these collective reexaminings and reworkings of nature and society can contribute much, we believe, to our ability as a global community to grapple successfully with social and environmental change. Indeed, such an understanding may prove even more important than advances in the science of climate itself. Global environmental governance, as presented in this book's accounts, functions by means of new, complex, hybrid forms of knowledge and power still being forged—and therefore still fragile, negotiable, and worthy of our most careful and creative attention.

Changing the Atmosphere

Throughout this book, we investigate three important connections between global environmental science and governance. First, we investigate how scientific research on the environment is changing the basic conceptions of nature that underlie international politics. In his ground-breaking work *The Idea of Biodiversity: Philosophies of Paradise*, David Takacs explores the ties between how people represent nature and how they value it. He suggests that scientists engaged in policymaking on biodiversity are attempting the Herculean task of reshaping basic human experiences of the environment. In their efforts to promote environmental protection, Takacs argues, these scientists seek to remake people's deeply held images of nature, and, with them, their environmental values: "Conservation biologists have generated and disseminated the term *biodiversity* specifically to change the terrain of your mental map, reasoning that if you were to conceive of nature differently, you would view and value it differently" (Takacs 1996, 1). By reconceptualizing nature as "biodiversity," Takacs argues, scientists hope to disseminate an image of nature in its global variety—and so to foster concern for environmental policies to protect nature on a global basis.

Making climate change into an international political issue has likewise involved efforts by scientists to alter the conceptual categories through which people understand and value nature. Until recently, *climate* denoted for most people merely the weather patterns characteristic of a particular locale. According to the *Merriam-Webster Dictionary*, for example, climate is "the average course or condition of the weather *at a place* usually over a period of years as exhibited by temperature, wind velocity, and precipitation" (emphasis added). The word *climate* itself derives from the Greek *klima*, meaning "sloping surface of the earth," linking a place to its weather through its elevation. Terms such as *tropical*, *desert*, *mountain*, *tundra*, and *temperate*, as common-language terms for particular climates, refer interchangeably to specific geographic regions and their typical meteorological conditions. The 1941 U.S. Department of Agriculture Yearbook *Climate and Man*—an early U.S. climate impact assessment report—defined this relationship succinctly: "The climate of a place is merely a build-up of all the weather from day to day" (U.S. Department of Agriculture, 1941, 4).

Contemporary climate science represents the climate very differently, however. Drawing on worldwide data-collection networks, computer models, and satellite images from space, scientists now view the climate as a set of integrated, world-scale natural processes linking the earth's atmosphere, oceans, land, and life. No longer do scientists in institutions such as the Intergovernmental Panel on Climate Change (IPCC) speak about climates of the world, in the plural, as separate objects of study. Instead, their discourses center on the dynamics of what they term the *global climate system* (or, even more inclusively, the *Earth system*). Drawing on the work of the IPCC, the text of the UN Framework Convention on Climate Change (FCCC) defines the *climate system* as "the totality of the atmosphere, hydrosphere, biosphere, and geosphere and their interactions." Today's scientific conception of climate thus connotes less the weather of any particular place than something more closely akin to the global environment: a natural object to be understood, investigated, *and managed* on planetary scales.

Second, this book explores the social processes by which scientists persuade other people to think about climate in global terms as they attribute meaning to events occurring around them. The meanings attached to climate and the weather are often highly "black-boxed" (i.e., they are complex, socially mediated concepts that are generally taken for granted). News media interviews of leading scientists often reveal the difficulties inherent in scientists' efforts to get their message across. One such interview took place on a local television news show after a scientific conference on El Niño, held in Hawaii during the 1997–98 El Niño event. Questioned by a reporter, former Assistant to the President for Science and Technology Jack Gibbons tried to use El Niño to illustrate why people should worry about climate change. El Niño, he argued, demonstrated that the atmospheric dynamics we perceive as weather are actually part of a worldwide climate system. The reporter was unfazed. If that is so, she persisted, what are the implications for Hawaii's weather? Mutual frustration quickly emerged as it became apparent that for Gibbons, the global climate system was paramount, while for the reporter (and, she presumed, her audience), the local weather mattered most. Her unwillingness to allow the interview to shift toward a more globally oriented discussion illustrates the conceptual gap that still remains between how most people perceive the atmospheric environment and how climate scientists would like them to see it.

Features of the new climatology only exacerbate the dissonance between expert and lay conceptions of the atmospheric environment. Existing climate models cannot predict exactly how climate change will affect the weather in any particular place (although some short-term, regional forecasts, especially those related to the El Niño/Southern Oscillation phenomenon, have begun to improve dramatically; see Cane 1997). Nevertheless, faced with the gap between global and local perspectives, some scientists and officials have tried to make the connections more explicit. Following the 1993 floods in the Midwestern United States, Vice President Al Gore toured the region, proffering the possibility that the disaster, while perhaps not a direct result of climate change, might be a foretaste of things to come. More recently, in an interview on the PBS *News Hour with Jim Lehrer*, IPCC Chair Robert Watson responded in careful phrases to a question about the increasing prevalence of powerful hurricanes such as 1998's Georges and Mitch:

This year was the hottest year on record, and the question is: are these just natural events that occur every so often, or is it possible that we humans are slowly but surely changing the Earth's climate, and by changing the Earth's climate, we are seeing more of these very extreme events? (Ponce 1998)

Third, the book investigates how people alter their values, behaviors, and institutions as they develop new understandings of nature. Representations of the environment are intertwined in countless human activities and relationships, and efforts to alter social and political arrangements often must go hand in hand with efforts to promote new images of nature. Of greatest concern for us in this book, expert depictions of the global environment have been used as the basis of calls for radical changes in global economic and political order. At the Second World Climate Conference in 1990, Mostafa Tolba, then Executive Director of the UN Environment Programme, observed:

The sum of research into the science and impacts of climate change makes it clear that nothing less than dramatic reductions in emissions of greenhouse gases will stop the inexorable warming of the planet. Nothing short of action which affects every individual on this planet will forestall global catastrophe. (Tolba 1991, 4)

Subsequent FCCC negotiations have challenged, if not yet over-turned, basic principles of international governance, including "the sovereignty of national governments over domestic affairs, the exclusive legitimacy of national identity as a basis for political mobilization in international forums, and the exclusive rights of national governments to participate in international legal agreements" (Miller, forthcoming). For many, such changes have become essential, if not yet inevitable, goals: "Critical threats to the Earth's habitability demand that humankind rise to the challenge of creating new and more effective systems of international environmental governance" (Young, Demko, and Ramakrishna 1996, 1).

Others, however, have contested the inevitability of the link between new global representations of nature and the necessity of transforming global governance. Climate scientists' projections of climate change have suffered frequent challenge from opponents offering competing interpretations of scientific evidence. Some scientists and policymakers argue that current models of the climate system are flawed, or that their results are too ambiguous to justify near-term action. Others claim that the global, systemic perspective presented by climate scientists cannot easily be reconciled with the practical requirements of politics. Like an overexposed photograph, globalism (from their perspective) washes out real differences between human communities around the world.

Writing in 1991, at the outset of the climate negotiations, Anil Agarwal and Sunita Narain of the Center for Science and Environment in India castigated those who would use "one-world" discourses about the environment to dilute responsibility for climate change by spreading it evenly among industrialized and developing countries (Agarwal and Narain 1991). Their work sought to counter what they viewed as biased estimates of individual nations' greenhouse gas emissions published by the World Resources Institute (see World Resources Institute 1990). Drawing on a model of moral responsibility that stressed overconsumption by the developed world, Agarwal and Narain accused WRI scientists of using environmental arguments to perpetuate limits on economic development in the South. (For a closer look at Indian perspectives on the epistemological and moral logic of climate change, see Jasanoff 1993.) Their critique attracted relatively little attention in the United States and Europe but was widely read among elites in developing countries.

Debates about the proper relationship between global environmental science and politics have also occurred during the creation of new, hybrid institutions for managing climate change under the FCCC. Experts' ability to represent nature in global terms has gained them privileged places in these new global institutions. In the IPCC and the Subsidiary Body for Scientific and Technological Advice to the UN Framework Convention on Climate Change Conference of Parties (SBSTA), experts—and even entire scientific communities—have been enrolled formally in global policymaking to a greater degree than in most other areas of international relations. This incorporation of experts directly into the policymaking apparatus extends to international organizations a century-long trend in the political transformation of liberal democratic governance (Ezrahi 1990).

At the same time, such bodies have enrolled political communities in the warranting of scientific knowledge. For example, one purpose of the IPCC is to conduct periodic assessments of the "state of the art" in climate science—at first glance an activity for which only scientists would qualify. Yet IPCC membership has always included governments and nongovernmental organizations whose representatives attend meetings, submit comments on draft IPCC documents, and vote on formal acceptance of IPCC scientific assessments. As Edwards and Schneider demonstrate in chapter 7, the IPCC's unique hybrid structure helps to enhance the credibility of its claims among diverse expert, policy, and lay audiences (see also Miller, forthcoming).

Not surprisingly, the construction of institutions like the IPCC and SBSTA has generated considerable conflict over the proper norms and practices of providing expert advice in international settings. Diplomats and scientists alike have wrestled with how much power to accord to scientists in making global policy and how much authority to accord to policymakers to shape processes of knowledge production, validation, and use. Both SBSTA and the IPCC have undergone long, complex negotiations over such questions as what standards they will apply to evidence, how they will organize peer review, who will be allowed to participate in their activities, and who will have the right to question their findings. Decisions on these matters influence not only the content and credibility of IPCC and SBSTA expert advice, but also their institutional legitimacy and long-term viability. These disputes go to the heart of what it means to use science legitimately to make decisions affecting, as Tolba (1991) put it, "every individual on this planet." What makes for legitimate knowledge when it is to be used for global policymaking? Who is entitled to speak for nature on the world stage? What rights should science be accorded in relation to other ways of knowing and other sources of authority in global politics? As we enter the twenty-first century, the ultimate resolution of these questions seems anything but clear.

Science Studies and the Politics of Technical Decisions

In recent years, many social scientists (and some scientists) have come to accept the idea that science and politics are deeply intertwined. The goal of this book is to go beyond this rather trivial observation to investigate in detail the actual processes by which this linkage is constructed in emerging institutions of global environmental governance. In the previous section, we argued for the need to attend carefully to the institutional, cultural, and political forces that shape the production and validation of scientific arguments as well as their uptake into individual and collective choices around the world. In this section, we review the literature on science studies and the politics of technical decisions as it applies to accomplishing this task.

Jasanoff and Wynne (1998) argue that science studies can contribute significantly to a deeper understanding of the human dimensions of global environmental change. (For other discussions of this and related literature, see Jasanoff et al. 1996; Yearley 1996a; Rayner and Malone 1998.) Science studies approaches focus attention on issues often left unexamined in other areas of social science theory, including the formation of new communities and social identities around new scientific conceptions of the environment, the resolution of scientific controversies, the institutionalization of expertise, and the stabilization (i.e., general acceptance) of knowledge. They seek also to recast global environmental debates in terms that display the tight linkages between the microsocial contexts in which knowledge about the environment is produced—for example, scientific laboratories, field experiments, and climate modeling centers—and the macropolitical and economic institutions that shape social and environmental change on global scales (e.g., Jasanoff, forthcoming, 1997b, 1996c, 1993; Miller et al. 1997; Edwards 1996b; Shackley and Wynne 1996, 1995, 1994; Wynne 1995; Zehr 1994).

This book offers a cross-section of this field of research, integrating the concerns of science studies, environmental studies, and policy studies. Like the broader literature from which it draws, much of the research presented in this book

can be broadly characterized as *interpretive*, because it emphasizes the significance of meanings, texts, and local frames of reference in knowledge creation; *reflective*, because it focuses on the role of reflection and ideas in building institutions; and *constructivist*, because it examines the practices by which accounts of the natural world are put together and achieve the status of reality. (Jasanoff and Wynne 1998, 4)

Recently, this approach has come under attack, particularly for its constructivist orientation and for placing too strong an emphasis on social and cultural factors in its accounts of science (Koertge 1998; Gross and Leavitt 1994). Such criticism largely misses the point, however. (For thoughtful rebuttals, see Jasanoff, forthcoming; MacKenzie 1999 and other contributions to the same volume; Hilgartner 1997.) Most scholars in science studies, including the authors writing here, do not attempt to evaluate the validity of specific knowledge claims. (To imagine that we could do this better than scientists themselves would be foolish indeed.) Rather, researchers in this field have sought to understand *processes* of scientific knowledge creation: how people in various cultural and historical contexts produce, validate, and use knowledge they consider scientific. Likewise, science studies research has sought to explain how scientific facts and images of science are taken up in society, as people incorporate them into collective norms, behaviors, and institutions. These concerns necessarily involve greater attention to social and cultural explanation.

In everyday scientific and policy practice, the vast majority of scientific facts enjoy considerable stability. Yet during scientific or policy controversies, scientists, public officials, and citizens frequently present competing interpretations and evaluations of scientific data, evidence, and theories (Nelkin 1992; Collins 1985). When newly developed scientific techniques alter fundamental ideas, or when social changes bring renewed critical attention to the facts of a particular issue, even well-settled scientific knowledge claims can be reconsidered and occasionally rejected, sometimes even en masse in the well-known phenomenon of "paradigm shift" (Kuhn 1962). Science studies has therefore sought to explain how new understandings of nature are stabilized—that is, *how scientific controversies are settled and why they remain so*, at least temporarily. Studies of scientific institutions, instruments, and communities have identified a wide range of practices by which scientists in various contexts articulate, validate, and resolve new claims to knowledge (e.g., Galison 1997; Kohler 1992; Kay 1993). Likewise, research on how science interacts with legislatures, courts, regulatory agencies, social movements, and other political institutions suggests that these institutions, too, can play important roles in helping to stabilize (and destabilize) new understandings of nature and new scientific claims (Cussins 1998; Jasanoff 1996b, 1990; Epstein 1996; Bimber 1996; Ezrahi 1990).

The stabilization of knowledge inevitably rests on the ability of various societal actors and institutions to create and maintain trust and credibility. Science studies research has therefore sought to explain how and why scientific knowledge acquires credibility among diverse audiences.

Studies in history, sociology, and comparative politics have pointed to the importance of trust and community as underpinnings of the persuasiveness of new knowledge (Shapin 1996, 1994; Dear 1995; Jasanoff 1986). For example, studies have shown that even experimental replication, the principal guarantor of scientific reliability, serves its intended function only when communities of researchers agree about what counts as an adequate replication of a previous experiment (Collins 1985). Such agreement, in turn, presupposes sufficient disciplinary or cultural cohesion among investigators for them to trust one another's competence and integrity (Shapin 1996, 1994; Collins 1985; Latour and Woolgar 1979).

Other studies have argued, further, that maintaining the public authority of science requires trust not only among experts but also between experts and other social groups (see, e.g., Gieryn 1999; Jasanoff 1996a, 1996b, 1995; Epstein 1996; Wynne 1995). Comparative studies of risk management and risk communication provide considerable insight into the role of political culture in building and maintaining (or corroding) trust among experts, government officials, industry and NGO representatives, and citizens (Wynne 1990; Jasanoff 1986; Brickman, Jasanoff, and Ilgen 1985). Likewise, historical work has illustrated the dynamic, shifting character of public trust in experts and expert knowledge, as well as of the forms of expert knowledge that command this trust (Porter 1995; Ezrahi 1990). In recent years, some social scientists have suggested that public confidence in experts is weakening. While views on this issue are hotly contested, many perceive the erosion of this once-strong confidence as a key contributor to the crisis of modernity (Beck 1992; Ezrahi 1990; Giddens 1990).

Finally, science studies research on expert advice highlights the pervasive influence and institutionalization of science in modern governmental decision making. Scientific knowledge and expertise have become enmeshed in the making of public policy, particularly (though not exclusively) in Western democracies. The ability to call on expert knowledge has become a key component of strategies for legitimating public policies and securing trust in public institutions. Polities frequently delegate considerable authority to science to identify and define issues of policy concern. Government, industry, and university laboratories, as well as scientific advisory committees, perform major functions in regulatory standard setting and risk assessment. Scientists also frequently take the witness stand as experts in the courtroom (Jasanoff and Wynne 1998; Jasanoff 1996c, 1990; Wynne 1982). Yet given the universalistic goals of scientific understanding, nations have institutionalized the policy roles of experts and expert knowledge in strikingly different ways. Many aspects of policy-relevant science, such as the conceptual framing of risks, the kinds of evidence deemed acceptable in policy discourses, and the norms and practices of expert advisory bodies, vary considerably from country to country. Such variation often reflects fundamental differences in political, social, and cultural organization (Jasanoff 1995, 1986; Wynne 1990; Brickman, Jasanoff, and Ilgen 1985).

Science thus emerges from recent science studies analyses as a powerful tool in the construction of modern political order, but one that acquires its credibility and solidity in policy settings at least in part from socially embedded, culturally specific norms and practices for warranting public knowledge (Shapin and Schaffer 1985; Jasanoff 1997b, 1995, 1986; Wynne 1990; Ezrahi 1990; Brickman, Jasanoff, and Ilgen 1985).

Toward a Science Studies Perspective on Climate Change

Deploying a variety of approaches informed by science studies, the essays in this book offer detailed, empirically grounded case studies of settings in which people make and interpret knowledge about the earth's climate and link that knowledge to political decisions. Chapters 2 through 4 study the creation of new scientific instruments, techniques, and laboratories through which scientists produce the knowledge underpinning current debates over climate policy. Chapters 5 and 6 examine the development of professional communities and organizations and their role in mediating between laboratory practices and broader aspects of political culture. The four final chapters investigate the uptake of policy-relevant knowledge into public decisions in expert advisory institutions, international negotiations, and popular culture.

None of these settings turn out to be either exclusively scientific or exclusively political. Rather, we found society being made inside the laboratory as well as out—and science being made outside the laboratory as well as in—although the nature of these constructions varied considerably according to the setting. Scientific research programs and political developments are intertwined not only *after* science-based issues become the subject of international diplomatic engagement (as international relations theorists have shown; see, e.g., Haas 1992) but often also *beforehand*. In other words, this book argues, creating and managing scientific research programs simply *is*, at least in part, an important way of working out social and political order.

As with all science, at some level knowledge claims and knowledge-making processes of the climate regime are value laden. This fact does not reduce the importance of science to global environmental governance. However, it does point to the need to develop new ways of understanding science's societal dimensions. The chapters in this book demonstrate that careful attention to the social, institutional, and cultural processes by which knowledge is constructed and taken up into policy can help illuminate how and why particular values come to be embedded in particular knowledge claims. In this way, the chapters help point toward ways of strengthening environmental governance by enabling policymakers to better identify, assess, and take into account the normative dimensions of expertise.

Building Worlds: Computer Modeling and Climate Politics

Chapters 2 through 4 begin our study of the links between climate science and politics with detailed investigations of the construction and use of computerized climate models as tools for producing knowledge of the earth's climate. They demonstrate that over the past several decades, the development of climate models has fundamentally altered both the science and the politics of global warming. Climate models have acted as a focus for community building and conflict both inside and outside science. Such models now constitute the principal tools for investigating and representing long-term variations in the earth's climate in policy deliberations as well as in science. They have played critical roles in establishing the credibility of the idea that human activities threaten the environment on planetary scales. In the process, they have come to underpin the authority of supranational organizations to set global policy (Miller, forthcoming).

These chapters form the basis for the principal argument of the book, namely, that a detailed understanding of scientific practice is necessary if we are to understand fully the dynamics of global environmental governance. They focus their analyses—respectively historical, philosophical, and sociological—on how scientists and others use computer models to construct and deconstruct claims to policy-relevant knowledge. In the process, they demonstrate that seemingly arcane epistemological debates—for example, concerning the relationship between simulations (models) and reality (observations) or between different national approaches to climate modeling—can carry fundamental normative and political as well as scientific implications.

In chapter 2, Paul N. Edwards charts the evolution of scientists' views about climate change from the nineteenth century to the present, describing how computer models became the only practical way to handle the "data deluge" created by post–World War II observing technologies (especially satellites; see Chervin 1990). The ability of computers to process these enormous data sets made possible, for the first time, the study of atmospheric dynamics on a global scale.

The epistemological dimension of Edwards's argument follows from his historical analysis of relations between models and data. In an important sense, he argues, computer models were necessary to create the first truly global data sets. The data record of climate, he demonstrates, is sparse, incomplete, and poorly fitted to modeling grids. Computer models are used to integrate, filter, smooth, and interpolate these data, building uniform, consistent global data sets that now form the basis of knowledge about climatic change over the last century. At the same time, computer-modeling techniques allowed scientists to create complex, dynamic simulations of long-term climatic changes, at sufficient levels of detail for comparison with the historical observational record. Today, using these models, scientists represent and investigate the dynamics of the earth's climate system on global spatial scales and decadal temporal scales.

These models of the earth's atmosphere are not simply scientific tools, however. Their claims to realism, global scope, and still-nascent, but ever-increasing ability to forecast short-term climatic shifts such as El Niño have made them significant tools in the inventory of those attempting to mobilize support for worldwide reductions in greenhouse gases. Climate models are essential to how scientists and others active in climate policymaking distinguish between human and natural climate change, evaluate policy options, and assess policy successes.

In turn, Edwards notes, the centrality of climate models to climate policymaking has shaped aspects of their development. Their prominence has made them targets for critical attention aimed at derailing policy action, particularly in the United States, where scientific controversies often become proxies for deeper social and political conflicts. Epistemological concerns—in the form of claims that climate simulations can (or cannot) be trusted to predict actual climatic changes—have often been the focal point for these debates (Edwards 1999; Edwards and Lahsen, forthcoming). Likewise, efforts to make climate models more relevant to perceived policy choices have prompted scientists to create new kinds of climate models, including integrated assessment models that link atmospheric and social science and Earth system models that couple atmospheric dynamics with those of the planet's oceans, land surfaces, sea ice, snow, and vegetation.

Models, Edwards concludes, have thus become tools for "world building": instruments by which scientists and other policymakers construct and articulate competing visions not only of today's world, but also of what they hope, and fear, for tomorrow's.

In chapter 3, Stephen D. Norton and Frederick Suppe develop another facet of our argument that a sophisticated analysis of climate science can shed important light on political conflicts in global environmental governance. In recent years, epistemological critiques of climate models have taken on considerable political significance. Testimony before the U.S. Congress has criticized the use of climate model results as a basis for political action on the grounds that models can never genuinely be validated by observations (Brown 1996; Edwards 1999), as have recent scientific and philosophical publications (e.g., Oreskes, Shrader-Frechette, and Belitz 1994).

Norton and Suppe challenge these arguments, questioning the validity of drawing epistemological distinctions between computer models and other scientific tools. Norton and Suppe argue that *all* scientific work, whether experiment, field observation, or numerical simulation, involves epistemologically similar modeling exercises (see Dear 1995, who makes a comparable argument about experimental and mathematical science). Before throwing out potentially invaluable tools for producing knowledge about the environment, they conclude, policymakers should take care to develop a nuanced understanding of how science actually works.

Norton and Suppe base their argument on a comparison of two approaches to generating knowledge about the atmosphere: satellite measurements and computer simulations. Their chapter examines, in detail, the actual steps by which patterns of light and electricity measured on satellite instruments get turned into "physical" measurements of atmospheric temperature. Using this material, the authors demonstrate that the measurement process depends on a host of assumptions and parameters that render the result contingent, uncertain, and dependent on theoretical considerations—that is, "model-laden" (see Edwards, chapter 2, this volume). No epistemologically important distinction can be drawn, they argue, between models and measurements of the sort required by global climate science.

This argument cuts both ways: it also shows that we can trust (some) computer models for the same reasons, and to the same extent, that we trust (some) techniques for observing nature. Norton and Suppe conclude that no philosophical grounds exist to exclude the use of computer models as a basis for responsible public decision making. Excluding simulations on the grounds that they are models, they argue, would similarly preclude reliance on most kinds of observations and experiments. Yet scientists, public officials, and citizens regularly and unquestioningly use the latter to advance policy goals. In contrast to popular images of science as the search for certainty, Norton and Suppe present a more tentative view of science that foregrounds the contingency of scientific claims. They encourage adoption of the more limited but achievable goal of making reliable and robust, albeit potentially less precise, claims sufficient for particular limited domains of action.

Viewing science thus as a kind of bounded rationality, Norton and Suppe emphasize careful attention to the assumptions embedded in all knowledge claims. In spelling out the implications of this conclusion, they highlight the responsibilities of scientists to reveal tacit assumptions that underlie their own work (see Bolin 1994). Admittedly, this may not always be practical given the complex, interdependent, and highly networked character of modern science. As an alternative, governments often structure expert advisory processes to promote the critical review of scientific evidence by diverse groups in society (see Miller, chapter 8, this volume). While such broader discussion of the complexities and uncertainties of knowledge creation may initially reduce public trust, in the long run it can lead to a more realistic and sophisticated public understanding of science (Jasanoff 1990).

In his study of the "epistemic lifestyles" of climate modelers (chapter 4), Simon Shackley builds the third aspect of our argument about the political import of a detailed understanding of the practice of climate modeling. He shows that fundamental differences in modeling "styles," and therefore in the kinds of knowledge produced by models, are intertwined with differing national and laboratory cultures. Norton and Suppe point out in chapter 3 that multiple approaches to modeling a problem may all yield "valid" conclusions depending on the assumptions made in constructing the model and the uses to which its results are put. For Shackley, this theoretical potential is concretely realized in the scientific laboratories of the United States and Britain. Shackley's chapter presents the results of an ethnographic survey of climate modelers, chiefly at the UK Hadley Centre, the NOAA Geophysical Fluid Dynamics Laboratory, and the U.S. National Center for Atmospheric Research. Shackley demonstrates that modelers—even those working in the same laboratory—can differ significantly in "epistemic lifestyle"—that is, in the strategies and assumptions they use to build and validate models of the climate.

Shackley describes three distinct approaches climate modelers bring to their work. He labels two of these *climate seers* and *model builders*, while the third is a hybrid, science-policy approach. Climate seers use models primarily to provide new insights into atmospheric dynamics; they tend to rely on relatively simple models that focus on a small number of key theoretical variables. Model builders, by contrast, tend to focus on improving the "realism" of atmospheric models by incorporating ever more comprehensive and detailed representations of physical processes. At a basic level, therefore, these approaches connect scientists' perspectives on how best to model the atmosphere with the particular problems they are trying to solve.

For Shackley, however, the more interesting observation is that the "epistemic lifestyles" of climate modelers vary systematically from country to country, in patterns that reflect important cultural and institutional dimensions of science. In the United States, for example, Shackley finds not only a plurality of modeling institutions, but also a plurality of epistemic lifestyles within each institution. In the United Kingdom, by contrast, climate modeling is dominated by a single institution, the UK Meteorological Office's Hadley Centre. Further, although Hadley Centre modelers exhibit a limited degree of diversity, the organization is much more hierarchically structured around a single model-building approach than its American counterparts. Systematic differences in the construction and use of policy-relevant knowledge from one country to the next are, of course, familiar from comparative studies of national regulatory policies (Jasanoff 1986; Brickman, Jasanoff, and Ilgen 1985). To date, however, the existence and implications of national differences in climate science have received little attention from either social science research or public policymaking. (For a handful of exceptions, see, e.g., Shackley and Wynne 1996; Rayner 1993; Jasanoff 1993. This topic is also taken up in our companion volume: Edwards and Miller, forthcoming.)

Taken together, these three chapters demonstrate that detailed examinations of scientific practice can reveal deep-seated connections between knowledge construction and macropolitical dynamics. Events transpiring in the laboratory or in computer modeling centers can be as influential in shaping the landscape of global environmental governance as the meetings of international diplomats.

Yet all three chapters also show that historical, philosophical, and sociological microanalyses of scientific instruments and laboratories cannot fully explain the intertwined character of scientific epistemology, laboratory life, and political culture. To better understand these connections, the rest of our book expands the scope of study beyond the laboratory to a broader array of institutions that link science and society: professional communities, scientific organizations, expert advisory processes, and public understandings of science.

Connecting Scientific Organization and Political Order

Chapters 5 and 6 argue that dynamics internal to the scientific community cannot, by themselves, explain the long-term trajectory of the atmospheric sciences. Rather, each demonstrates that changes in research directions over the past half century have gone hand in hand with changes in broader political order. They argue that these connections between science and society can best be understood in studies of the long-term historical evolution of atmospheric research programs, communities, and institutions from the early postwar era to the present.

In chapter 5, Chunglin Kwa investigates the history of U.S. research on deliberate modifications of weather and climate, detailing how the fortunes of this work paralleled broad-based changes in American attitudes toward the environment, risk, and technology. Through the 1960s, intentional weather and climate modification appeared to many Americans as a powerful technology for taming nature's fury. Extreme weather events—droughts, floods, hail, and hurricanes—perennially cause significant damage to U.S. agriculture, coastal communities, and flood control infrastructure. For meteorologists, public officials, and numerous corporations, weather control offered tantalizing possibilities for reducing or even eliminating this damage through such technologies as rainmaking and hail and hurricane suppression. In the early 1970s, however, funding and support for weather modification research dried up, and most weather-control companies left the business. What happened, Kwa asks, to change Americans' minds about this seemingly promising technology?

The conventional answer is that weather and climate modification simply did not work. Kwa challenges this interpretation, arguing that meteorologists' views of the potential viability of weather control changed only after support for the technology had largely fizzled among policymakers and the public. The fate of weather control was sealed, Kwa argues, by broad-based changes in American attitudes that inverted the traditional understanding of environmental risk. In the 1950s and 1960s, Americans saw nature as a source of risk, a violent force capable of damaging human communities. But by the early 1970s, in a remarkable reversal, Americans had come to view nature as itself at risk—and primarily from human activities. As part of this change, Americans came to view weather control as posing significant environmental risk, and once-significant research programs gradually died, despite sometimes-promising results.

Changes in American attitudes toward weather control provide a poignant lesson about the degree to which people's values regarding nature, risk, science, and technology are interconnected. Various scientists and public agencies continue, even today, to promote technological modification of the atmosphere to make rain, alter hurricanes, and even to counter global warming. Yet such suggestions for large-scale geoengineering have met nearly unanimous public opposition. The American public, it seems, remains committed to the values and problem framing developed since the 1970s: the atmosphere is not a system to be managed for human benefit, but rather a natural object to be protected from humanity's destructive capacity.

In chapter 6, Clark A. Miller shows how the ability of scientists to produce knowledge about global-scale natural phenomena rests on their ability to shape appropriate political as well as scientific institutions. He narrates the early history of the World Meteorological Organization (WMO), describing how its creation and operation linked a transformation of postwar meteorology to changes in the organization of international politics. Between 1945 and 1960, meteorologists succeeded for the first time in constructing a worldwide network of atmospheric observing stations that could illuminate questions about the atmosphere as a global entity. The construction of this network, Miller argues, entailed considerable social and political as well as scientific and technological work. Governments had to be persuaded to provide the necessary resources to construct, maintain, and operate the network. The network had to be integrated into acceptable norms of international political organization. Also, new states had to be convinced of the benefits of participating in an emerging international system.

To achieve these goals, Miller argues, meteorologists and American foreign policymakers collaborated in the two decades following World War II to create a transnational epistemic community (on epistemic communities, see Haas 1992, 1990a, 1990b). This community connected the creation of an international weather regime to the stabilization of postwar political order. In negotiating, organizing, and operating the

WMO, meteorologists drew on and helped shape three broad clusters of norms and practices—intergovernmental harmonization, technical assistance, and international coordination of scientific research—linking scientific cooperation to efforts to secure international peace and prosperity in the postwar era. These *modes of interaction*, as Miller labels them, helped provide meteorologists with the political legitimacy and authority to obtain the material and human resources to link up their far-flung observing stations into a global scientific network. More broadly, they helped establish the limits of legitimate scientific and political activity in international relations, thus contributing to the definitions of both "good science" and "good governance" that have characterized the operations of international organizations for much of the past half century.

On the surface, the interactions between atmospheric science and politics described by Kwa and Miller had little to do with worries about anthropogenic climate change. In fact, however, they were crucial to establishing conceptual and institutional frameworks that underpin contemporary concerns. In the 1960s, for example, the WMO refined its global observing network into today's World Weather Watch. In the 1970s, it sponsored the first global weather experiments. By the 1980s, the latter had evolved into the World Climate Program, which sponsored key meetings between scientists and politicians, such as the 1985 Villach conference (Davies and Ashford 1990). These programs not only enabled scientists to observe the atmosphere as a global system, but also established institutional pathways by which atmospheric science could speak authoritatively in global policy contexts.

Taken together, Kwa and Miller thus illustrate this book's argument that if we are to understand how and why global policymaking comes to be based on certain kinds of knowledge and not others, it is essential that we understand how scientific networks and institutions form, how they come to speak with a united voice, and how they acquire political influence—whether they take the form of epistemic communities, scientific research programs, or international organizations.

Expert Advice, Policy Uptake, and Public Understanding

The final four chapters of the book explore the uptake of science and science advice into political discourses and institutions dealing with the global environment. They concern how science achieves public credibility, how expertise is institutionalized, and how both are connected to norms of justice and equity in global environmental governance.

Trust, and its connection to the institutional norms and practices of global governing arrangements, occupy Paul N. Edwards and Stephen H. Schneider in chapter 7. They argue that formal procedural rules may take on far greater importance in international expert advisory institutions than they do in deliberations among small groups of scientists. Their study examines a recent controversy surrounding the IPCC. The IPCC's *Second Assessment Report* (SAR), released in the spring of 1996, concluded: "The balance of evidence suggests that there is a discernible human influence on global climate" (Houghton et al. 1996, 5). Skeptical scientists and energy industry lobby groups immediately attacked the report, asserting that IPCC participants had violated both scientific peer review conventions and the organization's own established procedural rules. These charges ignited a major debate, widely reported in the press, lasting several months.

Edwards and Schneider evaluate these charges, arguing that they rest on ambiguities in the IPCC's limited, underspecified rules of procedure. When the chapter authors behaved as if they were working according to the less formal norms typical of scientific communities, they opened themselves to unexpected attack in a mode typical of legal and political forums. Edwards and Schneider conclude that IPCC rules require revision to limit the organization's vulnerability to similar attacks in the future. (At this writing, they had just been so revised.)

Edwards and Schneider go on to explore this episode's ramifications for the role of formal review mechanisms in certifying scientific knowledge for policy contexts. How, they ask, can the IPCC and other institutions like it maintain scientific integrity in the face of intense political pressures and tightly constrained deadlines? Edwards and Schneider argue that the IPCC's widely inclusive, extremely intensive peer review process has opened the debate about climate change to a far wider range of actors than is usually consulted in science. By doing so, it has created a fairer, more thorough, and hence more powerful method for reaching consensus on the knowledge required for good public policy—serving ultimately as an important new model for the incorporation of expert knowledge into global governance. Edwards and Schneider explore the weaknesses and strengths of the traditional peer review process, concluding that peer review fails as a "truth machine" (largely since no human process could meet this impossible standard). Yet it succeeds in more important purposes related to trust, credibility, and the integration of science and politics. It helps to build collectively shared, publicly warranted knowledge and norms, and it maintains the democratic principle of "virtual witnessing" as a basic scientific tenet. These functions, and the principles they represent, remain the best hope of the slow and painful consensus-building process that global environmental governance necessarily requires.

In chapter 8, Clark A. Miller reaffirms our contention that constituting credible expert advisory processes in global environmental governance is a contingent, contested, and value-laden task that demands the careful attention of both scholars and policymakers. The central focus of his chapter is SBSTA, where government representatives negotiate the organization of scientific advisory institutions for global climate policy. Unlike the political systems of many Western nations, which have well-established, culturally specific systems of rhetoric and practice for warranting public knowledge, comparable international institutions are weak or nonexistent. Consequently, in settings such as SBSTA, negotiators have had to find other means of shoring up the public authority of science.

Miller's analysis of the evolution of science advice under SBSTA highlights the difficulties of avoiding unreflective reinscriptions of Western notions of the "right" relations between science and politics into the organization of global governing arrangements. As the history of developing country interactions with the IPCC illustrates, institutions that mirror too closely the assumptions of Western countries can encounter extraordinary difficulty in establishing credibility with far-flung developing country audiences. To overcome this and other challenges, SBSTA has sought ways of improving governments' opportunities to work out their political differences over science advice in incremental steps. SBSTA's strategies have included enabling participants to deconstruct proposed scientific advisory arrangements to reveal tacit, value-laden commitments and assumptions; finding tentative, partial, and highly responsive arrangements that can form the basis for institutional learning and confidence building; and tying advisory arrangements into a wide array of already-existent local, national, and international systems for warranting public knowledge and securing public trust. In this way, SBSTA has achieved some progress toward globally credible scientific advice even in the highly contested atmosphere of the ongoing climate negotiations.

In chapter 9, Dale Jamieson argues that despite the contributions of institutions such as IPCC and SBSTA, key moral dilemmas regarding science and global governance remain largely unresolved. As an example, Jamieson focuses on efforts to assign responsibility for global warming by counting greenhouse gas emissions. How we carry out this counting exercise, Jamieson points out, depends on how we value several important issues, including responsibility for past emissions and future equity. By failing to address such value choices explicitly, Jamieson believes, international negotiators have created bottlenecks to the inclusion of developing countries in the climate regime and, consequently, to the regime's long-term success.

To integrate developing countries successfully into the climate regime, Jamieson suggests, negotiators first need to agree on a broad normative framework for the next century. What kind of world, he asks, can countries from North and South both agree on as the place they would like to end up in a hundred years? Jamieson offers a principle: "Every person has a right to the same level of greenhouse gas emissions as every other person." From this follows a policy proposal: set permissible levels of greenhouse gas emissions on a national basis, but indexed to population (with every world citizen allotted the same permissible amount) and to a "baseline" year.

It may be that Jamieson's policy proposal has greater relevance as strategy than as substance. As Sheila Jasanoff has argued elsewhere, environmental problems sometimes find compelling solutions much more readily by addressing moral and political issues directly than by seeking scientific legitimation (Jasanoff 1997a). Now seems to be one such time. At stake in the ongoing negotiation of the climate regime is not merely the proper characterization of an environmental problem and its solution but also numerous questions about how we, as a global community, are going to live together on this planet. Jamieson argues that explicit discussions about the norms and values that should guide that community's collective decisions may have the power to move the debate forward. Negotiating the long-term "emission rights" of all world citizens, he suggests, might just be a good place to start.

The concluding chapter by Sheila Jasanoff brings the book back around to its central theme—an exploration of the shifting conceptual frameworks that underlie the emergence and dynamics of global institutions. Where earlier chapters investigated scientists' efforts to reconstruct understanding of the environment in global terms, however, in chapter 10 Jasanoff explores the role of visual images of the earth in the formation of global environmental consciousness among the lay public. Photographs of the earth suspended in space have achieved canonical status in environmental thought and action for their presumed power to compel a shared understanding of the fragility, finiteness, and indivisibility of the planet—one that might ground collective action to prevent its demise. The apparent simplicity and directness of this image and its message seems to underlie its broad appeal. It is, as Jasanoff puts it, the "vernacular" corresponding to the "high" scientific language of climate models and integrated assessments.

Yet through a detailed analysis of the reception of such images in American and international environmental policymaking, Jasanoff finds a more complex and contested reality behind this apparent clarity. Many groups have appropriated visual depictions of the earth in support of strengthening global environmental governance. Indeed, the view of the earth from space appears to underpin important new ethical discourses about equity and vulnerability around the world. For others, however, the image has an almost opposite import, signifying the dangers of "allowing environmental risks, and their control, to be globalized." Loss of sovereignty, bureaucratic inflexibility, neocolonialism, a disregard for the pervasive (but invisible from space) environmental problems faced daily by the poor—all appear to these observers as the potential upshot of narratives deriving power and authority from representations of the earth as "one world." Jasanoff thus reemphasizes a core concern of many of the authors in this volume: foundational aspects of global governance are at stake in how people choose to interpret and use competing representations of the earth's environment in their individual and collective choices.

Together, these four chapters illustrate the changing relationship between global environmental science and governance as we enter the twenty-first century. A few decades ago it seemed possible to encapsulate this relationship in terms of the activities of small, informal groups of scientists with access to key public officials (e.g., epistemic communities). Today, however, the interactions are broader, deeper, and more complex. Science's place in global policymaking is increasingly formalized, boosting its authority in policymaking processes but also subjecting it to new forms of political and legal oversight and review. International expert institutions such as the IPCC and SBSTA increasingly determine which knowledge counts and which does not, helping to shape crucial policy outcomes. At the same time, scientific ideas and images are pervading public consciousness, with equally powerful implications for how people make moral judgments about issues such as equity within and across generations. Only by investigating these broader processes of expert institutionalization and public understanding of science can we fully understand how the construction of new global governing arrangements is affecting the constitutional basis of global society.

Conclusion

Jasanoff's chapter reflects, in microcosm, the central argument of this book. Images of the earth, like general circulation models of the earth's atmosphere, have become embroiled in a much larger debate about how human beings should govern themselves—as a global society—in the face of persistent threats to the environment. No one has suggested that photographs of the earth in some way distort reality. Yet these images evoke very different responses, around the world, to the question of whether what we "see" in these photographs is the vision that should be foremost in our minds as we make the choices that will determine our common future.

Climate science has played, and will continue to play, a crucial part in the identification and analysis of global environmental change. In our rush to protect the earth, however, we cannot afford to ignore the broader, value-laden narratives about human agency and choice in which this science is unavoidably enmeshed. The issues for scientists, for scholars in science and technology studies, and for citizens alike regard not only how we can obtain reliable knowledge about the environment but also how we choose to frame that knowledge and to use it to build an environmentally and politically sustainable world for the twenty-first century and beyond.

2

Representing the Global Atmosphere: Computer Models, Data, and Knowledge about Climate Change

Paul N. Edwards

If the idea of a truly *global* environmental problem required a poster child, climate change would certainly top the list of candidates.

Political and scientific narratives of the last decade almost always frame climate change as a genuinely planetary risk. The UN Framework Convention on Climate Change (FCCC) defines the threatened climate system as "the totality of the atmosphere, hydrosphere, biosphere, and geosphere and their interactions" (United Nations 1992, Article 1). This way of framing the issue notes that the causes of climate change, such as fossil fuel combustion, deforestation, and rice and cattle agriculture, extend across the face of the planet. They are embedded in core sectors of modern economic systems and affect the daily activities of everyone on the planet. The global-risk perspective emphasizes that the projected consequences of climate change will implicate virtually all human communities and natural ecosystems. In the early twenty-first century, then, climate change can be put forward as the quintessential global environmental problem.

Yet when the human risks posed by rising atmospheric concentrations of carbon dioxide were first (re)discovered in the late 1950s and early 1960s, they were often understood primarily in local or regional, rather than global, terms. As with other environmental problems, such as earthquakes, urban smog, and drought, the consequences of climate change would, it was thought, affect some places more severely than others. In most places, on this view, people would simply adapt (as they always have) to such changes. Thus a 1966 report by the U.S. National Research Council (NRC) expressed complacency about climate change:

It is perhaps worth noting that, even in the more extreme estimates of the possible climatic consequences of increased atmospheric CO_2, the calculated temperature changes have been of the order of a few degrees, generally less than five or ten. From glacial-geologic data, it is known with some certainty that North America and Europe have, since the last maximum of the Wisconsin Glaciation, experienced climates that have averaged several degrees warmer than the present. As mentioned earlier, *although some of the natural climatic changes have had locally catastrophic effects, they did not stop the steady evolution of civilization.* (National Academy of Sciences 1966, emphasis added; see Miller, forthcoming, detailing the local framing of climate change and the subsequent transition to global framing)

Even two decades later, in its 1983 report *Changing Climate*, the NRC once again stressed that climate change not only could, but for pragmatic purposes *should*, be defined "flexibly" in local terms:

Viewed in terms of energy, global pollution, and worldwide environmental damage, the "CO_2 problem" appears intractable. Viewed as a problem of changes in local environmental factors—rainfall, river flow, sea level—the myriad of individual incremental problems take their place among the other stresses to which nations and individuals adapt. It is important to be flexible both in definition of the issue, which is really more climate change than CO_2, and in maintaining a variety of alternative options for response. (National Research Council 1983, 3)

Yet by the mid-1980s, most conceptions of climate change painted its risks almost exclusively in global terms.

How, and why, did scientific and political discourses about the human risks of climate change shift the emphasis from local to global concerns? To answer this question, this chapter will explore in detail the history of the competing, and contested, representations of climate through which the contemporary debate has been structured.

One clue comes from an important scientific paradigm shift. Historically, the science of climatology consisted primarily of record keeping and analysis of climate trends at a particular location (see Miller, forthcoming). Although a global conception of *weather* had developed by the early 1900s, climatology continued for decades along this rather separate, particularistic track. Even in the 1960s and 1970s, the primary scientific tools for representing climate remained the long-term statistical databases compiled by climatologists. But in the decade between about 1965 and 1975, the locally oriented climatologists were rapidly displaced by a new breed of global modelers. By the late 1970s, computer-based climate models—conceptually almost identical to the weather models in use since the 1950s, although used differently (see below)—had come to dominate climatological discourse. With this change, the very meaning of *climate* shifted from a local to a global understanding. The term *climatology* gradually fell from favor (although it is still used), replaced by the term *climate science*. This shift reflected the new model-based, globally oriented paradigm.

Today, climate models are essential not only for predicting future climates, but also for attributing the causes of climatic change in the recent past. Without a model of what would have happened *without* anthropogenic (human-caused) atmospheric change, scientists cannot separate out the effects of rising greenhouse gas concentrations from natural climatic variability. The inherent variability of weather makes it impossible to attribute individual storms, floods, droughts, or hurricanes to changes in the global climate. Only by coupling statistical analyses to climate modeling exercises have scientists been able to isolate and display the "fingerprint" of global warming in changing weather patterns around the world.

Because of its long-term, statistical character, even local climate change is difficult to grasp experientially. A few hot summers, an unusual spate of major storms, or even a decade-long drought can be elements of "normal," regional climate variation, rather than signals of long-term climate change. By the same token, *global* climate change cannot be grasped experientially at all. The most commonly cited figure in climate change debates—change in the average global temperature—has no correlate in anyone's actual living conditions. Thus, while public awareness and understanding of climate change have always depended on the work of the scientific community, they do so more now than ever before. Knowledge of climate change, in the contemporary sense of the term, comes *only* from science; in Bruno Latour's phrase, climate modeling has become an "obligatory passage point" for knowledge of climate change. Furthermore, knowledge about changes in the global climate system depends on very many sciences, from meteorology to oceanography to ecology. The theories, models, standards of evidence, and data sets of these various contributors are often quite different, not to mention contested. Modern climate science is a multidisciplinary and interdisciplinary field, rather than the specialty of statistical climatology.

The increasingly complex articulation between a science-based, descriptive understanding and normative climate politics—where the global climate is seen as a system in which political intervention could be both important and effective—has generated unprecedented interdisciplinary scientific collaboration and political coordination. These phenomena require historical explanation. At the same time, they present an opportunity to grapple with the central question of knowledge-power relationships at the interfaces between local, national, and global communities. How have scientists tied the wide-ranging strands of their work together? How have their efforts linked up with the formation of new transnational scientific and political communities? How have particular representations of climate linked multiple communities of researchers, government officials, and citizens? How have national efforts differed and conflicted, and how have these differences been handled in the international community? These are the kinds of questions to which a science studies theoretical approach may offer new and different answers.

This chapter begins to approach these questions, and also serves as an introduction to climate science concepts treated throughout this volume. To do so, it focuses on two key "boundary objects" in this enormous and enormously confusing arena: computer models of the atmosphere and global satellite data sets. Boundary objects are things, theories, symbols, or other entities used by multiple communities; although they may have different meanings and functions for each group, they provide conceptual and pragmatic links that bind the communities together (Star and Griesemer 1989). Computer models and global data sets play this role for many of the scientific and political groups focused on climate change.

Computer models are arguably the single most important tool of global climate science. They range in size from simple programs that can run on a desktop computer to ultracomplicated simulation models of the entire Earth system, which strain the capacities of even the most powerful supercomputers. Much of climate science could not exist without them, since planetary-scale processes cannot be studied by controlled laboratory experiments. Instead, climate science relies on global "experiments" performed on models to provide it with insights into the dynamics of the atmospheric system as a whole.

Satellite data—covering many facets of the atmosphere, and a smaller number of characteristics of the Earth's oceans, ice, snow, and land surfaces—are likewise central to contemporary scientific understandings of the entire planet. Although many other forms of data are collected on a worldwide basis (from surface stations, radar, weather balloons, ships, and so on), their coverage is far less uniform, less easily standardized, and less easily collected in a single location. The huge size of global data sets makes it impossible to process or understand them in any detail without the aid of computers (Edwards 1999). In fact, global data sets of the relevant density cannot even be collected without the aid of computerized interpolation models that mediate between raw instrument readings and usable data formats. I will return to this issue below.

I begin this chapter by sketching the history of atmospheric modeling and its relation to the development of global data networks. Second, I describe how modern climate models work and discuss some of the key problems faced by modelers. Third, I examine the extremely fuzzy boundaries between models and data in global climate science, and the major role of computer models in binding them into a coherent system of knowledge with a global, rather than a local or regional, basis. Finally, I explore some implications of the primary role given to computer models in representing the global atmosphere.

Climate Science: Concepts and Tools

Since most of the chapters in this book discuss climate modeling in one way or another, I will begin by describing briefly the scientific concepts on which the models are based and how these models work. I purposely ignore scientific controversies, since many of these are treated in detail by later chapters. In any case, at this level of generality there is little debate.

Scientific Principles

The principal sources of atmospheric science lie in various branches of physics: theories of the behavior of gases (pressure, temperature), the radiation absorption and emission characteristics of different gases, and turbulent fluid (gas) flows.

The earth is bathed in a constant flood of solar energy, all of which it ultimately reradiates into space. One key aspect of climate, the Earth's temperature, is therefore a matter of what climate scientists call *energy balance*: all the energy that goes into the system must, eventually, come out again. The atmosphere forms a blanket of gases (primarily nitrogen, oxygen, and water vapor) capable of absorbing and holding a great deal of this incoming energy as heat. The oceans, too, absorb and retain heat. They play a major role in the overall climate system, "damping" the system's response to change with their enormous heat-retention capacity (far larger than the atmosphere's). In theory, if the earth had no atmosphere its average surface temperature would be about −19°C. Instead, the heat retained in the atmosphere maintains it at the current global average of about 15°C.

Under the influence of solar heating and the earth's rotation, both the atmosphere and the oceans "circulate," carrying heat around the globe in currents of air and water. The *general circulation* refers to the motion and state of the entire atmosphere (or ocean); it is sometimes (more aptly) termed the *global* circulation. Ultimately, the circulation conducts heat from the equator, which receives the greatest amount of incoming energy from the sun, to the poles, where more heat is radiated into space than is received. Since weather moves freely around the globe and changes with considerable speed, only by modeling the general circulation can meteorologists hope to understand the evolution of weather over more than a couple of days.

The earth's wobble on its axis in relation to the sun, atmospheric and oceanic turbulence, and many other factors render circulation patterns highly complex. In the short term (hours to weeks), such patterns are experienced as weather: rain, dry spells, clouds, hurricanes. Long-term patterns (occurring over months to decades, and beyond) are known as *climate*, and include such phenomena as the seasons, with their regular annual changes in temperature and precipitation; prevailing regional climates (deserts, tropics, ice caps, and so forth); multiyear climatic variations (droughts, the El Niño/Southern Oscillation, and so on); and very long term climate changes such as ice ages.

Models

Climate models are mathematical simulations, based on physical principles, of these long-term atmospheric conditions. Although many discussions in this book (and in public debates) focus on the most complicated, supercomputer-based climate models, in fact there is a wide range of complexity, sometimes referred to by scientists as the *hierarchy of models*.

The simplest, "zero-dimensional" models rely solely on the principle of energy balance discussed above (and are called *energy-balance models*). Using measured values for such factors as solar radiation and concentrations of the atmosphere's constituent gases, they compute (for example) a single global average temperature, treating the earth as if it were a point mass. Models this basic may involve only a few equations and can readily be solved by hand. One- and two-dimensional energy-balance models also exist.[1] Another class of two-dimensional models, called *radiative-convective*, calculates the atmosphere's vertical temperature structure. In these models, temperature is computed as a function of latitude and either longitude (a second horizontal dimension) or air pressure (the vertical dimension). Many two-dimensional models remain relatively simple—typically, a few hundred to a few thousand lines of computer code—compared to the three-dimensional[2] models known as *atmospheric general circulation models* (GCMs, or AGCMs to distinguish them from OGCMs, which model the oceanic general circulation).

Contemporary atmospheric GCMs are typically expressed in some 30,000 to 60,000 lines of FORTRAN code. They represent the atmosphere as a three-dimensional lattice or "grid."[3] Typically, the grid resolution at the surface is 3°–5° latitude by 6°–8° longitude. (This translates roughly into squares or rectangles 300 to 500 km on a side.) Eight to twenty layers of varying depth represent the vertical dimension up to a height of 20 km or so, with more layers at lower altitudes, where the atmosphere is denser and most weather occurs. Equations of state compute the effect of various forces (radiation, convective heating, and so on) on the air masses and moisture (clouds and water vapor) within each grid box. Equations of motion compute the direction and speed of the air's movement into the surrounding grid boxes. AGCMs usually also include representations of certain aspects of the land surface, such as elevation and albedo (reflectance). In addition, they usually include some representation of the oceans, which may be as simple as a shallow, one-layer "swamp" ocean with fixed surface temperature. Today's most sophisticated models dynamically couple AGCMs to full-scale OGCMs of equivalent complexity. In addition, they may also include models of sea ice, snow cover, vegetation, agriculture, and other phenomena with important effects on climate; such models are sometimes known as *Earth systems models* (ESMs).

Such models demand enormous computational power. The first GCMs required twenty-four hours of computer time in order to simulate a single day of global circulation. By the mid-1970s, faster computers reduced the time to about twelve hours per simulated year. For a typical climate modeling run of twenty simulated years, a GCM still required as much as 240 hours—ten continuous days—of expensive supercomputer time. Although computer speeds continue to increase, these long run times have declined little since then. Modelers prefer, instead, to represent more variables, increase resolution, and carry out longer runs (Chervin 1990).

Modern weather forecasters also use GCMs. Weather forecasters use the highest possible model resolution, because their purpose is prediction and because their model runs are only a few days. Ideally, weather models must resolve relatively small-scale processes, such as the formation and motion of clouds. Grid cells as small as 60 km on a side are common in the best modern weather GCMs. They are initialized with observational data, such as temperature, humidity, and wind speed, from a wide range of sources, including surface weather stations, satellites, and radiosondes (weather balloons). The models then calculate the likely evolution of this observed initial state over short periods (hours to days).

Climate modelers, by contrast, use *coarse-grid* GCMs that cannot simulate clouds and many other atmospheric processes directly; climate scientists refer to such phenomena as *sub-grid-scale* processes. This necessitates *parameterization*, or representation of small-scale events by large-scale variables (Hack 1992; Kiehl 1992). In addition, when used for climate research, GCMs generally are not initialized with observational data. Instead, GCMs are autonomous simulations that generate their own "climates," starting—in principle, if not in practice—with only a few empirically derived inputs such as solar radiation, gas composition of the atmosphere, sea surface temperatures, and orbital precession. These simulated climates must be run until they stabilize, then compared with observed long-term trends and phenomena in the earth's climate system. This requires multidecade, even millennia-long time series (Manabe and Stouffer 1994, 1979, 1994; Wigley, Pearman, and Kelly 1992). These features of climate modeling lead to uncertainties and epistemological issues discussed by several contributors to this volume. I will return to them below, after sketching the place of general circulation modeling in the history of climate science and politics.

Computer Models and Global Data Networks

The concept of anthropogenic climate change has surprisingly deep historical roots, reaching back over 100 years. In this section, I discuss the history of climate modeling and its role in the emergence of climate change as a political issue.

Early Theories of Climatic Change

Scientific theories of climate change date to the mid-nineteenth century. In 1824, Jean-Baptiste Joseph Fourier hypothesized that the atmosphere retains heat, keeping the earth's surface temperature far higher than it would be if the earth had no atmosphere or if the atmosphere contained no water vapor or carbon dioxide (CO_2). Fourier likened the heating action of the atmosphere to a "hothouse," thus christening what we now call the *greenhouse effect*.

Physicist John Tyndall, in Great Britain, first calculated the radiative potential of CO_2 in 1863. His result paved the way for the Swedish scientist Svante Arrhenius, in 1896, to make the first calculation of the contribution of carbon dioxide to the earth's surface temperature. In 1900, another British scientist, T. C. Chamberlin, published a sweeping theory of climatic change over geological time scales, with CO_2 as the basic mechanism. Chamberlin argued that volcanic eruptions produce CO_2, warming the earth; the weathering of rocks absorbs the gas, accounting for glacial cycles. Chamberlin's idea that CO_2 was the *determining* factor in global climate change is no longer current. Orbital cycles, water vapor, ocean currents, and other factors are today believed to be more important. Yet Chamberlin did identify one of the basic mechanisms.

In 1903 Arrhenius went on to calculate that the CO_2 added to the atmosphere by human combustion of fossil fuels might eventually raise the earth's temperature substantially. In fact, he predicted that if the amount of CO_2 in the atmosphere were to double, the global average temperature would rise somewhere between 1.5° and 4°C. Even the most sophisticated modern climate models run on supercomputers still predict approximately this same range of probable change.

Arrhenius, perhaps since he lived in a very cold place, thought that such a global warming might be a good thing. But in his time, world consumption of fossil fuels remained low enough that this seemed merely an idle speculation, not a near-term possibility. Tellingly, Arrhenius also computed the effect of *decreasing* CO_2 on the atmosphere. He noted that natural factors might also produce such a change.

Although these scientists never thought of climate change as a political concern, in fact concern about anthropogenic climate change long predates the modern concept of greenhouse warming. James R. Fleming, Nico Stehr, Hans von Storch, and Moritz Flügel have uncovered numerous historical episodes of attribution of climatic changes to human causes. For example, in the Middle Ages climate anomalies were sometimes explained by the church as a divine response to human sin (Stehr, von Storch, and Flügel 1995), while Thomas Jefferson apparently believed that clearing land for agriculture altered the climate of the early United States in favorable ways (Fleming 1998). Similarly, Richard Grove (1997) has argued that some nineteenth-century colonial forest policies were predicated on a "dessicationist" theory of relations between deforestation and local, regional, and even continental climate change. According to Grove, these policies are the direct ancestors of some modern forest-conservation agendas. Grove also shows that colonial meteorologists in India and Australia developed early theories of global "teleconnections" (long-distance interactions of ocean currents and weather patterns, such as El Niño) by observing coincidence between Indian and Australian droughts.

The episode that perhaps most nearly resembles modern climate politics occurred in the 1890s, when German scientists Edward

Brückner and Julius Hann argued (separately) that human-induced climatic warming was already occurring in Europe and America as a result of deforestation and other causes. They based their speculations on trends derived from observations, rather than from physical theories of the atmosphere. Brückner thought that crop failures, economic crises, and epidemics would result, and argued that countries should undertake vast reforestation programs to counteract the trend. Prussia, Italy, and France established government reforestation committees as a result, while scientific societies debated the issue in the United States. In the end the issue fizzled, but it stands as a remarkable precursor to modern climate politics (Stehr and von Storch 2000).

Numerical Weather Prediction

Through most of this century, the history of climate politics is intimately linked with the history of numerical models of the atmosphere, initially created to forecast weather.

Toward the end of the nineteenth century, meteorology began to build theoretical foundations. By the early 1900s, the Norwegian meteorologist Vilhelm Bjerknes could argue that atmospheric physics had advanced sufficiently to allow weather to be forecast using calculations. He developed a set of seven differential equations, derived from basic physics, whose simultaneous solution would predict the large-scale movements of the atmosphere. Today these are known as the *primitive equations*.

Bjerknes proposed a *graphical calculus*, based on weather maps, for solving the equations. This analog technique made no attempt to treat the problem numerically, a feat far beyond the capacities of human or mechanical computers of the day. Although forecasters continued to use and develop his methods until the 1950s, both the lack of faster calculating methods and the dearth of accurate observational data limited their success (Nebeker 1995).

Richardson's "Forecast Factory" In 1922, the English mathematician Lewis Fry Richardson developed the first numerical weather prediction (NWP) system. His calculating techniques—finite difference solutions of differential equations in a gridded space—were the same ones employed by the first generations of GCM builders. Richardson's method, based on simplified versions of Bjerknes's equations, reduced the necessary calculations to a level where manual solution could be contemplated. Still, the task remained fantastically large. His own attempt to calculate weather for a single eight-hour period took six weeks and ended in failure.

His model's enormous calculation requirements led Richardson to propose a fanciful solution he called the *forecast-factory*. The "factory"—really more like a vast orchestral performance—would have filled a huge stadium with 64,000 people. Each one, armed with a mechanical calculator, would perform one element of the calculation. A leader, stationed in the center, would coordinate the forecast using colored signal lights and a telegraph system. Yet even with this impossible apparatus, Richardson thought he would probably be able to calculate weather only about as fast as it actually happens (Richardson 1922). Only in the 1940s, when digital computers made possible automatic calculation on an unprecedented scale, did Richardson's technique become practical.

Operational Computer Forecasting   The Princeton mathematician John von Neumann was among the earliest computer pioneers. Engaged in computer simulations of nuclear weapons explosions, he immediately saw parallels to weather prediction. (Both are nonlinear problems in fluid dynamics.) In 1946, von Neumann began to advocate the application of computers to weather prediction (Aspray 1990). As a committed opponent of Soviet-bloc communism and a key member of the World War II–era national security establishment, von Neumann hoped that weather modeling might lead to weather control. This, he believed, might be used as a weapon of war. Soviet harvests, for example, might be ruined by a U.S.-induced drought (Kwa 1994; Kwa, chapter 5, this volume; von Neumann 1955). On this basis, von Neumann sold weather research to military funding agencies.

Under grants from the Weather Bureau, the Navy, and the Air Force, he assembled a group of theoretical meteorologists at Princeton's Institute for Advanced Study (IAS). If regional weather prediction proved feasible, von Neumann planned to move on to the extremely ambitious problem of simulating the entire atmosphere. This, in turn, would allow the modeling of climate. Jule Charney, an energetic, visionary young meteorologist, was invited to head the new Meteorology Group.

The Meteorology Project ran its first computerized weather forecast on the ENIAC in 1950. Although not identical to Richardson's, the group's model followed his in representing the atmosphere as a grid, calculating changes on a regular time step, and employing finite difference methods to solve differential equations numerically. The 1950 forecasts, covering North America and part of the surrounding ocean, used a two-dimensional grid with 270 points about 700 km apart. The time step was three hours. Results, while far from perfect, were good enough to justify further work (Charney, Fjörtoft, and von Neumann 1950; Platzman 1979). Anticipating future success, Charney and his colleagues convinced the Weather Bureau, the Air Force, and the Navy to establish a Joint Numerical Weather Prediction (JNWP) Unit in 1954.

In December of that year, an independent effort at the Royal Swedish Air Force Weather Service became first in the world to use computer models for routine real-time weather forecasting (i.e., with broadcast of forecasts in advance of weather), using a model developed at the University of Stockholm (Bergthorsson et al. 1955; University of Stockholm Institute of Meteorology 1954). Routine computer forecasting began in the United States in mid-1955 (Nebeker 1995).

General Circulation Modeling
As late as the 1970s, the weather models used by forecasters were still regional or continental (vs. hemispherical or global) in scale, and they made no attempt to look ahead further than a few days. Calculations for numerical weather forecasts were limited to what could be accomplished in a couple of hours on then-primitive digital computers.

Yet for theoretical meteorologists, more interested in causal patterns than in real-time forecasting, general circulation modeling rapidly became a kind of holy grail. By mid-1955 Norman Phillips had completed a two-layer computer model of the general circulation (Phillips 1956). Despite its primitive nature, Phillips's model is now often regarded as the first working GCM. Like other early GCMs, this model employed major simplifying assumptions, modifying the equations to reduce the number of variables and calculations. As computer power grew, the need for simplifying assumptions diminished (although as we will see, it has hardly vanished). Between the late 1950s and the early 1960s, a number of separate groups in the United States, England, and Germany began—more or less independently—to build many-leveled, three-dimensional GCMs based on the full Bjerknes/Richardson primitive equations.

The General Circulation Research Section of the U.S. Weather Bureau became the first laboratory to build a continuing program in general circulation modeling. It opened in 1955, under the direction of Joseph Smagorinsky. Smagorinsky felt that his charge was to continue with the final step of the von Neumann/Charney computer modeling program: a three-dimensional, global, primitive-equation general circulation model of the atmosphere (Smagorinsky 1983). The lab—renamed the Geophysical Fluid Dynamics Laboratory (GFDL) in 1963—moved to Princeton University in 1968, where it remains.

Beginning in 1959, Smagorinsky and his colleagues, especially the Japanese émigré Syukuro Manabe, proceeded to develop a nine-level primitive-equation GCM, still hemispheric (Manabe 1967; Manabe, Smagorinsky, and Strickler 1965; Smagorinsky 1963). Other important general circulation modeling groups formed during the 1960s at UCLA, the U.S. National Center for Atmospheric Research, the United Kingdom Meteorological Office, and elsewhere.

By the end of the decade, general circulation modeling was firmly established as a basic research tool of meteorology and climate science. The European Center for Medium Range Weather Forecasts (ECMWF), founded in 1975, was the first to employ a *global* general circulation model in operational weather forecasting, beginning in 1979. Today, most weather forecasting centers around the world get global and regional weather data from the ECMWF and a few other large centers; they may use these data as is, or plug them into their own regional models. GCMs have not replaced simpler models altogether; in fact, many climate modelers play off one- and two-dimensional models against GCMs in the course of their research. But GCMs' sophistication and their apparent realism—as the highest-resolution theoretical representations presently available of the general circulation—have earned them a possibly inordinate prestige. Perhaps ironically, this is somewhat less true in the modeling community than among the "consumers" of model results in politics, the media, nongovernmental organizations—and academic communities, such as the one responsible for the book you are presently reading—working on climate science and climate change.

Data for Numerical Models

Modeling is, of course, only one part of the story of meteorology and climate science. The collection and processing of weather data (and its derivative, climate data) is the other. Once meteorology had provided theories of the general circulation, scientists confronted the problem of acquiring data commensurate with the models' needs and capabilities. As computers became the tools of choice for this purpose, this task evolved in unprecedented ways that fundamentally transformed the atmospheric sciences.

As Miller and Edwards note in chapter 1 of this volume, international agreements to share weather observations date to the 1878 founding of the International Meteorological Organization (IMO). Long before World War II, standard coding systems had been worked out to facilitate transfer of this information. In 1951, under the United Nations, the IMO became the World Meteorological Organization (WMO). Both organizations developed and promoted data distribution systems and standardized observational techniques.

Computerized weather prediction brought vastly intensified needs for data and for ways to handle those data. Weather models required information about the state of the atmosphere from ground level to very high altitudes. Ideally, data would be collected from locations as near as possible to the points on the models' three-dimensional grids. But in the 1950s, such data were simply unavailable. Observation stations, concentrated in urban regions, provided only very scattered and irregularly spaced coverage of the world's oceans and sparsely populated land areas.

In addition, the mere acquisition of weather information was only one step toward real-time numerical forecasting. The computer programs were useless without well-structured, reliable data *in digital form*. Most weather data, at that time, were collected by analog instruments (e.g., mercury thermometers or barometers). A human instrument reader converted them into numbers (digits) and charted them on maps, interpolating intermediate values, by eye or with simple calculating aids. The time delays inherent in this analog-digital data conversion were magnified by delays in long-distance communication. All of this imposed limitations on the scale of operational weather forecasting and left the global general circulation out of reach of predictive models.

In 1954 "gathering, plotting, analyzing and feeding the necessary information for a 24-hour forecast into a computer [took] between 10 and 12 hours," with another hour required for computation ("Long-Range Weather Forecasts by Computer," 1954). Well into the 1960s, even in the industrialized world, much weather data was hand-recorded and hand-processed before being entered into computers (Collins 1969). Data distributed in potentially machine-readable form, such as teletype, often arrived in a Babel of different formats, necessitating conversions (World Meteorological Organization 1962). Much of the available data was never used, since the time required to code it for the computer would have delayed forecasts beyond their short useful lifetimes.

Meteorologists had always engaged in "smoothing" data (eliminating anomalous data points). Another standard practice was the interpolation of intermediate values from known ones. To feed the grids of computerized weather models, this activity became a central element of meteorological work. By the 1960s, it was being automated. The methods did not really change, but their automation required explicit, computer-programmable theories of error, anomaly, and interpolation. The effect was simultaneously to render this data "massage" invisible (Filippov 1969).

By the early 1960s, atmospheric scientists realized that the core issue of their discipline had been turned on its head by the computer. In the very recent past, through the data networks built for World War II, they had acquired far more data than they could ever hope to use. But now, already, they did not have enough of it—at least not in the right formats (standard, computer processable), from the right places (uniform grid points), and at the right times (on the uniform time steps used by the models). The computer, which had created the possibility of NWP, now also became a tool for refining, correcting, and shaping data to fit the models' needs.

Global Data Networks and Anthropogenic Climate Change

The International Geophysical Year (IGY), a UN-sponsored program of global cooperative experiments to learn about the earth's physical systems, including the atmosphere and the oceans (see Miller, chapter 6, this volume), began in 1957. As it opened, the theory of carbon dioxide–induced global warming was finding renewed scientific attention. Suess had already concluded that fossil fuel combustion was producing so much carbon that some of it remained in the atmosphere, causing a continual rise in the atmospheric concentration of carbon dioxide (Suess 1953). Plass aroused new interest in carbon dioxide as a factor in climate change (Plass 1956).

Suess and Revelle predicted that fossil fuels might soon induce rapid changes in world climate. They wrote, in 1957, that humanity was conducting, unawares, "a great geophysical experiment" on the Earth's climate (Revelle and Suess 1957). To track the "experiment's" progress, Revelle proposed a monitoring station for atmospheric CO_2 at Mauna Loa, Hawaii, as part of the IGY. The Mauna Loa station, along with another station in Antarctica, has since documented a steady annual rise in the atmospheric concentration of CO_2, due primarily to human activities.

The IGY's meteorological component focused most of its attention on the global general circulation problem. Three pole-to-pole chains of atmospheric observing stations were established along the meridians 10°E (Europe/Africa), 70°–80°W (the Americas), and 140°W (Japan/Australia). Dividing the globe roughly into thirds, these stations coordinated their observations to collect data simultaneously on specially designated "Regular World Days" and "World Meteorological Intervals." An atmospheric rocketry program retrieved information from very high altitudes. Data from all aspects of the IGY were deposited at three World Data Centres, major repositories of climatological information to this day (Comité Spécial de l'Année Géophysique Internationale 1958; Jones 1959).

The IGY efforts represent the first global data networks conceived on a scale to match the developing atmospheric models (see Miller, chapter 6, this volume). Nevertheless, even these covered the Southern Hemisphere only sparsely. They marked the start of a trend toward global programs such as the World Weather Watch (WWW) and the Global Atmospheric Research Program (GARP). WWW, coordinated by the WMO, eventually linked global data collection from satellites, rockets, buoys, radiosondes, and commercial aircraft as well as conventional observing stations. This was necessary, according to one participant, because "currently conventional methods . . . will never be sufficient if the state of the atmosphere over the whole globe is to be observed at reasonable cost with the time and space resolution which can be used with advantage in computer-assisted research and forecasting" (Robinson 1967, 410). GARP's roots lay in a 1961 U.S. proposal for "further co-operative efforts between all nations in weather prediction and eventually in weather control," backed by President John F. Kennedy (Robinson 1967, 409). Among GARP's chief organizers was Joseph Smagorinsky, founder of GFDL and builder of the first primitive-equation GCM. With the participation of scientists from the United States, the Soviet Union, and many other nations, GARP eventually sponsored a series of regional and global observations and experiments. At the height of the Cold War, the IGY and these successors marked a new era of international cooperation in meteorology.

Despite the interesting links between climate change concerns, the IGY, and the construction of global digital data networks, the concrete issue of climate change had little, if any, effect on data networks until the 1970s. Instead, the early impetus came from the desire of both weather forecasters and theoretical meteorologists to achieve better coverage of the globe. The former were motivated by practical concerns of forecasting. But the latter hoped to achieve a grander goal: modeling the detailed dynamics of the global atmosphere.

By 1969, the WMO had called for extending the global atmospheric data network to monitor pollutants that might change the climate, such as CO_2 and particulate aerosols. Anthropogenic climate change began to become a public policy issue within the U.S. government (and, soon afterward, in some other industrialized nations; see the seven country studies in Edwards and Miller, forthcoming). However, it was almost two decades before global warming became a genuine public concern at the level of mass politics. Meanwhile, attempts continued to build models capable of "realistically" representing the global climate, and to construct data sets accurate enough to distinguish "signals" of long-term climatic change from the "noise" of natural climatic variation. These goals were tightly intertwined. Without global data sets, modelers could neither validate nor parameterize their models. Without computers and satellites, uniformly gridded global data sets could not even be created, much less manipulated. Without NWP models and GCMs, these data could not be understood.

The UN Conference on the Human Environment, held in Stockholm in 1972, approved plans for an extended global data network "with little discussion" (Hart and Victor 1993). During the rest of the 1970s, increasingly sophisticated data collection and processing networks developed in tandem with GCMs and weather models at NCAR, ECMWF, the U.S. National Meteorological Center, and numerous other locations (virtually all in the industrialized world).

Models and Global Atmospheric Politics Scientific concerns about climate change allowed climate science to ride the 1970s political wave of environmentalism. By no means was this a cynical attempt at self-promotion; for many, it represented genuine alarm aroused by significant scientific results. Nevertheless, these worries provided practical reasons for accelerating expensive research programs. But in the 1970s, the immaturity of climate science and climate models made consensus on anthropogenic global warming impossible. Climate theorists did not agree on the relative roles of such factors as solar variability, sunspots, and cloud feedbacks in climate change; global *cooling*, some argued, was also a possibility (Edwards and Lahsen, forthcoming). Chaotic qualities of the climate system added to the uncertainty over model results.

Nevertheless, by the late 1970s climate change had begun to generate an ever-broadening circle of scientific and policy concerns. At the First World Climate Conference in 1979, WMO scientists established the World Climate Programme to coordinate and develop climate research and climate data. The "nuclear winter" issue of the early 1980s and the Antarctic "ozone hole" discovered in 1986 were the first events to elevate the general issue of anthropogenic atmospheric change to the level of front-page news (Edwards and Lahsen, forthcoming; Morrisette 1989; Schneider 1989). Both issues created an awareness that human actions were capable of causing sudden, potentially catastrophic changes in the atmosphere not just regionally, but on a global scale. The Cold War fizzled to a close, leaving a sort of "apocalypse gap" in popular political consciousness, which was readily filled by global warming scenarios. As the basis of key scientific results, computer models played substantial—even decisive—roles in the nuclear winter and ozone depletion issues as well as in climate change.

In 1985, at Villach, Austria, an influential climate-science conference recommended policy studies of climate change mitigation techniques, including international treaties. Ozone depletion concerns resulted in a near-comprehensive international ban on chlorofluorocarbons, completed in 1990 after just five years of negotiation. In 1988, the WMO and the UN Environment Program formed the Intergovernmental Panel on Climate Change (IPCC). The group, consisting of experts on climate, ecology, and environmental and social impacts from around the world, was asked to serve as the scientific adviser for international climate negotiations. In 1990 the organization released its first report, designed as input to the Second World Climate Conference held later that year. The report noted a qualified consensus on two points. First, greenhouse gas concentrations were rising rapidly due to human activities. Second, if this trend continued, global average temperatures were likely to rise somewhere between 1.5° and 4°C by about 2050 AD.

The IPCC played a crucial role in the 1992 UN Conference on Environment and Development (UNCED). The UNCED produced the landmark FCCC. Signed by 165 nations, the Framework Convention entered into force in early 1994. It sets voluntary goals for stabilizing greenhouse gas emissions. More important, the FCCC requires signatories to prepare national greenhouse gas emission inventories and commits them to ongoing negotiations toward an international treaty on climate change (Bodansky 1993). The IPCC continues to provide scientific input to the periodic Conferences of Parties to the FCCC. An era of global atmospheric politics had dawned, with computer models at its very core (Bodansky 1994; Donoghue 1994).

Associated with the political arrival of the climate change issue was a trend toward ever more comprehensive global models, from two directions. The first, *Earth system* or *Earth systems* models (ESMs), was a direct extension of natural-science efforts to couple oceanic and atmospheric general circulation models. The goal is to couple models of other climate-related systems (land surface, sea ice, and so on) to an

OAGCM, eventually capturing all of the major elements of the total climate system—including anthropogenic effects such as agriculture and artificial greenhouse gas release (Schneider 1994; Trenberth 1992; Turner, Moss, and Skole 1993). In general, sophisticated models of human socioeconomic activities have been last in line for integration into ESMs, which focus most of their effort on natural systems. Most of these models descend from existing GCM efforts. The second type, *integrated assessment models* (IAMs), aims to simulate the impacts of climate change on human society, and the costs and benefits of possible mitigations (Alcamo 1994; Dowlatabadi and Morgan 1993; Hope 1993; Rotmans 1990, 1992). IAMs typically do not incorporate GCMs directly. Instead, they rely either on selected and aggregated GCM outputs or on much simpler energy-balance climate models. Their purpose is to allow rough, rapid analysis of the possible effects of various politicoeconomic scenarios on climate change. IAM developers generally spend much more of their energy than climate system modelers do on the social, political, and economic elements of their models, relying for the natural-systems side on outputs from other efforts based in the natural sciences.

IAMs incorporate empirically derived trends, heuristics, and unproven or qualitative theories into their modeling techniques far more freely than do climate and Earth system models. Their goal is comparison of policy scenarios and forecasting of trends, not prediction at statistically significant levels; this is the point of the term *assessment*. Not all IAM outputs are global in scope—for example, the first IMAGE model focused primarily on the Netherlands (Rotmans 1990). Many IAM builders hope that their models—unlike the hypercomplex, supercomputer-based ESMs—will be simple, transparent, and portable enough that policymakers, or perhaps their staffers or administrative agencies, can engage with the models directly. If so, they could observe for themselves, on a desktop computer, the differential effects of various politicoeconomic scenarios, such as carbon taxes, population stabilization, or reforestation efforts, on global change. The idea is to offer policymakers an effective way to learn a set of *heuristics*—a quasi-intuitive "feel" or rule of thumb based on, yet not fully determined by, data-driven analysis—for global change policy options. Table 2.1 compares a first-generation

Table 2.1
Comparison of the GENESIS Earth system model and the IMAGE integrated assessment model

GENESIS	IMAGE
• Origin: previous NCAR climate models	• Origin: Ph.D. thesis
• Orientation: natural/physical science	• Orientation: policy analysis tool
• Based on sophisticated, high-resolution atmospheric GCM; other models added later	• Simple, one- or two-dimensional atmospheric models; modular; built as integrated unit
• Parameterization: moderate, relatively model-specific	• Parameterization: extreme, literature-based
• Approach to terrestrial biosphere: potential vegetation, to be modified later by agriculture model	• Approach to terrestrial biosphere: actual land use; mosaic of natural and human-altered landscapes
• Typical experiments: regional climate change due to CO_2 doubling; paleoclimate	• Typical experiments: impacts of IPCC scenarios on Dutch coastal defenses; emissions scenarios
• Technology: supercomputers	• Technology: PCs
• Architecture: partially modular, model-specific, but publicly available and user-manipulable within limits	• Architecture: highly modular, open, links easily with other models, user-manipulable
• Institutional context: Interdisciplinary Climate Systems Group, Climate and Global Dynamics Division, NCAR, USA	• Institutional context: National Institute for Public Health and Environmental Protection (RIVM), Netherlands
• Audience/accessibility: climate science community	• Audience/accessibility: climate science community, terrestrial ecosystems community, policymakers, educators
• Funding: EPA; NSF	• Funding: RIVM, European Economic Community

Earth system model (GENESIS) with a first-generation integrated assessment model (IMAGE).

ESMs and IAMs have become focal points in a relatively new, very broad effort to integrate results and methods from many different sciences. Social, behavioral, economic, and policy sciences are part of this mix, albeit more so in IAMs than in ESMs. Doing this kind of modeling means that each discipline—often extending to members of policy communities, such as regulatory agencies—must ultimately embody its data and principles in computer code that can "talk" to the model's other modules (i.e., perform "intermodel handoffs"). Thus IAMs and ESMs are increasingly the foci of an emerging *epistemic community*.

This is Peter Haas's term for a knowledge-based professional group that shares a set of beliefs about cause-and-effect relationships and a set of practices for testing and confirming them. Crucially, an epistemic community also shares a set of values and an interpretive framework; these guide the group in drawing policy conclusions from their knowledge. Its ability to stake an authoritative claim to knowledge is what gives an epistemic community its power (Haas 1990a, 55–63; 1990b). In the arena of global-change science, where wholly empirical methods are infeasible, computer modeling has become *the* central practice for evaluating truth claims. It lies at the center of the epistemic community of global change science. Other, roughly equivalent ways of describing the fundamental role of models in climate science/policy communities would be as supports for climate-change discourse (Edwards 1996a) and as boundary objects in a knowledge exchange system (Star and Griesemer 1989).

Whether or not they are ever used directly by policymakers, ESMs and IAMs in fact contribute substantially to the basis of global change politics, in the important sense that they serve as a central organizing principle for a large, growing, epistemologically coherent community. This community shares the crucial belief that *global* natural systems may be significantly affected by human activities—a belief to which few would have subscribed three decades ago. It also, in general, shares the values that such systems are worth preserving and that rational political decision making can be achieved, at least to some degree, which could preserve them (Jasanoff and Wynne 1998). Integrated model building contributes directly to this base of common assumptions, to a scientific macroparadigm that accepts computer simulation as a substitute for (infeasible) traditional forms of experimentation, and to a network of individuals, laboratories, and institutions such as the U.S. Global Change Research Program and the IPCC. The models help to create a public space, including shared knowledge, shared values, and access to common tools and data, for consensus building on global change issues. In this very important, entirely nonpejorative sense, comprehensive model building—as the core representation of the global atmosphere—is simultaneously scientific and political (Edwards 1996b; Jasanoff 1990).

Techniques and Problems of "Global" Representation

As climate change became a major public issue in the last decade, an acrimonious debate about the relationship between models and data moved from the scientific arena into the mass media. IPCC "consensus" opinion met intense opposition from a small but vocal "contrarian" group, especially in the United States. These skeptics raised many objections to models, from poor parameterization of cloud feedbacks to differences between the observed warming to date (about 0.5°C) and model calculations showing a 1°C warming for the same period (White 1990). Recently, GCMs incorporating particulate aerosol effects have aligned more closely with observations; the IPCC's most recent report states that "the balance of evidence suggests a discernible human influence on global climate" (Houghton et al. 1996, 5). But despite the appearance of scientific consensus and moves toward a binding emissions-limitation treaty at the international level, debate continues into the present, with skeptics arguing that models cannot be trusted without higher levels of observational confirmation (Edwards and Lahsen, forthcoming).

These debates are simultaneously scientific, political, and epistemological. They go to the heart of the question of what we know about the world and how we can know it. At the same time, by projecting the extent and impacts of climatic change in the future, and by locating it in data about the past, they set the stage for policy choices. In this section, I explore some of the epistemological issues behind these debates. While I will focus on GCMs, my conclusions apply equally to ESMs, IAMs, and other kinds of models as well. The surprising upshot of this discussion is that *all* important knowledge and choice about climate change depends fundamentally on modeling.

Model Resolution and the Computational Bottleneck

GCMs recompute the state of the entire atmosphere every fifteen to thirty simulated minutes. This process is extremely computationally intensive. At each time step, hundreds to thousands of complex calculations must be performed on each of the tens of thousands of grid boxes. This consumes vast quantities of supercomputer time; a typical state-of-the-art GCM currently requires tens to hundreds of hours for a full-length "run" of twenty to a hundred simulated years. In principle, climate modelers could achieve far better results with high-resolution NWP models. But the number of model calculations increases exponentially with higher resolutions. This creates a computational bottleneck, forcing GCM builders to make trade-offs between a model's resolution and its complexity.

"Complexity" here refers to two related things: the number of phenomena simulated, and the level of detail at which they are modeled. Existing models do not directly simulate a vast number of basic atmospheric events. The most important of these is the formation of clouds, which form typically on scales of a few kilometers or less. Clouds are believed to play many key roles in climate, such as trapping heat at night or reflecting it back into space during the day. These phenomena are notoriously difficult to study empirically, and their role in climate remains controversial. Clouds are not yet perfectly modeled even with NWP techniques. Other phenomena not well captured at GCM resolutions are the activity of the planetary boundary layer (the layer of air nearest the earth's surface) and many factors relating to the land surface, such as its roughness and elevation. (For example, many current models represent the entire region between the Sierra Nevada range in California and the Rocky Mountains as a single plateau of uniform elevation.)

Their low resolution is one reason for the high levels of uncertainty surrounding climate models. Techniques for getting the most out of these low-resolution models have improved them, but have also been intensely controversial. The next section reviews some of these techniques and the associated problems and controversies.

Parameterization and Tuning

Most of the major issues in climate modeling stem from the problem of scale described above. All sub-grid-scale processes must be represented parametrically, or *parameterized*. For example, rather than represent cloud formation in terms of convection columns, cloud condensation nuclei, and other direct causes, a GCM typically calculates the amount of cloud cover within a grid box as some function of temperature and humidity. This approach embodies what is known as the *closure assumption*. This is the postulate that small-scale processes can ultimately be represented accurately *in terms of the large-scale variables available to the models.*

Parameterization is controversial, and its effects on the activity of models are not entirely encouraging (Shackley et al. 1998). For example, some cloud parameterization schemes in early GCMs resulted in cloud "blinking," an oscillation between the presence and absence of cloud cover in a given grid box at each time step when certain variables happened to be just at the critical threshold. Real clouds do not, of course, behave like this. The question is whether and how unrealistic behavior of this sort in one element of the model affects the quality of overall model results.

Another example of a parameterized function is atmospheric absorption of solar radiation, the energy driver for the entire climate system. Atmospheric molecules absorb solar energy at particular frequencies known as spectrographic "lines."

The contribution of each narrow absorption line must be accounted for to model the transfer of radiation. . . . There are tens of thousands of such lines arising from all the absorbing gases in the atmosphere. Thus, to include all lines in a parameter of absorption would require an explicit summing over all lines at each model level and horizontal location. These types of calculations can be performed on present day supercomputers and are called line-by-line models. (Kiehl 1992, 338)

But such modeling is too computationally expensive. Instead, absorption is represented in GCMs by coefficients that implicitly integrate all the absorption lines.

In an ideal model, the only fixed conditions would be the distribution and altitude of continental surfaces. All other variables, such as sea surface temperature, land surface albedo (reflectance), cloud formation, and so on would be generated internally by the model itself from the lower-level physical properties of air, water, and other basic constituents of the climate system. To say that current GCMs are far from reaching this goal is a vast understatement. Instead, "virtually all physical processes operating in the atmosphere require parameterization" in models (Kiehl 1992, 336). Generating these parameters is therefore the largest part of the modeler's work.

Climate modelers do this partly by reviewing the meteorological literature and observational data to try to determine how small-scale processes and large-scale variables might be related. When they succeed in finding such relations, they call the resulting parameters *physically based*. Often, however, they do not find direct links to large-scale physical variables. In this common case, modelers invent ad hoc schemes that provide the models with the necessary connections. For example, one method of cloud parameterization represents all the cumulus clouds in a given region as a single "bulk" cloud (Yanai, Esbensen, and Chu 1973). In addition, observed patterns exist that can be mathematically described, but whose physics are not understood. These, too, are represented in the models as parameters.

Another, very important part of modelers' work is known as *tuning* the parameters. *Tuning* means adjusting the values of coefficients and even, sometimes, reconstructing equations to produce a better overall model result. "Better" may mean that the result agrees more closely with observations, or that it more closely corresponds with the modeler's judgment about what one modeler called the *physical plausibility* of the change. In some cases parameters fit relatively well with observed data. In others—as in the case of cloud parameterizations—the connection is so uncertain that tuning is *required*. Such parameters are said to be "highly tunable." Since many parameters interact with others, tuning is a complex process. Changing a coefficient in one parameter may push the behavior of others outside an acceptable range.

Flux Adjustment

Today's most sophisticated climate models couple atmospheric general circulation models with general circulation models of the oceans. The latter operate on principles much like those of atmospheric GCMs. These "coupled" models, known as OAGCMs, must somehow provide for the exchanges or "fluxes" of heat, momentum (wind and surface resistance), and water (precipitation, evaporation) between the ocean and the atmosphere. Empirical knowledge of these fluxes is not very good, but their values have profound effects on model behavior.

Most OAGCMs include ad hoc terms, known as *flux adjustments*, that modify and correct the overall model results to bring them more closely into line with observations. Without them, the models' climates drift out of line with observed values and patterns (Meehl 1992). These adjustments are "nonphysical" model terms, in modelers' language, although they are also characterized as "empirically determined" (Houghton et al. 1995, 237, 34); they are an excellent example of "highly tunable" parameters. Recently the National Center for Atmospheric Research introduced the first OAGCM that does not require flux adjustments (Kerr 1997).

Parameterization and tuning are, in effect, scientific art forms whose connection to physical theory and observational data varies widely. As one modeler told me in a confidential interview,

Sure, all the time you find things that you realize are ambiguous or at least arguable, and you arbitrarily change them. I've actually put in arguable things, and you do that all the time. You just can't afford to model all processes at the level of detail where there'd be no argument. So you have to parameterize, and lump in the whole result as a crude parameter.

Common Code: GCMs as a Family

One final issue about GCMs concerns their relationships with each other. Because of their complexity and expense, the total number of atmospheric GCMs is not large—probably fewer than fifty worldwide. Many of these models share a common heritage (Edwards 2000). Typically, one modeling group "borrows" another group's model and modifies it. This avoids unnecessary replication of effort, but it also means that the "new" models may retain problematic elements of those from which they were created. Several modelers told me that substantial segments of the computer code in modern GCMs remain unchanged from the original models of the 1960s. This may be one reason for the fact that some systematic errors in GCMs are common to virtually all extant models (Boer 1992; World Meteorological Organization 1991).

Data-Laden Models

Simulation models are typically described as theoretical constructs, deriving their results from equations representing physical laws (Oreskes, Shrader-Frechette, and Belitz 1994). In this conception—shared by most modelers—models use basic information about key physical variables only as a starting point (what modelers call *initial conditions*.)[4] However, as the foregoing discussion has shown, the reality of climate modeling practice is at best an approximation of this goal.

Many of the basic physical laws governing atmospheric behavior are well understood and relatively uncontroversial. Modelers call these the *primitive equations*. But the huge range of spatial and temporal scales involved—from the molecular to the global, from milliseconds to millennia—makes it impossible to build models from these principles alone. Schneider notes that

> even our most sophisticated "first principles" models contain "empirical statistical" elements within the model structure. . . . We can describe the known physical laws mathematically, at least in principle. In practice, however, solving these equations in full, explicit detail is impossible. First, the possible scales of motion in the atmospheric and oceanic components range from the submolecular to the global. Second are the interactions of energy transfers among the different scales of motion. Finally, many scales of disturbance are inherently unstable; small disturbances, for example, grow rapidly in size if conditions are favorable. (Schneider 1992, 19)

Hence the necessity of parameterization, much of which can be described as the integration of observationally derived approximations or heuristics into the model core. Schneider sometimes refers to parameters as "semiempirical," an intriguingly vague description that highlights their fuzzy relationship with observational data. For the foreseeable future, all GCMs will contain many of these "semiempirical" values and equations. Thus we might say that GCMs are *data-laden*.

I use this phrase symmetrically with the well-known observation that data are "theory-laden" (Hanson 1958; Popper [1934] 1959). In one sense there is nothing odd about this, since theory in the physical sciences always includes constants (such as the gravitational constant or the sun's energy output) derived from empirical measurements. However, physical-science practice normally attempts to explain large-scale phenomena as an outcome of smaller-scale processes. The "data-ladenness" I describe here refers to the inclusion of large-scale, empirical statistical data in models, which necessarily goes against the reductionist imperative of the physical sciences.

Model-Filtered Data

Global climatological data sets are deeply problematic.

Some of the reasons are obvious. Many kinds of measurements, from many different instruments, are necessary to make up a data set that covers the entire global atmosphere in three dimensions and over many years. These measurements are taken under a vast variety of conditions, which differ for reasons that are not only physical (e.g., Antarctic vs. temperate zones), but social (differing levels of understanding, technical skill, and experience in different countries) and historical (changes in techniques, instrumentation, and so on over time).

Fairly good records of land and sea surface meteorology exist for the last hundred years, but changes over time in instrument quality, location, number, and measurement techniques create many uncertainties. For example, most thermometers are located on land and clustered in urban regions, where "heat island" effects raise local temperatures above the regional average. Meteorological records at sea tend to be drawn from shipping lanes, ignoring the globe's less traveled areas. For the last several decades, records from the atmosphere above the surface have been drawn from increasingly extensive commercial aircraft, radiosonde (weather balloon), and rawinsonde (radar-tracked radiosonde) networks, but these too are concentrated in particular areas. Coverage in the tropics and in the Southern Hemisphere is particularly poor. Heroic efforts continue to purify these data by estimating and correcting for systematic errors (Houghton et al. 1996, 133–192). For example, satellite data are being used to estimate the effects of urban heat island bias on global surface temperature data (Johnson et al. 1994); historical sea surface temperatures have been corrected for the effects of different kinds of buckets used to draw water samples (Folland and Parker 1995); and problems with rawinsonde data are being addressed by comparisons with satellite data and corrections for various sampling errors (Parker and Cox 1995).

Among the chief tools of this data-filtering process are what we might call *intermediate models*. These include models of instrument behavior, interpolation techniques (for converting actual observations into gridded data), techniques for automatic rejection of anomalous data points, and many other methods (Christy, Spencer, and McNider 1995; Hurrell and Trenberth 1997; Jenne 1998; Karl, Knight, and Christy 1994). Recently, a number of laboratories have used computer models to produce "reanalyses" that attempt to calibrate, correlate, and smooth data from multiple sources into long-term, internally consistent global climatological data (ECMWF Re-Analysis Project 1995; NASA Goddard Space Flight Center 1998; National Oceanic and Atmospheric Administration 1999).

Satellites and Global Data Sets   Unlike all others, satellite data have the signal advantage of being genuinely global in scope. Weather satellites overfly the entire globe at least twice every day. This total coverage makes satellite data extremely attractive to climate modelers. "We don't care about a beautiful data set from just one point," one modeler told me. "It's not much use to us. We have one person whose almost entire job is taking satellite data sets and putting them into files that it's easy for us to compare our stuff to."

Yet satellite data are also problematic. Satellites provide only proxy measurements of temperature; these may be distorted by optical effects. In addition, their lifespans are short (two to five years) and their instruments may drift out of calibration over time. A number of scientists, including one responsible for satellite data analysis at a major climate modeling group, told me that the quality of these data was not very good. One said that their main practical value has been for television weather images. Nevertheless, the satellite data are generally regarded as the most reliable global observational record.

Here too, the solution to problems in these data is a suite of intermediate models. Statistical models filter out "signals" from noise; models of atmospheric structure and chemistry are used to disaggregate radiances detected at the top of the atmosphere into their sources in the various atmospheric layers and chemical constituents below. In addition, models are used to "grid" the data and to combine them with other data sources. Among the most important data sources are the twice-daily atmospheric analyses of the U.S. National Meteorological Center and the European Centre for Medium-Range Weather Forecasting. These atmospheric analyses "incorporate observational data from both the surface and from satellites into a 4-D data assimilation system that uses a numerical weather prediction model to carry forward information from previous analyses, giving global uniformly gridded data" (Kiehl 1992, 367–368). Thus the twice-daily periods of actual observation are transformed into twenty-four-hour data sets *by computer models.*[5]

Conclusion: Modeling as World Building

The model-data relationship in climate science is thus exceptionally complex. Models contain "semiempirical" parameters, or heuristic principles derived from observations. Meanwhile, global data sets are derived from direct observations by modeling. Since the problems of scale that create this situation are present in all global dynamic processes, the same could be said of all world-scale models. These facts about data and models have a number of important but rarely noticed consequences for climate change concerns.

First, it is *models*, rather than data, that are global. They make inaccurate, incomplete, inconsistent, poorly calibrated, and temporally brief data *function as* global by correcting, interpolating, completing, and gridding them. The sheer size of global data sets makes it unlikely that much could be learned from them without the computer models that make them comprehensible. Furthermore, global uniformly gridded data would never have been generated in the first place without the models that required data in that form. The dynamics of the earth's atmosphere could not be understood without them—at least not at a level of detail that would confer the ability to make long-term projections.

Second, the structure of knowledge about the past (data) and knowledge about the future (model projections) exhibits a surprising symmetry in the climate change field. Models, often the same ones used for future projections, are required to produce global data in the first place (Suppes, 1962, called these *models of data*). The youth of the climate field means that data sets remain in flux, not only because new data are acquired but because intermediate models continue to evolve. Yet detecting climatic change in the historical record depends on comparing a present state against a past baseline—which in this case remains a moving target. Because detection requires a long historical record, there is no alternative other than model-based reanalysis. Forecasts of future climate change rely on comparisons with the same (shifting) baseline.

Third, even if the historical record were absolutely firm, so that future trends could be projected "directly" from the data, forecasts of climatic change would still necessarily rely on modeling. This is the case because in a highly complex system with multiple feedbacks, there is no a priori reason to suppose that any given historical trend will continue on the same path. In effect, extrapolating directly from data trends would itself be a model of atmospheric behavior—but one without any basis in physical theory. The point here is that without *some* model of atmospheric behavior—even this primitive and almost certainly false one—the exact shape of the curve of global climate change could not be projected at all.

Fourth, modeling is necessary to separate human from natural contributions to climate change. For example, major volcanic eruptions, such as those of El Chichón (1982) and Mount Pinatubo (1991), can inject vast quantities of particulate aerosols into the stratosphere, causing cooling near the surface and warming in the stratosphere that can last several years. To understand the human role in global climate change, the effects of these and other natural events must be extracted from the global data. This can only be done through modeling.

Finally, *models offer the only practical way to discern the effects of policy choices about climate change.* As a thought experiment, imagine that a strong, comprehensive climate change policy (regulating, say, not only greenhouse gas emissions and energy efficiency, but agriculture, forestry, population, and economic development) were somehow instituted tomorrow and continued for fifty years. At the end of that time, how would we measure its success or failure? The only way to do so would be to compare the historical record with models of what would have happened had the policy never been introduced. This point is important, since it indicates one potential role for models that has rarely been highlighted.

For all these reasons, computer models are, and will remain, the historical, social, and epistemic core of the climate science/policy community. Without them, we would know little if anything about the causes and possible future consequences of climate change. At least as importantly, from a science-studies viewpoint, the global epistemic community that now surrounds the climate change issue would have lacked a fundamental organizing principle.

As I have shown, this point has sociological and historical ramifications as well as epistemological ones. Climate science communities developed around the practice of computer modeling. By its nature—resource-intensive "big science"—this practice limited the number of expert groups and focused them on a common strategy of "parameterization" and model-based experimentation. With the growth of computer power and the expansion of political interest in (and funding for) climate science, especially in the last decade, related sciences began to cluster around the models, using them as a common language for scientific integration. At the same time, policy communities came to depend on (and at least in part to trust) models for advice. Thus global modeling does not merely represent, but in a social and semiotic sense *constructs*, the global atmosphere.

Notes

Background materials for this chapter include published literature, archival sources, and interviews with the following people: Akio Arakawa, David Baumhefner, John Bergengren, Bruce Callendar, Robert Chervin, William Clark, Curt Covey, Richard Davis, Anthony Del Genio, Carter Emmart, John Firor, Jim Hack, Milt Halem, James Hansen, Bill Holland, Anthony Hollingsworth, Sir John Houghton, Geoff Jenkins, Roy Jenne, Tom Karl, Akira Kasahara, Jeff Kiehl, Andrew Lacis, Cecil E. "Chuck" Leith, Richard Lindzen, Jerry Mahlman, Syukuro Manabe, Linda Mearns, John Mitchell, Roger Newson, Bram Oort, David Pollard, Robert Quayle, David Rind, William Rossow, Peter Rowntree, Robert Sadourny, David Schimel, Stephen Schneider, Gus Schumbera, Starley Thompson, Kevin Trenberth, Joe Tribbia, Warren Washington, Richard Wetherald, Tom Wigley, and Austin Woods. The author acknowledges the National Science Foundation for its generous support of some of this work, under grants SBE-9310892 and SBR-9710616. He also wishes to thank Stephen Schneider and

Clark Miller for helpful comments, and John Anguiamo, Katherine Bostick, Amy Cooper, Margaret Harris, Robb Kapla, and Yuri Tachteyev for research and administrative support.

1. Zero-dimensional models compute energy balances as if the earth were a single point in space rather than a volume. One-dimensional models usually compute according to latitudinal bands (without vertical depth). Two-dimensional models add either longitude (horizontal east-west) or air pressure (vertical) dimensions to produce a grid (horizontal or vertical). Three-dimensional models extend the grid either longitudinally or vertically to produce a gridded volume.

2. The designation *three-dimensional* is slightly misleading. Most GCMs are really four-dimensional, the fourth dimension being time.

3. The most popular modern modeling technique, the spectral transform method, does not use grids in this simple Cartesian sense. Spectral models represent the atmosphere as a series of interacting waves. They are mathematically complex and difficult to grasp intuitively, but for my purposes here, this simple description is adequate.

4. This sense of the term applies mainly to *mathematical* models; it is worth pointing out that this is not the only important sense of the term. Analog models, in which one physical system is used to model another (by "analogy"), may be largely nontheoretical. Early experiments in climate modeling sometimes used analog models, such as dishpans (representing the earth) filled with fluids (representing the atmosphere) rotating above a heat source (representing the sun) (Hide 1953). Today, analog models have virtually disappeared from the field, although one might argue that climate studies of other planets serve a similar purpose.

5. Long-term, contemporary data sets are not the only ones against which to test climate models. The seasonal cycle provides a well-known, reasonably well-understood benchmark. Paleoclimatic (prehistoric) data from a variety of "proxy" sources, such as tree rings, ice cores, and fossilized pollen, are also available. Model inputs can be set to the different conditions (orbital precession, trace gas concentration, and so on) of past periods and evaluated by how well they simulate the paleoclimatic record. Naturally, the level of detail in paleoclimatic data is far lower than in contemporary instrumental observations.

3
Why Atmospheric Modeling Is Good Science

Stephen D. Norton and Frederick Suppe

Far better an approximate answer to the right question, which is often vague, than an exact answer to the wrong question, which can always be made precise.
—John Tukey (1962, 13)

Computer models are absolutely essential in the efforts of atmospheric scientists to represent the earth's climate and its possible evolution. As Edwards argues in chapter 2, computer modeling constitutes a form of world building, where controlled experiments impossible to perform physically can be run in virtual environments. Without such models, we would be unable to understand the climate system as a single, integrated whole and discern the effects of policy decisions.

It is not particularly surprising, therefore, that the issue of atmospheric computer model validity has become an explicit point of contention in the politics of global warming. Reliability of model projections has come under serious attack, especially in the United States, with its adversarial policymaking culture and tendency to couch complex political issues within artificially and inappropriately construed frameworks. In such an atmosphere, the efforts of global-warming skeptics to discredit theories of climate change have sought to impeach all conclusions based on complex global climate models in two ways: first, by exploiting uncertainties associated with modeling and, second, by denying that models are capable of yielding objective knowledge. Their enterprise is to stigmatize modeling as inferior science on philosophical grounds.

This chapter challenges the basic philosophical position underlying this argument. We begin with a detailed examination of the practice of modeling in contemporary scientific research. We then refute contrarian claims by showing that modeling is central to all empirical science—not only computer-based investigations, but also traditional experiments with laboratory controls. The epistemological issues faced by climate modeling are no different in kind than those encountered by traditional experimentation. If contrarian objections had merit, they would impeach all of science, not just complex modeling, forcing the absurd conclusion that science is in principle incapable of producing knowledge.

It is important to note that we do not address normative questions raised by the use of computer models in public policymaking. The decision to use climate models as the basis for far-reaching social and economic changes subsumes moral as well as technical choices. The choice between harm-based regulatory standards and more precautionary approaches, for example, entails differing roles for computer models in policy; in the first case, evidence of actual damages might be required before mitigating actions could occur, while a precautionary approach might rely heavily on model projections of future climate change. Likewise, the choice between cost-benefit analyses and ecological-protection approaches to environmental damage might involve choices between models focusing on human consequences and models geared to the analysis of ecosystem effects. Individual, institutional, and cultural values will inevitably color how particular communities decide to employ climate models in their policy deliberations. Although our conclusions have implications for these important issues about how models should be used, we do not address them explicitly.

Instead, we focus on the more fundamental questions of whether model-based claims constitute scientific knowledge, the extent to which such claims can be relied on (if indeed one chooses to use models in policy deliberations), and their relationship to traditional experimental and observational scientific knowledge. We argue strongly, *pace* recent suggestions, that communities should not in principle dismiss model-based claims as epistemologically unreliable. The first section of the chapter discusses digital computers' transforming impact on science. The second section examines models of data and their essential role in reliable data production, undercutting claims that mere use of models is epistemologically deleterious. In the third section, we treat simulation models, showing they can function as probe instruments yielding real-world data that face essentially the same epistemological issues as models of data. The fourth section examines those issues in atmospheric simulation contexts and techniques developed to confront them. In the final section, we summarize the findings and examine some epistemological blunders underlying skeptical attacks on scientific modeling.

Modeling in Computationally Intensive Sciences

Computational capability limits scientific data analysis. A basic unit of scientific computation is multiplication. Prior to digital computers, skilled operators, even aided by mechanical calculators like the Friden, could sustain about one multiplication per minute (Burks and Burks 1981; Goldstine 1972). This effectively limited data points used in establishing hypotheses to tens or hundreds. Even with Tycho Brahe's approximately 70,000 observations, 200–300 which were of Mars, computational limitations forced Kepler to average these down to 20 data points to determine Mars' elliptical orbit.

Digital computers changed things. By any measure, amounts of data collected or used to evaluate models and hypotheses have increased exponentially. Only computationally intensive modeling and visualization techniques prove capable of rendering intelligible today's massive data sets and allow investigation of climatic processes (Kaufmann and Smarr 1993). Indeed, many complex weather and climate simulation runs perform more computations than did humans collectively before digital computers. John von Neumann viewed numerical weather prediction as the ideal problem for digital computers—"one that would challenge the capabilities of the fastest computing devices for many years" (Thompson 1983: 757).

Models of data bring into relief structures hidden in masses of raw data and make possible proper interpretation. Simulation models evaluate models of data, allow nonanalytic solution of difficult equations, and are alternative experimental data sources. Suitably deployed, both kinds and their hybrids function as probe models that reliably detect real-world phenomena.

Modeling Climate Data

Neither raw data nor raw sensory experience carry their own interpretations. To be properly interpreted and deployed, data must be modeled. As one researcher put it, "If you ask a scientist for her data, she just gives you a model" (George Philander, professor of geoscience, Princeton University, personal communication, 1993). This section characterizes models of data, discusses inverse problems, and examines the roles of interpolation and smoothing in data sets for climate research. Epistemological issues surrounding model construction, interpretation, and reliability are explored.

Models of Data

Science does not study phenomena in their full complexity. Simplified versions are studied to bring key aspects into bold relief. Scientific experiments create instrumented environments where simplified phenomena can be observed and interpretable data collected, errors corrected, and data transformed into forms allowing unambiguous interpretation. Key aspects are experimental control, instruments, calibration, and data reduction.

Low-speed wind tunnels are paradigmatic of traditional experimentation. Wind is blown at known regulated speeds past airplanes attached to pylons in chambers, causing planes to move pylons as a function of lift, pitch, roll, and yaw. Six principal flight components are separated and transmitted via mechanical linkages to separate balances measuring forces.

Test chambers are controlled environments where flight is simplified so that air velocity from just one direction determines performance. Factors like crosswinds, updrafts and downdrafts, pilot maneuvering, and engine variations are removed in this simplified environment. Such experimental control yields raw data measurements that are relatively uncontaminated reflections of principal flight components.

However, the measurements are systematically inaccurate. Experimental control introduces artifacts—systematic errors—into the data. Unlike real flight, test chambers are closed environments. Walls introduce turbulence affecting airflow past models, causing systematic airflow variations at different chamber locations. Careful chamber design, including sufficient model-to-wall separation and unusually smooth wall surfaces, reduces but does not eliminate distortions. The only resolution is measuring airflow variations at different chamber locations, producing calibration tables indicating characteristic distortions at location, then correcting measurements to remove distortions. Calibration removes some but not all data distortions. Systematic errors remain after calibration corrections because the pylon mount introduces other distortions making measurements of each principal component systematically wrong. Mathematical models are applied to correct data.

Random errors remain after correcting systematic data errors. Instruments have ranges they can detect with accuracy (absolute correctness of measurement) and precision (repeatability and stability of measured values over time). Instrument resolution determines the smallest effect it reliably can detect. Imprecisions manifest themselves as random deviations from true values. Random instrument error or "white noise" limits instrument precision. Filtering often increases data precision by separating real effects from random disturbances. So, too, can combining repeated measurements. Excessive filtering may introduce new artifacts.

Even after correction and filtering, data still are of limited accuracy and precision. A well-tweaked tunnel produces principal-component measurements, x, precise to one part in 100,000. All you can know is that the true value probably lies somewhere in the interval from $x - 0.00001$ to $x + 0.00001$. Thus, all measurements are vague and cannot support claims more precise than measurement errors and instrument resolution allow.

Once corrected data are produced, performance characteristics like laminar-flow disruption usually will not be apparent. So, principal-component data are embedded into more complex models relating them to performance characteristics. Outputs are transformed using logarithmic or other conversions that make graphed performance characteristics obvious. Calibration, transformation, model embedding, and data display comprise data reduction—the process of converting raw data into forms useful for interpretation and understanding phenomena. Models of the data are the end product of data reduction (Suppes 1962).

Computerization allows complex instrumentation involving perhaps 100,000 detectors producing huge data quantities. The largest atmospheric data sets contain hundreds of terabytes.[1] They are collected mostly by satellite-based instrumentation under circumstances precluding traditional experimental control. Suitably done, enhanced computer modeling of data introduces vicarious control every bit as good as, sometimes superior to, traditional experimental control.

Such trade-offs between physical control and vicarious control by modeling data also occur in traditional experiments. Wind-tunnel mechanical linkages separate independent force components, transferring them to six balances. Mechanical separation is imperfect and each component measurement may contain small influences from other components. Early control attempts employed counterforces to negate confounding influences but were unsatisfactory. Physical control is not attempted today. Instead one simply measures interdependencies and removes them. Physical control could be obviated by measuring every unwanted influence and removing their effects during data reduction. Whether achieved by physical or computational means, the points of experimental control and data reduction are to produce data about uncontaminated real effects by removing artifacts introduced by instrumentation, observational environment, and extraneous influences.

Even raw data involve modeling built into the instrumentation. An instrument consists of probe, modifier, and output transducer. Probes are input devices that typically convert physical quantities into electrical signals. Modifiers then alter signals through amplification, shaping, filtering, and digitizing. Output devices convert modified signals into non-electrical display forms like meter readings or printouts. For example, a thermocouple probe is two dissimilar metal wires silver-soldered together and inserted into a heat source. Joined dissimilar metals produce currents that are a function of ambient temperature. Circuitry modifies them to produce outputs which, when connected to a voltmeter with °C scale, accurately indicate probed-region ambient-temperature raw data.

The modifier physically realizes or implements a mathematical model correlating thermoelectric voltages with temperatures. Because the model contains two magnetic permeability parameters specific to the joined metals that must be specified, it is a parameterized model. Another approach inputs thermal currents into a computer whose data-reduction process incorporates the parameterized model. Whether physically or computationally realized, all data collection from instruments involves modeling. Thus raw data also are models of data. Therefore, there is no important epistemological difference between raw and reduced data. The distinction is relative.

Satellite-borne radiometers—a principal source of temperature data for climate models—collect thermal emissions from planetary surfaces or atmospheric gases. The intensity of those emissions at wavelength is the target brightness temperature. Thermocouple-like input devices convert these emissions to electric currents, then modifiers convert the signal to measures of target brightness temperature.

A blackbody emitter (such as a large body of water) is a 100 percent efficient emitter of physical temperatures. Most surface and atmospheric materials have less than 100 percent efficiency, with their emission efficiency at a wavelength characteristic of the specific materials. Target emissivity is the anomaly of actual emittance compared to a perfect blackbody emitter. When the targets are not perfect blackbody emitters, brightness temperatures are not the physical temperatures of the target region that produce the emittance. To obtain measurements of target region physical temperatures, the brightness temperatures must be divided by the emissivity (Rees 1990, 35–42, 126–127). Thus radiometer temperature measurements require recourse to actual or virtual realizations of blackbody-emitter models.

There is a second way blackbody-emitter models are used in radiometers. The collectors that focus emissions onto the thermocouple-like input devices introduce complex distortions (analogous to wind-tunnel wall effects) that are corrected by calibration adjustments. Periodically turning collectors on a physical blackbody of known temperature produces brightness-temperature readings different from what perfect collectors and modifiers should produce. These anomalies indicate correction factors for brightness temperatures. Radiometer data involves both computational and physical realizations of blackbody models when the former are used to obtain physical temperature data and the latter for calibration.

Many physical measurements are anomalies from idealized models. In 1881 A. A. Michelson performed experiments designed to measure

Earth's motion relative to the ether—a substance nineteenth-century scientists believed necessary for propagation of light. The ether was thought to pervade all space and be absolutely at rest. Calculations indicated the earth's motion through the ether around the sun would produce an ether drift or difference in the speed of light between summer and winter observation times. Michelson built his famous interferometer to detect these light velocity variations by measuring fringe-shifts anomalies. His initial experiments failed to detect fringe shifts of the predicted magnitude (Michelson 1881). An error in the original calculations was later discovered indicating fringe-shift magnitudes much lower than originally predicted. Michelson's instrument was not designed to detect these smaller effects. Thus, the lack of fringe-shift anomalies proved an artifact of limited instrument resolution. The null result was not a detected real effect (Michelson 1882; Shankland 1964).

Crucial to both traditional experimentation and today's computationally intensive satellite-based observations is determining what are real effects in data versus artifacts of experimental or observational setups, instrumentation, and data reduction. Much experimental design concerns identifying artifactual effects and removing them through improved experimental control. Other artifacts are corrected for during data reduction. Either way, determination of artifacts involves modeling: First, one has to model potentially confounding influences and their impact on real effects of interest. This model is implemented in experimental design or data reduction. Second, one varies model parameters to determine how sensitive real effects are to identified potential confounders. For physical control, one varies potential influences and sees what effects are produced. Only confounders making significant differences will be controlled for in the final design. Similarly, when control is vicarious, variant modeling (simulation) is done to determine real-effect sensitivity to potential contaminators by varying the intensity of confounders in the model. When real effects are insensitive to potential confounders they do not have to be measured or controlled.

Michelson redid his interferometer experiment in 1887 with Edward Morley. Proper data analysis and a more refined experimental setup were used, but fringe-shift anomalies were not detected at magnitudes ether theories required, although a small anomaly was, which they interpreted as noise (Michelson and Morley 1887).

Dayton C. Miller suspected this "noise" might be a real effect, albeit smaller than expected (Miller 1933). Miller commenced careful interferometer experiments requiring unusually severe physical controls for contaminating influences due to radiant heating, magnetic influences, altitude, wind effects, and so on. His controls included insulating instrument parts with cork, variant nonferromagnetic instrument versions, and testing at different elevations. Whatever he did, fringe-shift anomalies were obtained. Eventually he became convinced the fringe-shift anomalies were real effects, not artifacts (Miller 1922, 1925a, 1925b). His results could not be replicated by others using alternate instrumentation (Miller 1928).

In 1955, Robert Shankland reanalyzed Miller's data using a contaminating-influence computer model, finding the fringe-shift anomalies were a combination of statistical fluctuations in the readings and subtle but systematic temperature disturbances (Shankland 1955). Shankland used data modeling as a surrogate for the physical control Miller could not achieve to demonstrate that the fringe-shift anomalies were artifacts that could be removed, producing models of data displaying real null effects.

There is no guarantee that reduced data will allow unambiguous discernment of real effects. Effects close to instrument resolution cannot be reliably interpreted—for example, interpretation of geological features smaller than 8×8 pixels is unreliable in radar images (Suppe, forthcoming c, chap. 7). Sometimes data are incomplete. One cannot always collect the kind of data needed for unambiguous interpretation. In such cases, assumptions are used in lieu of missing data. Reduced data are embedded into more complex interpretative models of data, where additional structure comes from assumptions in lieu of data. For example, interpolation assumptions produce complete data sets that fill data gaps.

The main finding of this subsection is that models permeate scientific data, whether it be in traditional experimentation or satellite-based observations, and that models play the same roles in both cases. No epistemically significant wedge can be driven between them due to reliance on models. If one wishes to impeach atmospheric data merely in virtue of their model dependence, consistency demands rejecting all experimental and observational data. This amounts to claiming science is incapable of yielding any knowledge about the world.

Temperature-Profile Retrieval

Detecting anthropogenic greenhouse effects requires extensive data modeling. Results have been inconclusive due to uncertainties regarding the strength and structure of the signal, noise of regional-scale climatic patterns, and incomplete knowledge of global variability (Santer et al. 1995). Adequate temperature records help mitigate many difficulties.

However, surface-based air and sea data sets are plagued with problems, most prominent being lack of global coverage. Satellite remote-sensing devices typically provide near-complete Earth coverage twice daily, helping mitigate this deficiency. This subsection examines how modeling satellite radiometer and spectrometer measurements produces atmospheric temperature profiles.

Temperature profiles are continuous distributions of temperatures with respect to altitude or pressure over a specified surface location (figure 3.1). Temperature-profile retrieval from thermal-emittance measurements is not straightforward. Instrument design, calibration, orbital drift over time, atmospheric structure, and chemical composition must be corrected for. Even if controls are appropriately implemented and corrections adequately done, radiometric emittance measurements do not uniquely determine temperature profiles. Infinitely many profiles are compatible with any set of measurements.

An idealization illustrates the issues (Twomey 1977, 1–2). It assumes, correctly it turns out, that each atmospheric layer emits thermal radiation that is related in a known way to the layer temperature. It further assumes, falsely it turns out, that each layer emits radiation of one specific wavelength emitted by no other layer (figure 3.2). Given these two assumptions, satellite radiometers could uniquely determine any layer temperature by measuring its characteristic wavelength (figure 3.3a). Furthermore, ideal radiometers could measure the temperature of every atmospheric layer, uniquely determining vertical temperature profiles for given atmospheric slices.

The actual situation is not so accommodating. First, real radiometers have limited resolution (ability to isolate any given wavelength).

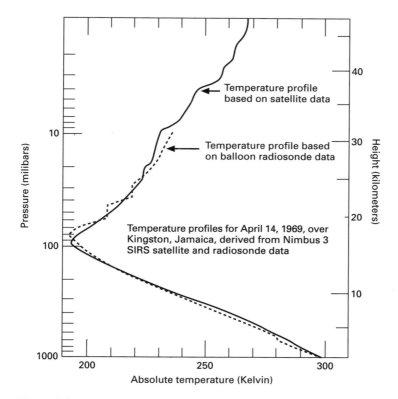

Figure 3.1
The first satellite temperature profiles. *Source:* Modified NASA image.

Radiometers designed to measure CO_2 emissions register intensities primarily due to radiation of the desired wavelength but also include contributions from wavelengths within a certain range around the target wavelength. The range of wavelengths detected and measured is determined by the inherent resolving power of the instruments.

Second, actual radiometers only measure a small number of channels or wavelengths. Limited resolution and number of channels allows assigning average temperatures to only a few atmospheric layers. Even if one could uniquely determine temperatures at radiometer resolution, infinitely many profiles still would be compatible with measurements.

Third, observed wavelength intensities are composites of signals from many different levels, not emissions originating from single layers.

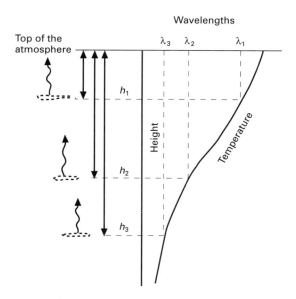

Figure 3.2
Idealized composite temperature-wavelength dependency and unique wavelength-height correlation. *Source:* Based on Twomey 1977.

Signals actually measured will be blurred. Primary contributions are localized around individual atmospheric layers with contributions from other layers (figure 3.3b). Primary contributions to signal at wavelength actually can only be localized to a 10–15 km height range (figure 3.4).

Temperature-profile retrievals from remote-sensing measurements are inverse problems. Inverse-problem solutions estimate the distribution of some quantity from a set of proxy measures—in this case, layer temperatures from remote-sensing emittance observations. Each measurement aggregates unknown temperature distribution values, and the problem is to arrive at a solution by disaggregating them. In their simplest versions inverse problems amount to reconstructing individual summed items from just the sum. Inverse problems are underconstrained by data or ill-posed when available measurements alone are insufficient for determining unique profiles even when measurement error is neglected (Tikhonov 1963).

Ill-posed inverse problems superficially resemble what philosophers of science call the underdetermination of theory or models by data. This

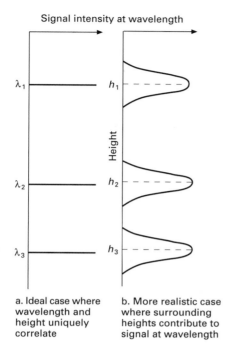

Signal intensity at wavelength

a. Ideal case where wavelength and height uniquely correlate

b. More realistic case where surrounding heights contribute to signal at wavelength

Figure 3.3
Idealized vs. more realistic relationships between height and wavelength. (*a*) In the idealized case, contributions to measured wavelength intensities are zero everywhere except at one height. (*b*) Measured wavelength intensities actually include contributions from surrounding heights. *Source:* Based on Twomey 1977.

view assumes that data analysis is a curve-fitting problem, where data are represented by points on a plane and hypotheses by curves drawn through each point. Since interpolation between the data points is unconstrained, there will always be infinitely many mathematical curves exactly fitting the data with no objective empirical basis for eliminating any, let alone selecting just one. For temperature-profile retrieval, this view entails that profile selection will be arbitrary.

This conclusion is at odds with science's ability to produce well-defined, plausible solutions for many classes of inverse problems including temperature-profile retrieval. Indeed, since they involve solution of inverse problems (Katz 1978; Suppe 1997a), standard CAT scans and MRI medical-imaging techniques would be impossible if this philosophical underdetermination thesis were correct.

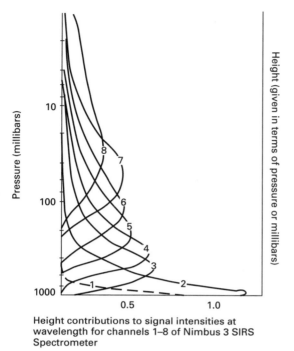

Height contributions to signal intensities at
wavelength for channels 1–8 of Nimbus 3 SIRS
Spectrometer

Figure 3.4
Actual contributions of height to wavelength as measured by Nimbus 3 SIRS
Spectrometer. Unlike figure 3.3b, signals from different heights overlap in real
radiometers. *Source:* Adapted from Houghton et al. 1984, 83.

Where philosophers go wrong is in their reliance on the curve-fitting
analogy. Mathematical proof exists that most geophysical inverse problems are not equivalent to, and cannot be reduced to, unconstrained
curve-fitting problems (Parker 1994, 58). Such philosophical considerations are irrelevant to much real science, including atmospheric research
(Tarantola 1987, xiv).

Quixotic philosophical concerns with underdetermination are due
partially to widespread acceptance of mistaken views about scientific
data and data analysis. Measurement data are the result of applying
dimensional scales, and thus consist of a numeric quantity plus a unit of
measurement having additional mathematical properties. For example,
a measured acceleration of 30 km/sec/sec consists of the numeric

quantity 30 plus a unit of measurement (km/sec/sec) having rich mathematical properties including being the second derivative of the velocity of the object. Only curves so related to velocity can be fitted to acceleration data points. But mere curve fitting allows any arbitrary intersecting curves. By construing model fits to data as simple curve fitting to numeric values, the underdetermination thesis divorces measurement data from their observational contexts, instrumentation, and the wider body of scientific knowledge. As Laudan (1990, 267) notes, the underdetermination doctrine incorrectly takes "the logically possible and the reasonable to be coextensive."

In the philosopher's curve-fitting problem, the solution space includes every logically possible fit to numeric data values. In contrast, solution spaces for inverse problems are highly constrained due to richer mathematical structures associated with the measures used to obtain data and added information or assumptions. Such mathematical properties often exclude many possibilities allowed by corresponding curve-fitting problems (Rodgers 1977, 119). This fact allows meteorologists and climate researchers to develop practical means for solving inverse problems that regularly generate reliable knowledge.

Data are not mere numeric values. Data result from measurement. Measurement scales and their units endow them with mathematical properties that inter alia define what mathematical operations can be performed on the data (Stevens 1946). For a model to "fit" data it must accommodate their measure's full mathematical properties, not just numeric values. This severely constrains a solution space. For example, dimensional analysis sometimes identifies the general form all solutions must take (Stull 1988, chap. 9).

Curve fitting presumes finite-dimensional representations for finite data sets, whereas most geophysical inverse problems require infinite-dimensional spaces to fully represent finite numbers of data points. In practice those representations are such that either properties like upper bounds are common to all solutions or else optimization to an additional parameter yields a unique solution when measurement error is neglected (see Parker 1994, 58, 73).[2]

Such mathematical constraints allow constructing specific solutions from data, but may not produce unique solutions. Atmospheric inverse problems often remain ill-posed even when measurement properties are exploited. No matter how much data are available, they never suffice to solve ill-posed inverse problems analytically, even neglecting measurement error. In temperature-profile retrieval "the only way of solving the problem is by making use of some kind of extra . . . information" where "the source of . . . information may be the physics of the problem, statistics of other measurements, arbitrary restrictions, prejudice, etc." (Rodgers 1977, 119). In short, "Assumptions can always be substituted for data" (Woodward, Boyer, and Suppe 1989, 105).

Additional information allows deriving well-posed extensions of inverse problems. Climate researchers typically create well-posed extensions by casting inverse problems as optimization problems where the goal is finding a model solution that minimizes or maximizes some "fitness" function. The specific formulation depends on both the details of the problem and available data. Whatever method is used, data together with assumptions must have sufficient structure that parameters can be accurately determined. A common procedure in profile retrieval associates observations with pertinent atmospheric or surface parameters to generate a database of profiles (Chedin and Scott 1983). Once prepared, correspondence between atmospheric parameters and radiometer measurements is established through statistical regression by selecting a particular database profile and fitting to the data by minimizing the mean difference between measured and fitted values, then iterating the process until the difference is within specified error limits—a process known as *statistical regression.*

Unlike philosophical curve-fitting stories, inverse-problem augmenting assumptions either produce unique solutions when measurement error is neglected or else constrain the set of acceptable curves. For example, one set of assumptions added to temperature data concerns general atmosphere structure. Curves fit to temperature sounding data must conform to the structural requirements of these assumptions; otherwise, they are rejected as unreasonable physical approximations.

Augmenting assumptions need not be true. Modelers routinely use soft, unsubstantiated, highly idealized or even false assumptions. Assuming the atmosphere is composed of a small number of discrete layers is

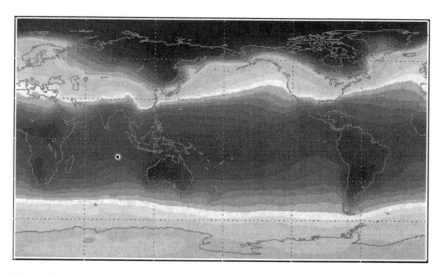

Figure 3.5
Mercator model of Earth surface temperatures. (Created using NOAA-CIRES Climate Diagnostic Center visualization tool for GFDL data: www.cdc.noaa.gov/cgi-bin/DataMenus.pl?dataset=gfdl.)

clearly false, yet reasonable given limited radiometer resolution and need for tractable algorithms to solve inverse equations.

Many philosophers believe recourse to false or unsubstantiated assumptions introduces artifacts and that mere presence of artifacts precludes models from producing knowledge (Cartwright 1983; Oreskes, Shrader-Frechette, and Belitz 1994). Simple cartographic considerations show the conclusion unwarranted. Maps are models of data used in creating climatological models. All Earth maps create data representations with artifacts due to projecting spherical data onto flat surfaces. Though no map faithfully represents all surface features, various aspects are accurately represented. Mercator projections accurately represent locations but distort areas. Thus, a surface-temperature Mercator model accurately represents temperatures at given latitude and longitude even though total surface areas at given temperatures are distorted (figure 3.5). By contrast, sinusoidal or Eckhert equal-area projections correctly represent areas but distort distances and shapes.

Objective knowledge can be retrieved from map models if claims are suitably adjusted to account for distorting effects. Similarly, the finite-layer assumption creates features not literally true. But, as with map models, if claims are selectively qualified to mask artifacts and reflect instrument resolution, modeled data reliably do support objective knowledge claims.

More generally: Even when inverse problems have nonunique solutions, in practice they have restricted solution spaces. Some properties or relationships may be robust in the sense that they are common to all allowed solutions. We do not need unique fits to obtain knowledge of those robust effects. Knowledge claims should be vague as to which solution fitting the data is correct, but precise as to nonartifactual robust findings.

The inverse-problem discussion has neglected uncertainties due to measurement error and focused on idealized solution spaces. Limited resolution, precision, and accuracy of instruments means actual measurements are valid only within certain error limits. Measurements should be reported with error bars or other error-limit indications.[3] When measurement error is taken into account, there are no unique data points to fit. Each set of values in the vicinity of measured values constrains an idealized solution space, and the set of all idealized solution spaces comprises the actual solution space. The latter contains the error-free idealized solution space and is thus a kind of smearing out of it.

Nevertheless, when measurement error is accounted for, solution spaces still are constrained and some nonartifactual effects may remain robust. As before, knowledge claims should be tailored to be insensitive to artifactual aspects of models and precise about real effects.

These considerations show that posing the underdetermination problem in terms of unique fits to data is wrong. Obtaining knowledge from models of data only requires constrained solution spaces with nonartifactual robust effects, and scientific knowledge claims that are suitably qualified and vague (Rosenberg 1975).

Diagnostic Analysis
Once modeling of data has provided satisfactory temperature profiles, additional modeling is required to discern complex climate trends includ-

ing enhanced global warming. A key step is providing a series of synchronous and spatially homogeneous data representations at suitable grid resolution and time intervals. In practice, surface- and satellite-based temperature measurements and profiles are asynchronous, inhomogeneous, incomplete, and of insufficient or excessive resolution.

The processes used to correct these problems collectively are known as *diagnostic analysis* and use interpolation to alter data points. Gridded data are generated from observations by hand plotting or automated procedures involving function fitting, methods of successive corrections, and statistical interpolation schemes (Daley 1991).

Polynomial function fitting once was the interpolation technique of choice (Panofsky 1949; Gilchrist and Cressman 1954).[4] Polynomial-fitting assumptions need careful evaluation. Consider Charney's (1955b) geostrophic adjustment that differentially weights observations with respect to quality. Improved weather predictions are obtained at higher latitudes, but the assumption introduces nonrandom errors, making lower-latitude predictions unsatisfactory unless geostrophic artifacts are corrected (see the section "Lumping Artifacts" below).

Areas of the globe exist for which data are sparse or nonexistent. Missing data must be supplied by interpolation. Sparse data means less accurate atmospheric representations since larger data voids increase interpolation error and uncertainty. The situation can be significantly improved if interpolations are constrained by preliminary estimates of wind directions and temperature gradients made via numerical weather-forecasting models.

Philip Thompson (1961) developed such an iterative method for generating more accurate atmospheric representations. A preliminary analysis produces a gridded field that is more reliable in the observationally rich areas surrounding the voids. First, approximations for missing data are used as initial conditions in a numerical predication model. The model is run and then solved for one time step, with the effect that observationally dense area information propagates into voids, providing better data-void approximations. The simulation is solved for the next observation time period. Another analysis is performed with new observational data and forecast approximations for the data voids. Iterating the procedure achieves progressively better results than the original analysis.

Asynchronicity of satellite data results from measuring along successive orbital swaths while Earth rotates beneath the instrument. Data-collection times between grid points can vary from twelve to twenty-four hours. Conversion to synchronous data involves interpolating grid-point temperatures between successive actual measurement times, picking a standard reference time, then replacing each grid point's temperature measurements by the estimated interpolated values for that reference time.

Correcting asynchronous, missing, or insufficient resolution data is done by interpolating values. Excessive resolution is addressed by smoothing data—a filtering process involving replacement of actual measured values by interpolated values. For example, wind measurements contain information about motions on many different scales. Some of these motions have little or no influence on weather patterns. If appropriate corrections are not made, these motions propagate throughout the gridded data and during subsequent analysis produce noise levels obscuring signals of interest.

Filtering smoothes data. When data are gridded, smoothing is driven by polynomial equations used to interpolate between data points. The appropriate level of smoothing depends on targeted phenomena. Some smoothing generally is desirable since it eliminates white noise or random measurement errors. Smoothing can also remove modeling-assumption artifacts.

Summary

Both traditional experimentation and remote-sensing observation employ models to obtain data. They rely on models in exactly the same ways. Recourse to models marks no fundamental epistemological difference between traditional experimentation and today's computationally intensive observations.

Epistemologically important features of data modeling have been identified: Under certain conditions data and assumptions are interchangeable. Addition of weak, unsubstantiated, and even false assumptions can structurally enhance poorly constrained data into models displaying coherent structures. These structures may either reveal real relationships in data or be assumption artifacts. Further modeling of the data often enables determining which structures are real and which are artifacts.

Control of artifacts and undesirable effects often is made possible by modeling. Although model fits to real data are not unique, in practice allowed fits often are so highly constrained that nonartifactual robust effects are common to all solutions. Objective knowledge results when claims about robust real effects are suitably tailored to model constraints and uncertainties.

Probing the Climate

Simulations often are alleged to be only heuristic or ersatz substitutes for real experimentation and observation. This will be shown false. Properly deployed simulation models are scientific instruments that can be used to probe real-world systems. Thus, simulation models are just another source of empirical data.

Simulation Models

Traditional experimentation collects data on a small number of parameters $p_1, \ldots, p_n$ at discrete time intervals, and uses experimental control to isolate the phenomena from other influences. The analog, in simulation modeling, is the base model B consisting of states that are simultaneous values for $p_1, \ldots, p_n$ that change in discrete time steps. Base models are thus simplified versions of real-world phenomena, and their possible state transitions correspond to what could be observed experimentally under perfectly controlled circumstances. State transitions can be deterministic, indeterministic, stochastic, or chaotic in form.

Simulation uses a simpler lumped model L to investigate B. For example, base model states might be gridded daily precipitation levels. The simulation might investigate average annual regional precipitations using a model where daily grid-point values are "lumped" together as regional annual averages. In a good simulation, lumped-state transition sequences allowed by L's equations mimic B's possible state transitions.[5] The simplifications associated with lumping sometimes allow stochastic B to realize deterministic L, deterministic B to realize stochastic L, and so on (Burks 1975).

A programmed computer C "runs" L. Properly done, C also is a realization of L. Thus, for valid simulations, mapping relations between the

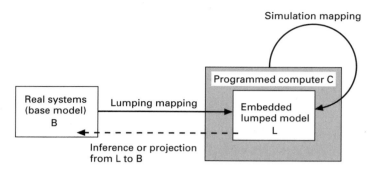

Figure 3.6
General simulation modeling situation. Both base model B and programmed computer C realize lumped model L, which is embedded into C to produce a real-world system.

systems must be such that both base model B and programmed computer C simultaneously realize the same lumped model L (Zeigler 1976, 27–49).

Programmed computers realize lumped models by embedding them into the computer (figure 3.6).[6] L embedded into C—a simulation model M—is a real-world physical incarnation of the lumped model that can be run and observed. Since it is a real system, observations of it are epistemically creative in the sense that new knowledge of the world can thereby be obtained. This is why simulation models are more powerful than mere suggestive analogies and their use not merely heuristic.

The circumstances under which behavior observed in valid simulation models can be projected back to base model systems depends critically on the extent to which the B ⇒ L and C ⇒ L mappings hold. Mappings between systems preserve behavior and/or structure. What legitimately can be determined about B from observations of L depends essentially on the extent of behavior preserved under these mappings.

Fairly strong theorems have been proven indicating conditions that must be met for simulations to yield data or knowledge about B (Suppe, forthcoming a). Roughly, they are: (i) the lumping and simulation mappings hold, (ii) equivalence classes of states of B are mapped onto the states of L, and (iii) the base system is isolated from influences not identified in defining the states used in the B ⇒ L mapping. Much

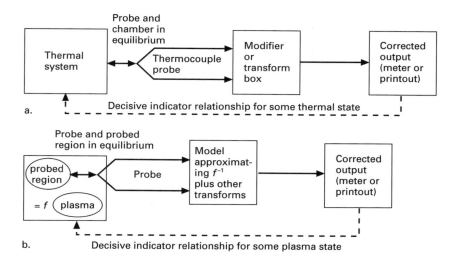

Figure 3.7
Probe models. The basic case involves placing an instrument directly into the thermal region (*top*). When that is not feasible, the probe is inserted into some other systematically associated region and thermal region temperatures are derived by solving inverse problem $f - 1$ (*bottom*).

experimental work and modeling goes into determining whether or not, and to what degree, these conditions hold (see below).

Probe Models

When simulation model M of B is valid and conditions (i)–(iii) are met, M can be used as an instrument for probing or detecting real-world phenomena. Empirical data about real phenomena are produced under conditions of experimental control.

Probes are inserted devices used to explore or measure phenomena by interacting with them. A thermocouple input device connected to galvanometer output device through a Wheatstone bridge modifer is a paradigmatic probe instrument. These probes may be inserted directly into heat sources (figure 3.7a), but that is not always possible. For example, plasma jets will vaporize any temperature probe. In plasma physics one measures something else (temperatures removed from the plasma jet), then uses thermal inverse models to convert those measurements into plasma jet measures (figure 3.7b). What counts is that observed outputs

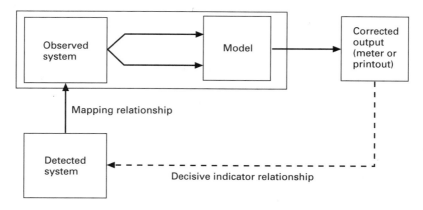

Figure 3.8
The general probe situation. The direct and indirect probings in figure 3.7 both instantiate this more general scheme.

correctly and unambiguously represent the actual state of affairs—that they be decisive indicators. This condition can be met even if there are systematic errors in output that must be corrected through data reduction, calibration, solving inverse problems, and interpretive modeling.

Instruments embody models, as we saw earlier. When probed regions are associated with, but are not, the phenomena of interest, additional data modeling is required to convert readouts to derived measures. Such modeling can be computationally intensive. Indeed, inverse modeling is a particularly important example of probe modeling where probed aggregate effects are projected back to their source regions, as in CAT scans (Suppe 1997a).

Simulation models can function as probes (figures 3.8 and 3.9). Indeed, when a simulation meeting conditions (i)–(iii) runs, the embedded lumped model functions as a probe of either the actual base model or a physically possible variant of the actual world under conditions of perfect experimental control (Suppe 1997c, forthcoming a). Simulation models probe real-world physical possibilities.

Outputs of simulation models functioning as probes are models of the data. Thus, the general discussion of models of data earlier in the chapter carries over to simulation-model probes. Specifically, they are instruments where experimental control is completely vicarious. Epistemically,

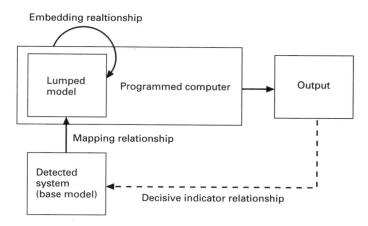

Figure 3.9
Simulation models functioning as probe instruments. The general probe scheme in figure 3.8 is instantiated in a manner reminiscent of indirect thermal probing in figure 3.7 (*bottom*).

simulation is more powerful in that it can probe situations where neither experimental nor statistical control for contaminating influences presently is feasible using traditional techniques.

Determining whether the $B \Rightarrow L$ and $C \Rightarrow L$ mappings hold and conditions (i)–(iii) are met amounts to determining whether the simulation probe outputs are decisive indicators of real-world conditions—that is, determining whether the instrument is functioning properly. As with any other instrument, simulations may have systematic errors in their output, and must be calibrated. Typically this is done by tuning model parameters to get close agreement with empirical data, or inserting models of data into the lumped model. Simulations may embody false or unsubstantiated simplifying assumptions and still detect real effects. One has to determine which aspects of simulations are real effects versus artifacts and tailor claims to the reliability and limitations of the data-model output.

Just as with experimentation and models of data, the ability to control for contaminating influences and artifactual effects is essential for proper interpretation of simulation results. Isolation condition (iii) embodies this fact. It is functionally the same as physical control of experimentation and vicarious control exploited by models of data. As with traditional experimentation, if appropriate circumstances are realized and relevant contaminating influences controlled for, then well-calibrated simulation results can be extrapolated back to base systems to yield reliable knowledge of the real world.

Simulation modeling is just another form of experimentation, and simulation results are nothing other than models of data. Thus, the epistemological challenges facing simulation models are, at bottom, identical with those of any other experimental situation involving instrumentation.

Simulating Climate Change

This section examines validity threats that climate modeling encounters and techniques used to meet them.

Lumping Artifacts

The heart of scientific experimentation is creating simplified versions of phenomena through experimental control. When simulation probes are used, control is obtained by lumping variables.

Lumping limits simulated effects to simplified replicas of real-world phenomena. In fact, "Limitation is implicit in every model, and is the real purpose underlying the use of a model" (Platzman 1967, 529). Lumping makes possible the study of climate processes by bringing into relief their main features and excluding masking secondary effects. This is not unique to simulation. Gridding data is a form of lumping where local effects are averaged at regular spatio-temporal locations. Diagnostic analysis alleviates many gridding problems, as noted earlier.

Lumping often introduces artifacts. Sorting real effects from artifacts in climate simulation outputs is essential for assessing model reliability. Indeed, much controversy surrounding climate simulation concerns which lumped-model effects are artifacts.

General circulation models (GCMs) are essentially low-resolution numerical weather prediction (NWP) models used to study climate. Both correct lumping artifacts in similar ways. For instance, different atmospheric air densities are lumped together by assuming constant or pressure-dependent density. This allows atmospheric motions to be modeled hydrodynamically. The hydrodynamic state equation, Newton's laws of motion, mass conservation, and the thermodynamic energy equation are the foundation of all weather and climate models. These primitive equations lump together atmospheric wave motions from different scales, introducing significant artifacts into simulations.

The primitive equations are nonlinear and lack analytic solutions. Various approximations make them tractable. The first numerical weather prediction relied on the hydrostatic approximation (Richardson 1922). This lumping treats the atmosphere as if it were hydrostatically balanced—meaning that the force of gravity and pressure acting on each atmospheric particle is zero—exploiting the fact that actual atmospheric deviations from this condition are negligible. Sound and other high-frequency waves tend to obscure or severely distort meteorological phenomena and need to be removed. The hydrostatic approximation eliminated sound but not high-frequency gravity waves. Consequently, Richardson's solution produced a catastrophic lumping artifact of short-term pressure changes 145 times larger than those actually observed.

Charney (1949) identified the artifact source: the primitive equations represent the entire spectrum of waves from small scale sound waves to mesoscale eddies. Study of synoptic processes requires filtering small-scale motions, which are meteorological noise relative to large-scale perturbations. This noise excited Richardson's spurious high-speed fronts (Charney, Fjörtoft, and von Neumann 1950).

Charney proposed isolating meteorological patterns from unrealistic high frequencies by additional lumping. He assumed horizontal winds speeds to be in near balance with the Coriolis force and the pressure gradient (i.e., nearly or quasi-geostrophic). Charney's refined lumped model, the quasi-geostrophic model, successfully controlled for the high-frequency contaminating effects, but introduced systematic errors including far-northward displacement of depressions and preventing frontal discontinuities from developing (Charney 1955b).

Seeking to avoid these errors, Charney later returned to the primitive equations and used filtered initial conditions. This geostrophic adjustment of initial-condition observational data blocked spurious high-speed-wave solutions of the primitive equations (Charney 1955b).

More sophisticated initialization procedures have since been developed, making primitive-equation models robust tools for investigating atmospheric phenomena. (See the earlier section on "Diagnostic Analysis.")

Phillips performed the first general circulation experiment (Phillips 1956). It was a simple two-layer quasi-geostrophic hemispheric model. The resulting three-week baroclinically unstable flow (equator-to-pole transport of warmer air resulting in large meandering, westerly flowing air mass) agreed well with observed atmospheric behavior. However, it also encountered some serious instabilities. Phillips later identified their source as aliasing errors (Phillips 1959). Inadequate wave-frequency sampling caused the energy of high-frequency waves to appear as false enhancements of related lower frequencies—that is, they were aliased into the lower frequency signals. Model refinements eliminated aliasing errors by periodically damping (correcting) low frequencies to suppress the instability—that is, selectively eliminating high-frequency components. Arakawa later developed an alternate lumping for horizontal transport variables that completely eliminated spurious transfer of energy to small scales, thus avoiding artificial numerical damping (Arakawa 1966).

In summary, early NWP models and GCMs lumped variables in ways that introduced distorting artifacts. Initial control attempts introduced new artifacts. Eventually, lumping techniques were developed that did not create significant artifacts. Once that was done, simulation models became probes—instruments that could be used to investigate selected aspects of atmospheric phenomena under controlled simplifying circumstances. The model-refinement process was no different than refining conventional instruments into reliable measurement devices or shaking down experimental designs to produce better control. Success is not guaranteed. Just as with experiments, the adequacy of lumped models is assessed by the extent to which artifacts confounding interpretation can be corrected, removed, or controlled for.

Since the primitive equations cannot be solved analytically, approximate numerical solutions are derived. This requires discretization—a form of lumping that replaces continuous equations with discrete approximations at grid points. Continuous changes in variables are approximated by differences over finite space and time intervals. The discrete equations can be solved using an iterative method of successive approximation.

Finite difference and other discrete approximation techniques can affect simulation accuracy: truncation errors are the differences between primitive equations and discrete approximations (Logan 1987, 490–496). This can be understood via polynomial functions that can be expressed as infinitely long summations (Taylor-series expansions). Only a finite number of expansion terms can be computed. Excluding more terms introduces larger truncation error. Absolute or statistical error bounds can be calculated based on the number of expansion terms used. This indicates deviations of the discrete approximation from the primitive equations (Logan 1987, 500). Errors from other sources such as improper boundary- and initial-condition discretization can be similarly evaluated.

Early terms in Taylor-series expansions often provide basic "first-order" characterization of phenomena (and subsequent higher-order terms produce embellishments). Modelers may elect to detect just first-order effects. Or the phenomena of interest may involve higher-order effects. Deciding how many terms in the Taylor series to model is a lumping decision. Excluding terms amounts to imposing a filter to remove effects that are irrelevant or ignored.

Discretization of equations is tied to spatiotemporal gridding. Spacing affects whether numerical solutions are well behaved. For example, numerical solutions are unstable when time-step increments are less than is required for waves to travel from one grid point to another (Courant, Friedrichs, and Lewy 1928). Diagnostic analysis ameliorates other gridding problems, as noted in the diagnostic-analysis section of the present chapter.

Effects close to instrument resolution are white noise and cannot be reliably interpreted. For simulation probes, instrument resolution is grid resolution and white noise is grid-scale "turbulence." No matter how fine the grid scale, effects close to grid resolution are noise, potentially manifesting themselves as aliasing errors. Diagnostic analysis provides some recourse, but the ultimate solution is restricting knowledge claims to effects well above grid resolution.

Climate is importantly influenced by many sub-grid-scale processes that cannot be explicitly represented in climate models. They must be estimated by parameterized models correlating below grid-resolution effects with gridded variables. For example, energy-balance climate models need to account for equator-to-pole heat transport. Large-scale eddies are responsible for transport but below model resolution. A "diffusion approximation" is therefore used that relates energy flow directly to the latitudinal temperature gradient—a lumping that mimics real-world behavior. That the parameterized model does not faithfully represent actual underlying mechanisms is epistemically irrelevant so long as effects at model resolution are preserved and the results interpreted with nuance.

Parameterized models can be tuned to improve estimates. For example, clouds influence climate via feedback processes affecting temperature. GCMs can resolve large-scale condensation and convection processes, but not cloud-formation processes. Instead, model parameterizations relate grid-box cloud coverage to condensation and convection effects like relative humidity. The models are tunable in the sense that better agreement with observed statistical correlations is obtained by adjusting correlations between cloud fractions formed for any grid box and the critical relative humidity at which fractions are formed. Such adjustment is just calibration of the simulation probe.

Just as models of data embodying false or unsubstantiated assumptions can yield reliable knowledge, so too can parameterized simulation models. For example, an early version of NCAR's GCM cloud prediction scheme called for 95 percent cloud coverage when relative humidity exceeds 80 percent (Schlesinger and Mitchell 1985). This captures neither true cloud-formation mechanisms nor actual humidity/cloud-coverage correlations, and introduces "cloud blinking" artifacts: when grid-box humidity fluctuates around some critical value, clouds may form at one time step, then disappear the next. This blinking artifact is more problematic for weather prediction than climate modeling. Long-term statistics characterize climate variables. If blinking does not affect these statistical properties, GCM results will be reliable.

The essence of scientific experimentation is detecting simplified versions of phenomena at some limited resolution under circumstances of actual or vicarious control. Therein lies the epistemic power of science.

Lumping is the primary control process for simulations; grid resolution is their probe-instrument resolution. Suitably done, simplification via lumping and limited grid resolution are virtues, not vices. By itself, introduction of artifacts is unimportant. What counts is whether the artifacts interfere with collection of reliable and interpretable data. This is so for any scientific instrument. Producing reliable simulations is epistemologically on par with shaking down experimental design and instrumentation.

Robustness

GCMs are among the most complex simulations. Despite lumping, course gridding, parameterization, and other simplifications, single runs push computational capabilities of the largest supercomputers. Good science seeks not unique model fits, but rather constrained solution spaces with robust effects. Sensitivity analysis explores model robustness by investigating whether solutions are driven by choices of parameter and initialization values about which there is uncertainty.

When models are linear and parameter values well circumscribed, variant modeling can be used to explore entire solution spaces by modeling extreme cases (end-member modeling). Linearity ensures that high/low extreme cases will bracket the solution space in the sense that all possibilities are intermediate. Effects common to extreme cases are robust over the entire solution space. Thus comparison of models with extreme parameter values provides intercomparison of all models.

This is not so when models are nonlinear. For example, intercomparison of GCMs using plausible parameter values indicates temperature increases between 1.5° and 4.5°C for CO_2 doubling. But, unlike linear models, such intercomparisons are not guaranteed to bracket the actual situation.

More generally, traditional sensitivity methods are not feasible for atmospheric models. First, model complexity limits one to relatively few runs with selective variant initial and parameter values. Second, primitive equation nonlinearity means parameter influences cannot be exhaustively assessed with standard end-member techniques. Changes in parameter values cause unpredictable results through complex feedback processes, making it difficult to isolate individual sensitivities.

Adjoint sensitivity analysis (ADA) overcomes these limitations. Simulations are modified so that after model runs, the dependence of specific effects on individual parameters can be diagnosed (Cacuci 1981a, 1981b). For example, a single calculation with a radiative-convective model determined individual contributions to surface-air-temperature sensitivity for all 312 model parameters when CO_2 concentrations doubled (Hall 1982). These sensitivities, in conjunction with actual parameter-value uncertainties, are then used to assess model uncertainties. For example, knowing the sensitivity of CO_2-induced temperature increases to surface reflectivity (albedo), it can be shown that a one-σ standard deviation (10 percent) of uncertainty in the known surface-albedo value creates little uncertainty in the radiative-convective results (Hall 1982, 2045). Robust claims result when they are suitably qualified to reflect such model uncertainties.

Another method for investigating robustness exploits complexity to constrain solution spaces and remove ambiguities. Simulation outputs tracking individual- or single-variable variations, such as changes in the earth's average surface temperature, may be compatible with the data signals from various sources. When the same signal is expected from different sources, these univariate analyses remain nonrobust. However, simultaneously tracking multiple-variable changes can create "crucial experiments" that eliminate all plausible alternate interpretations by tightly constraining the solution space. (In effect, no plausible alternative model could produce the same signal.) Detecting such multivariate signals allows robust, unambiguous data interpretation.

Greenhouse fingerprint studies illustrate the process. Until recently, enhanced-greenhouse detection typically employed model-derived univariate signals of expected effects—for example, average temperature change. Even if enhanced greenhouse effects are present and the probe signals accurate, detecting such signals in observational records does not decisively indicate anthropogenic global warming since univariate signals cannot be unambiguously distinguished from natural fluctuations. Fingerprint methods are more powerful. They seek to derive multivariate time-varying signals that can be used as "fingerprints" or decisive indicators of enhanced global warming (Barnett and Schlesinger 1987). The more variables used, the more constrained the solution space, the less likely the detected signal is artifactual, and the more robust the interpretation.

Correct scientific claims express effects that are robust relative to experimental/observational limitations. Model complexity, nonlinearities, and imperfect understanding of phenomena complicate determining such robust effects. Multivariate ADA techniques are means for determining sensitivity, hence robustness, in the face of such challenges. GCMs introduce no epistemological challenges different in kind from what scientific experimentation routinely confronts.

Epistemological Confusions about Scientific Modeling

Seven considerations have emerged from examination of modeling data, simulation modeling, and their use in probe modeling. These lay the epistemological basis for impeaching pessimistic dismissal of all global climate models.

Summary of Findings

The most basic finding is that there are no epistemically significant differences in how models are used in traditional experimentation and today's computer-intensive remote-sensing atmospheric sciences. In each case, models are intrinsic to the collection, reduction, and interpretation of data. More specifically, the following epistemological principles apply to both traditional experimentation and remote-sensing-based climate modeling.

1. Simulation Models as Real-World Probes   Some phenomena do not allow collection of good data under controlled natural circumstances. This can be due to difficulties or impossibilities of introducing physical or vicarious control or due to instrument limitations. In such circumstances, simulation models provide an alternative technology for probing real-world phenomena. When done properly, simulation probes yield data about actual or possible real-world systems. Thus simulation models are another source of experimental data. Determining whether simulations yield reliable data involves essentially the same considerations as for instruments and data in traditional experimentation.

2.   Interchangeability of Data and Assumptions   Raw data and observations are noisy and often indecipherable. Only if stable, coherent structures are discernible is it possible to interpret relationships inherent in data. Practical and theoretical considerations limit both data quantity and quality. This is especially so in global change research, which relies on a spatially heterogeneous global observation system employing many different instrument types. Thus, scientists must model their data by adding assumptions that serve as surrogates for unobtainable data. The basic insight here is that under certain conditions data and assumptions become interchangeable.

3.   Structural Escalation   Assumptions are added to bring structural coherence to existing data, much as adding a stain to a microscope slide brings into relief features otherwise unseen. This use of assumptions is analogous to what Mallows and Tukey call procedures of structural escalation "in which an apparently weaker a priori structure is conjoined with the actual distribution of data sets to generate an apparently stronger a posteriori structure" (Mallows and Tukey 1982, 126). The assumptions employed need not even be true. However it is achieved, the goal of structural escalation is to bring into relief interpretable structures inherent in data.

4.   Real vs. Artifactual Effects   Appropriately augmented and transformed data displays structure, which may be inherent or artifactual. Thus results of structural escalation can be either inherent real effects or artifacts of augmenting assumptions.

5.   Differentiating Inherent vs. Artifactual Effects via Modeling   Differentiating between structures that are real effects inherent in data and those that are merely artifacts of assumptions is of utmost importance. Further modeling often provides the only effective means for making these determinations. One looks for effects that are robust when assumptions are varied. Some aspects of structure revealed in data may be robust while others are not.

6. Controlling for Undesirable Effects Observational instruments detect signals composed of real-effect information and random (white) noise. Unwanted influences can introduce systemic (black) noise (Suppe 1985). Physical control attempts to isolate phenomena from contaminating influences. When appropriate controls cannot be physically implemented, measurements often contain real effects combining information both relevant and irrelevant to phenomena of interest. Data reduction imposes vicarious control by filtering, calibration, removal of contaminating-influence effects, and interpolation. Vicarious control often is superior to physical control.

7. Suitably Tailoring Knowledge Claims Scientific claims are about robust real effects. They convey interpretations of data that are themselves models of data (Suppe 1997c). Thus, investigating the robustness of model structures is central to interpreting data and establishing results—regardless of whether data come from naked-eye observations, traditional controlled experiments, remote-sensing measurements, or simulation modeling.

The basic job of interpreting data is discerning real effects inherent in data. Data are inherently restricted in their precision and accuracy. Thus, there is an inevitable vagueness to scientific findings. In the best of circumstances, proper scientific data interpretations never advance a single model of data, but rather claim that the true situation falls within some family of models indistinguishable from the offered model in relevant respects.

Philosophical Confusions

Philosophers concerned with underdetermination of models and theories have been exercised by the belief that valid scientific knowledge must rest on finding a unique model or theory that "fits" the data. But the very nature of data precludes unique fits in this sense. Uniqueness of fit to data is irrelevant to scientific knowledge. What is relevant is that scientific claims be carefully tailored to express robust features common to the family of nonunique models. Thus, whether based on direct observation, controlled experiment, remote-sensing measurements, or the generation of data via simulation modeling, claims carefully crafted to model resolution and suitably tailored to the constraints and uncertainties of the modeled data are at the heart of scientific knowledge.

Though models can in fact reliably support knowledge claims, there are those who continue to deny this possibility. Standard philosophical doctrine is that models are mere analogical or heuristic devices useful only for suggesting scientific hypotheses, theories, or explanations, and that because their truth cannot be uniquely established they cannot function as vehicles for scientific knowledge. Such philosophy is reflected in the views of many scientists, politicians, and other intellectuals who challenge the legitimacy of basing any climate, environmental, or other policy decisions on computationally intensive modeling. The findings in the previous section directly refute such claims.

A recent and lamentably influential paper in *Science*, written collaboratively by two geoscientists and a philosopher, exemplifies the philosophical blunders leading to such incorrect and pessimistic epistemological assessments. Oreskes, Shrader-Frechette, and Belitz (1994) claim that because models employ assumptions that are either false, soft, or underdetermined by data, they are not reliable sources of knowledge. In their view, reliability is intimately tied to truth: if unqualified truth cannot be established with certainty, then reliability remains in question. And since unqualified truth of models can never be established with certainty, neither can their reliability.

They conflate the underdetermination of curve fitting with nonuniqueness of inverse problems (Oreskes, Shrader-Frechette, and Belitz 1994, 642) and utilize the underdetermination thesis to emphasize the intrinsic lack of certainty associated with models. No model can be uniquely selected by any conceivable amount of data. If more than one model fits the data, one is uncertain as to what is true. Hence they say fitting a model to data yields no knowledge.

From this they conclude that "the primary value of models is heuristic" (Oreskes, Shrader-Frechette, and Belitz 1994, 641)—that is, as aids to discovery. Models indicate where future work needs to be done, reveal flaws in other models, and help generate evidence to more firmly establish scientific claim. But, they conclude, however helpful models may be, they are not vehicles of reliable empirical knowledge.

Their first blunder is failure to distinguish the underdetermination of simple curve fitting from the nonuniqueness typical of modeling. This conflation reveals a commitment to the specious idea that uniqueness is of central epistemic import for scientific model selection. This has been shown false. Modeling involves marshaling relevant constraints to circumscribe the range of physically reasonable solutions. Typically no unique solution is possible, but constrained solution spaces yield reliable knowledge when model-based claims are crafted to be insensitive to model uncertainties. What is essential to reliable model-based knowledge claims is that they be no more precise than their robust model features allow. Uniqueness of model fits is irrelevant.

Their analysis commits a second, more fundamental blunder: reliance on certainty as an epistemological criterion for distinguishing modeling from other more allegedly reliable scientific practices. All empirical propositions, not just model-based claims, are uncertain in their sense. Hence, if the existence of uncertainties undermines the epistemic status of modeling, then all science is equally suspect. Consequently, on Oreskes et al.'s analysis it follows that there can be no scientific knowledge.

This conclusion is absurd. It reveals that their criterion of certainty is epistemologically irrelevant to actual scientific knowledge acquisition.

If models are epistemically disadvantaged with respect to other scientific products, it cannot be because their veracity is always open to question. For this is true of all empirical claims and practices. If there is an epistemological wedge to be driven between modeling and other allegedly more reliable scientific practices, it must be found elsewhere.

Some critics of global warming seem to suggest that such a wedge exists. In political debates over global climate change, some contrarians attack conclusions based on complex simulation models because of the associated uncertainties. But all models, whether simple or complex, have uncertainties. Certainty is an unreasonable requirement to impose on scientific results.

The appropriate response to uncertainties associated with models is to craft one's claims carefully, limiting them to robust real effects. Indeed, the standard argumentative form of a scientific paper is to rebut uncertainties surrounding one's claims, showing they are the only plausible interpretations fully consistent with the data (Suppe 1998). In political debates, many of the problems surrounding climate change stem from attempts to draw inferences from complex climate models that go far beyond their robust/nonartifactual findings. This has nothing to do with limitations on the knowledge that can be obtained from models. It is simply misuse of science.

Conclusion

There are no defensible epistemological grounds for challenging reliance on climate modeling results in global-change policy deliberations. Models of data and simulation models are not mere heuristic devices. Models of data are intrinsic to the nature and collection of data both in today's remote-sensing modeling science and in traditional controlled experimentation. If climate change modeling is to be challenged legitimately, it will have to be on the basis of how good specific models are, and whether their application to the debated issues is appropriate.

Notes

We thank Arthur W. Burks, George Philander, John Suppe, John Tukey, Ching-Hung Woo, and Bernard P. Zeigler for crucial insights; Tony Dahlen, Michael Dickson, Owen Gingerich, and students in Suppe's various modeling seminars for help; and Paul Edwards and Clark Miller for patience. Partial support was provided by the National Science Foundation and the University of Maryland General Research Board. Much of the general analysis of modeling and many nonatmospheric examples are drawn from Suppe 1985, 1997a, 1997b, 1997c, 1998, 1999, forthcoming a–c.

1. A terabyte is 1,000,000 (million) megabytes or 1,000,000,000,000 (trillion) bytes.

2. Other examples: In particle physics measures frequently embody holomorphic functions such that representations of real-valued data in complex spaces generate unique model solutions. Symmetry groups also constrain solution spaces. If relative-frequency data are construed as empirical probabilities, Borel spaces are required to represent frequencies and so only models having Borel space compactness properties qualify as solutions. When measurement error is taken into account, unique solutions are not obtained via these methods, though highly constrained solution spaces still do result.

3. Standard error reporting practice is to list measurement error as $x \pm s$, where x is a measurement and s is one *standard deviation* (the range within which 68 percent of measurements lie).

4. Polynomials describe complex curves as weighted sums of variables raised to powers. Such curves are generated by taking a specific polynomial form and fitting it to data by adjusting the *values of the weights* (polynomial equation coefficients) through some procedure such as a least-squares method.

5. B then is a *realization* of L. More precisely: A system S_2 is said to be a *realization* of another system S_1 when there is a many-one behavior-preserving mapping from the states of S_2 *onto* the states of S_1.

6. Intuitively, a model B of A is *embedded* into A just in case the mapping from A to B partitions complete states of A into equivalence classes that become the complete states of B. Whenever A undergoes a complete state transition from s to s^* in B the equivalence class containing s will be succeeded by the equivalence class containing s^*. See Burks 1975 for formal characterizations of modeling and embedding in terms of one system realizing the other.

4

Epistemic Lifestyles in Climate Change Modeling

Simon Shackley

Climate models allow scientists to combine insights from many disciplines, helping to integrate the understanding of complex natural systems. At the same time, they enable scientists to portray climate as a global system, and human interventions as potentially global in their effects. Climate modeling can thus be portrayed as a form of "world building" (see Edwards, chapter 2, this volume).

In building global climate simulations, scientists make numerous epistemological choices and assumptions. These choices affect their final outputs in a myriad of ways, some trivial, others quite significant. The modeling strategies employed by individual modelers, modeling centers, and even national research and funding communities often vary considerably. This chapter explores several approaches to modeling represented among American and British climatologists, tying their systematic variation to a concept I term *epistemic lifestyles.*

In 1993 I worked for several months at the Hadley Centre, the United Kingdom's premier climate modeling center, as part of a sociological study of climate change science and policymaking. The degree of unanimity and consensus among the scientists concerning the Hadley Centre's aims and ambitions struck me at the time. The scientists there did not appear to disagree very much and argued in rather subdued and implicit ways. A year later I visited several American climate modeling centers, where I encountered much greater dissension and argument (within and between centers) over priorities and use of limited resources, methodologies, evaluations and interpretations of model results, and appropriateness of tools, and so on. The American scientists seemed to enjoy argument and debate in a relatively open fashion.

In this chapter I explore and analyze the different approaches to climate modeling I witnessed in the U.S. research community and consider why a similar demarcation was not evident among the British climate modelers. The main source of material for this essay is a set of detailed, semistructured interviews with climate modelers and other scientists, conducted in the United States and United Kingdom in 1993–1995 and followed up by further private communications. I start with a detailed case study of differences between two U.S. climate modelers. I go on to identify different styles of climate modeling and explore the set of institutional, organizational, policy, and scientific elements that lie behind them. I should stress, however, that I am describing only a limited segment of the world of climate modeling, namely, that most directly involved in developing, evaluating, and conducting climate change experiments. Such work represents only the "tip of the iceberg" of climate modeling, and relatively few scientists actually do such work.

Two Visions of Climate Modeling: An Extended Story

In the summer of 1993, a climate scientist from the United States attended an international meeting on climate change. I will call him Scientist B. B later distributed his report on the meeting to selected members of the U.S. climate change research and policy communities. Scientist C, a climate modeler who had been invited to present one of the plenary lectures at the meeting, was one recipient of the report. B described C's presentation of the results from a coupled atmosphere/ocean general circulation model (AOGCM). In an attempt to reproduce the low-frequency natural variability of the climate system, C and his team had run the coupled model as a "control" (i.e., with fixed inputs based partly on observations) for 1,000 years. They then ran the model with an increase in carbon dioxide (CO_2) of 1 percent per annum to the end points of a doubling and quadrupling of CO_2.

In his report, B commented on C's presentation as follows:

What was striking compared to later talks at the meeting was how coarse this resolution [was] and how crude the physics were. . . . (AGCM: R-15, 9 [vertical] layers; OGCM: 4° × 4°, land: bucket hydrology). . . . [Scientist C] argued that the coarseness allowed him to carry out extended runs; however, others are typically running T-42 with detailed physics representations, etc. and are doing much more detailed diagnostics studies. What is not clear is the value to be gained from running a model that is no longer state-of-the-art. [Scientist C] argued on his behalf that one cannot trust complicated models—or maybe even simplified models, but not including the result of the many process studies underway seems a strange way to proceed (even if it saves computer time). (Scientist B, report, August 11, 1993)

The atmospheric model at "R-15" translates to a spatial grid of approximately 4.5° latitude by 7.5° longitude, while "T-42" translates to a grid of approximately 2.8° latitude by 2.8° longitude. B reflected a viewpoint, widely heard in the climate modeling community, that the spatial resolution of climate models should aim to be as fine as possible, this being because more of the features of the general circulation in the atmosphere, such as large-scale storms, cyclones and anticyclones, and equivalent (but smaller-scale) eddies in the ocean, are thereby directly resolved by the model. B also reflects a similarly widespread preference for more detailed physical representations of those climate-related processes that are not directly resolved by the model and instead have to be "parameterized," or represented by a parameter related to some large-scale variable that is resolved in the model.

B's comments attracted considerable attention because he was involved, at a senior level, in designing the research policy of the U.S. Global Change Research Program (USGCRP). His report contained a plea for more central coordination in the research efforts of the USGCRP. He seemed particularly concerned that despite the enormous resources being devoted to the USGCRP, the climate modeling effort in the United States was in danger of losing out to other countries, especially in Europe. For example, he identified a large U.S. research investment in process studies and observations, but noted that the amount dedicated to climate and earth systems modeling had leveled off or even declined. As he put it in a subsequent letter to C: "I have been working in Washington these past months and have seen them adding millions to observation programs and millions to process studies . . . and support for systems modeling is shrinking!" (Scientist B, letter, October 19, 1993)

This situation concerned B because he saw systems modeling as the one activity that could unify the disparate process studies and observations and hence that had a large "value added." B proposed much more coordination between researchers in the United States—for example, through development of "consortia" with the primary climate modeling centers (such as the National Center for Atmospheric Research (NCAR) and the Geophysical Fluid Dynamics Laboratory (GFDL)) at the hub. B had been very impressed by the Max Planck Institut für Meteorologie (MPI) in Germany and the Hadley Centre (HC) in the United Kingdom. Both countries had rapidly developed highly organized and coordinated climate-modeling centers, which were increasingly at the forefront of building complex, comprehensive climate models and providing scientific insights to the Intergovernmental Panel on Climate Change (IPCC) and Framework Convention on Climate Change (FCCC). How could the United States be in danger of lagging behind, given that it had pioneered the climate-change field from the start—and still had many of the leading research scientists, and far larger research resources, than Germany and the United Kingdom put together?

The answer to that question for B appears to have resided in the European countries' very centrally managed and coordinated research effort, and their achievement of broad consensus on how to conduct climate modeling and interpret its outputs. B perceived the United States, by comparison, to have a very fragmented and diverse research effort, with multiple centers, multiple funders, and different and sometimes conflicting scientific and research policy objectives and management approaches.

Not surprisingly, C did not agree with B's assessment of his climate model. He wrote a letter in response, again widely circulated, which initiated further bilateral correspondence. C argued that his group's model compared favorably with others. His choice of a coarse resolution partly reflected the need to "perform many exploratory integrations prior to the successful execution of the main experiment." In other words, the final run written up as a paper and presented at conferences rested on many other, shorter, exploratory runs needed to tune the model, test computer code, study the model's drift, and so on. C's model run was distinctive in integrating the model as a control run for a very long time (1,000 years) in order to simulate low-frequency natural variability or "noise." From this baseline, he could then ask how likely it is that the observed climate change of the past 100 years is due to anthropogenic influences rather than natural variation. Priority was given, and resources devoted, to producing a long control run rather than to increasing the resolution or making parameterizations more complex.

C conceded that "obviously, we would have preferred to use a coupled model with higher computational resolution. . . . The choice of the computational resolution of a model should be made by carefully balancing the scientific requirements and available computer resources" (Scientist C, letter, August 23, 1993).

B's interpretation of this statement was that C's group had had to "make a choice so based on limitations of computer resources that, I would argue, you were doing significantly less than what you are capable of and what you 'obviously' preferred to do." The solution for B was to increase available resources. But he implied that in return, funders would require greater coordination of the climate modeling research community. Lack of computer resources had become increasingly serious, according to B:

The answers [to questions about human-induced climate change] may not yet be certain, but they must be the best we can provide, or we must say clearly that resource limitations are preventing us from doing this. . . . The systems modeling community is experiencing such serious constraints that we do not now have the resources needed to assure that we have a set of system models under development and being used that is at the forefront of current understanding of what could and must be done. We, as a nation, need a range of system models from conceptual to exploratory to definitive that embodies all that we understand as being important to getting better answers; and, when we provide our best answers for important near- and long-term decisions valued in the many billions to trillions of dollars, we need to be incorporating all that is necessary in the models to assure we have the best answers we can provide. (Scientist B, letter, October 19, 1993)

B's concern is related therefore to the perceived needs of climate policymaking and to the collective effort required by the climate science community to meet them. C outlines a different interpretation than B of what are "the best answers we can provide," as explained below:

I am not comfortable with the suggestion that a model with the highest possible resolution and the most detailed parameterizations should always be used for an experiment just because nature is infinitely detailed. . . . I am not convinced that many current parameterizations successfully incorporate the information from field experiments as you implied. This is because it is difficult to fully understand the details of the process involved and how to aggregate the information from a local field experiment to the grid scale of a climate model. . . . In my opinion, a parameterization should be as simple as possible in order to improve our understanding of model behavior, facilitate the tuning of the model, and reduce the computational requirement of the model.

... Although a Manhattan Project of climate modeling, such as what you seem to have alluded to ... can yield a very detailed model, such a model alone may not be very effective in gaining a "predictive understanding" of future climate change. To get predictive understanding, it is also essential to conduct a large number of numerical experiments by using not only the most detailed models but also simpler models and then comparatively assessing the results from these experiments. Whether a model is "state of the art" or not depends upon the objective for which it is used. (Scientist C, letter, August 23, 1993)

C is concerned that centrally coordinated model development and application would reduce diversity in climate modeling. Hence C also writes: "In dealing with such a complicated issue as global change, multiple approaches and strategies are essential."

B responded to C's preference for simple parameterizations as follows:

While we can never expect to be fully as detailed as nature, I would hope and expect that we could aim to realistically represent the full set of critical aspects and interactions. The USGCRP is investing several hundred million dollars each year ... to improve understanding of processes and to provide the basis for improving parameterizations. I believe that much progress has been and is being made (and if this is not the case, that is if all this process research is not improving on the simple parameterizations that have been around for 15 years we really need to rethink our research philosophy). ... The [model] results will be much better received if the most advanced and best tested (not necessarily the most complex) parameterizations are included; clearly there will need to be tradeoffs when very significant increases in computer resources gain virtually no advance in accuracy, but we should not easily give up on requiring accuracy—if modelers are expected to produce more and more accurate results, then we need to demand the resources needed to produce such results. Thus I feel that resources must be insufficient when your simulation is using bucket hydrology, convective adjustment, average diurnal radiation, etc.[1] Further, I do not see the logic of simplifying parameterizations that are working well just because other parameterizations are not working. (Scientist B, letter, September 17, 1993)

In B's account, promotion of more complex parameterizations becomes entwined with the institutional objectives of the USGCRP and its stated ambition of improving parameterizations in climate models (see, e.g., National Science and Technology Council, 1994). B argues that some parameterizations must become more physically realistic to improve accuracy, even if that improvement is incremental and piecemeal, and involves revisiting other parts of the model. While not explicitly disagreeing with B's comments above, C continues to express a preference for relatively simple parameterizations.

It is very important to quantitatively calibrate a parameterization by use of large scale observation (i.e. satellite observation of radiance, runoff from a major river basin, etc.). Because it is very difficult to uniquely determine a large number of parameters from relatively few variables available from large scale observation, a parameterization should be as simple as possible. . . . A prediction of future climate, which is not based upon satisfactory understanding of model behavior, is indistinguishable from the prediction of a fortune teller. . . . It is not obvious that the use of an improved model automatically reduces uncertainty. (Scientist C, letter, October 18, 1993)

Here C is alluding to the danger that parameterizations will become more complex than is warranted by the data or theory available for confirmation.

On the desirability of a high model resolution, B and C are in closer agreement. In particular, both say that higher resolution improves the accuracy with which the hydrodynamic equations (based on classical physical theory) are computed (letter from C, September 23, 1997). Yet B's requirements vis-à-vis the model resolution are somewhat more stringent than C's. B maintains that climate models used in making predictions should resolve the major dynamic features of the circulation in the atmosphere and ocean:

Even for exploratory runs, I am not at all comfortable that using coarser resolutions will produce meaningful results (as, for example, in an experiment coupled with a dynamic ocean) in that the major atmospheric circulation systems that drive the ocean are not well located or of proper magnitude. . . . [A] resolution of 3 degrees [in the ocean model] may be adequate if all one wants is the global meridional [equator-to-pole] heat transport . . . but if one wants to represent the largest contributor to ocean variability (at least on interannual scales), namely the ENSO [El Niño Southern Oscillation[2]] cycle, it seems clear that resolution must be less than a degree. In addition, with bottom water formation apparently a small scale phenomenon, it would appear that fine resolution is likely needed to represent processes that may play a role in abrupt climate change.[3] In that your study was looking at variability, the 4 degree resolution just seems to be an unsatisfactory simplification. (Scientist B, letter, September 17, 1993)

It is clear that B wants C's team to be going full-steam ahead on "improving" their model: making the parameterizations more complex, increasing model resolution to improve the representation of dynamics, including new suites of diagnostic variables, and so on. Hence, B wrote: "that [your center] (like some of the other centers) has to make a choice between making simulations and improving the model aggressively is unfortunate."

C proposes a somewhat different "triad" strategy for model improvements, consisting of

1. Reliable monitoring of thermal forcing (i.e., greenhouse gases, aerosols, solar irradiance, and so on) and basic variables of climate (including temperature and salinity in oceans)

2. Prediction of climate and climate change by models of various complexities

3. Comparison of predicted and actual climate change (including exploring the natural variability to detect a signal from the noise)

C notes that:

> Although it may take a very long time, I believe this may be the only way to slowly but steadily reduce uncertainty in our climate prediction. It is, therefore, very urgent to develop a comprehensive plan for the reliable, long-term monitoring of climate with sufficient accuracy and to implement it as soon as possible. Without such monitoring, it would be impossible to demonstrate the credibility of climate models and prediction of future climate. (Scientist C, letter, October 18, 1993)

This plea for appropriate monitoring should be read not so much as a comment on the USGCRP as on NASA's Earth Observation System (EOS), which many climate modelers have felt to be inadequate in terms of providing relevant, long-term data on climate forcings and feedbacks (Hansen, Rossow, and Fung 1993).

Epistemic Lifestyles

In the rest of this chapter, I explore different ways of being a climate modeler, or different epistemic lifestyles. By *epistemic lifestyle* I mean the set of intellectual questions and problems, and the accompanying set of practices, that provide a sense of purpose, achievement, and ambition to a scientist's work life, as well as the more mundane sense of carrying out those activities necessary to "getting the job done."[4] Additionally, an epistemic lifestyle includes the social networks and connections through which scientists organize their individual and collective work. We can ask, for example, to what extent a scientist is within a well-bounded organization and a highly structured social grouping (see Douglas 1986).

The factors that give rise to different sorts of epistemic lifestyles include many of those identified in sociology of science: disciplinary concerns and practices; institutional culture, structures, and processes; policy and "user" relationships and support; funding sources; peer-group concerns; career trajectories; and so on. The word *lifestyle* indicates the immersion of scientists in an intellectual-organizational setting and trajectory that is only partly consciously chosen and not easily exchanged for something different. Such lifestyles are, of course, only "ideal types," evident to different degrees in any individual modeler or organization.

Climate Seers and Climate Model Constructors

What are the principal motivations and knowledge ambitions of climate change modelers? We can arguably distinguish between the following two groups of modelers:

• Those conducting model-based experiments to understand and explore the climate system, with particular emphasis on its sensitivity to changing variables and processes, especially increasing greenhouse gas concentrations. Such scientists can be termed *climate seers*.

• Those developing models that aim to capture the full complexity of the climate system, and that can then be used for various applications. Such scientists can be termed *climate model constructors*.

The climate seers are most clearly identified with model studies of human-induced climate change, though their interest is more widely on the functioning of the forcings and feedbacks within the climate system and their sensitivity to change (whether induced by CO_2, sulfate aerosols, changes in solar energy input, the effects of a nuclear war, deforestation, desertification, or other factors). The climate seers use climate models because they are regarded as the most appropriate and useful tools for exploring sensitivity to change, including for producing projections of future climate change, and for answering the more fundamental scientific questions about how greenhouse gases (GHGs) change climate processes and with what effects. The seers are usually driven by a desire to understand the scientific processes and feedbacks that will occur in a climate system due to changes in climate "forcing," and to predict the effects of a given change in forcing—that is, something similar to C's "predictive understanding."

The model constructors are much more interested in advancing the actual climate models "for their own sake." The models are an important end in themselves, not merely a means to answer a prior scientific question. The model constructors see their main task as building better climate models of the atmosphere, ocean, sea-ice, and land surface. This, for them, involves higher resolution; better, more physically based, and hence (it is thought) more "realistic" parameterizations; performing better diagnosis; and so on. The assumption behind such development work is that a better simulation of the *current* climate will result. The models can then be coupled to one another without the need for correction factors or excessive tuning (regarded as "fudges"). The model constructors' assumption is that a "best possible" simulation of reality is feasible and desirable. The model that produces it can then be used for a whole range of applications, whether it be to study atmospheric chemistry, climate interactions, land-surface effects, paleoclimatology, or the enhanced greenhouse effect.

Herein resides an important distinction between the seers and developers. For the seer, *which model is "state of the art" depends on the model's intended application* (as C expressed it above). For the model constructor, by contrast, *a single state-of-the-art model exists irrespective of its application.* It is defined as one whose control simulation is close to the observed climate, that uses the most up-to-date observations as inputs and for verification, and that contains the most detailed, most physically realistic parameterizations (sentiments expressed by Scientist B).

Seers, Constructors, and Parameterizations

The difference between seers and constructors is illustrated well by how each group views parameterizations. As we saw above, C, who approaches the ideal-type climate seer, views parameterizations not as realistic representations of natural processes, but rather as means to create models that are effective for answering questions about (human-induced) climate change. This is achieved through tuning the model in accord with the modeler's own judgment of what is a good climate simulation, while keeping the parameterization sufficiently simple that its influence in generating the representation of physical processes, and the response of the model to perturbation, can be understood. Climate seers do not negate the importance of better parameterizations, but rather see little room for improving on them in the medium-term future. For a model constructor, on the other hand, parameterizations should attempt to represent real physical processes as closely as possible. Hence the relatively simple parameterizations developed in the 1960s (convective adjustment, bucket hydrology, and so on) are viewed as inadequate for the contemporary task of building better models. In summary, while for climate model constructors the sole purpose of parameterizations is the representation of physical referents, for climate seers physical referents are only one consideration to take into account alongside other factors related to the model's intended application (tractability, computer power requirement, interpretability, and so forth).

The representation of clouds provides a good example of the difference between these contrasting approaches. The simplest parameterization for clouds entails prescribing the amount of cloud, zonally or globally, according to climatological means. More complicated schemes involve predicting the amount of cloud based on other model-resolved variables. In many early models, for example, whenever the relative humidity in a grid box reached a given quantity (e.g., 95 or 99 percent), clouds immediately formed within that box. In many respects this parameterization is unrealistic. For example, clouds can form at much lower relative humidities than 95 percent, even down to 60 percent in the case of cirrus clouds. Climate seers at GFDL therefore began to use a somewhat more complicated cloud parameterization that takes account of the change of relative humidity with height. Model constructors, however, also wish to represent the influence of different cloud types and their highly complex radiative and microphysical properties; their cloud parameterizations can be quite elaborate (McGuffie and Henderson-Sellers 1997; Rasch and Williamson 1990).

All climate modelers agree that the climate system is very sensitive to relatively minor changes in the radiative and microphysical properties of clouds (such as changes in the ratio of liquid water to ice). But climate seers have been skeptical of parameterizations that aim to predict cloud optical and microphysical properties, because of the difficulty of confirming and evaluating such formulations. Even one of the groups that developed more complex formulations warns that "although the revised cloud scheme is more detailed, it is not necessarily more accurate than the less sophisticated scheme" (Mitchell, Senior, and Ingram 1989, 132). Some climate seers have expressed a concern that complex, but unverifiable, physical parameterizations of clouds are in danger of confusing analysis and understanding of model behavior. "Solving" the problem of clouds is seen by them as a task that will take several decades (interview, April 25, 1994). One seer even likened attempts to represent the physics of clouds in climate models to Don Quixote tilting at windmills!

The dilemma of the climate-model construction approach is that complexity does not increase in neat increments. One scientist expressed this as follows during an interview:

> You go through at least two or three stages of model development. One is where you do produce something very simple and that just has one or two constants which you set and you're done. And then you come back and you can put in something much more complicated, a lot more physical basis to it, but it's now got 27 different numbers you've got to set. And you don't know how to set them, because those numbers depend on more things you haven't done yet. So you've just got to set them somehow. At the next stage you push the problem back a level because now you're modeling the things the numbers are based on. So now you take out the bunch of arbitrary numbers in the first parameterization, but you have arbitrary numbers in what you just put in. So there's a stage where you go from something simple, where you only have one or two numbers that you pick, to something that is complicated, with a bunch of arbitrary numbers in it, and then you go beyond that to predicting what you need in order not to have a bunch of arbitrary numbers any more. . . . At this point . . . we're in that second stage of adding a bunch of arbitrary numbers. (Scientist D, interview, April 15, 1994)

Increased complexity frequently has the counterintuitive effect of reducing the quality of the control climate simulation. This is because the model's parameterizations have been tuned *to each other*, so that if one is changed, the other parameterizations are no longer "in sync" and the quality of the simulation suffers. The solution in the model-construction approach is to open up the other parameterizations and to improve them by making them more physically realistic as well. The belief is that a better representation of reality in all major parameterizations will eventually come to simulate the real processes of the climate system.

Climate seers tend to see such model developments as a luxury they can ill afford. One compared the different purposes of models to the different purposes of cars in Formula One racing and everyday driving (interview, April 27, 1995). Formula One vehicles have very high performance capabilities but are notoriously unreliable. Like the climate model constructor's ideal model, they have state-of-the-art technologies and incorporate the latest insights, but for routine everyday use they are not so reliable. By contrast, climate models used in CO_2 perturbation experiments need to be reliable enough to operate for very long time periods. For example, some long simulations take many months to compute; hence climate seers have to be reasonably confident that their models are not going to "crash" and can be easily fixed if necessary. What is then required is more akin to the family car, which starts every day on turning the ignition switch and can be easily and cheaply maintained and repaired.

For similar reasons, the climate seers also perceive model development quite differently from the model constructors. Seers tend to be more cautious about changing a model that "works" and is reliable. An incremental reductionist strategy is adopted in changing the model, with the influence of each model component analyzed separately. Only after the model is well understood is it appropriate (in the climate seers' opinion) to add complexity. Additional elements of complexity are then added one at a time, and their implications for the rest of the model are analyzed. This is a much less open-ended and more time-consuming strategy (relative to how much the model changes) than the many-pronged approach of model constructors. Using a simpler model also means that it is possible to conduct multiple runs for the purpose of model testing and diagnosis. Furthermore, climate seers argue that there is little point in introducing a level of complexity in one model component that is negated by simplicity elsewhere in the model.

An example of the latter condition is the parameterization of convection (the movement upward of warm air masses). Climate model constructors are critical of simple convection schemes (such as the moist convective adjustment scheme in the GFDL model) and have developed more physically based approaches. However, these new parameterizations were initially used in models without a diurnal cycle of radiation. The influence of day-night heating on convection far exceeds that arising from the difference between moist convective adjustment and the more complex schemes. The value of the added complexity in a model without diurnal heating was questioned by climate seers (interview, April 27, 1994). The model constructors would instead perceive the case for including diurnal heating in the model to be strengthened by the adoption of the more complex parameterization of convection.

The Heat and the Wind: Thermodynamicists and Dynamicists

Climate is commonly understood by scientists as the product of the interaction of thermodynamics (the movement of heat) and dynamics (the atmospheric and oceanic circulation). Yet some climate modelers appear to give more emphasis to thermodynamics, while others give more emphasis to dynamics.[5] According to the thermodynamically oriented, the climate system is driven primarily by changes in heat fluxes—such as those caused by increased concentrations of greenhouse gases—and associated thermodynamic feedbacks: the ice-albedo feedback, cloud changes (influencing the amount of incident or reflected radiation, and so on) and water vapor feedback.[6] As Scientist E put it: "To a first order, if you think of the whole world as a box—how much heat goes in and out of the box—that [dynamical] stuff doesn't matter" (interview, April 25, 1994). E went on to explain:

Scientist E: My experience with heat has been that if you get the global mean right, you get the rest of the places right too, more or less, as long as you have sea-ice at about the right places.

Interviewer: I was talking to Peter Stone at MIT about some of the analyses they've been doing in which they inferred the heat fluxes in the ocean [from AGCMs], and they seem to think that they're often quite in error, with the ocean not transporting enough [heat].

Scientist E: But the atmosphere will adjust to it pretty fast. You might not get the split right between the atmosphere and the ocean. But if your polar-equator temperature gradient gets a bit too strong, the winds will blow a bit harder and the atmosphere will transport more heat.

Interviewer: But wouldn't that change the sensitivity of the system to perturbation?

Scientist E: It's conceivable. . . . My view of the world is much more thermodynamic. It's much more important to get the sea-ice edge in the right place, and to have about the right temperature distribution, than it is to worry about what the winds are. It's just my bias. As long as the circulation looks something like the observed . . . I forget it.

If you think the transports are what makes everything go then you have to worry about model resolution, because the higher your resolution, the better the transports you're getting because you're doing all the finite differencing better and all that kind of stuff. And if you're a thermodynamicist you worry about where the sea-ice edge is, what your cloud albedos are, stuff like that.

. . . The dynamicists are more mathematical and therefore more rigorous in their proofs of things, and therefore more theoretical. The thermodynamicists tend to think more physically based, and [their] arguments are all physical. They don't use equations—they wave their hands a lot more when they talk. (Scientist E, interview, April 25, 1994)

E described how his model transferred a heat imbalance at the top of the atmosphere almost exactly to a heat imbalance at the interface between the atmosphere and ocean. The dynamical properties of the atmosphere between the ocean-atmosphere interface and the top of the atmosphere do not appear to change the heat balance, providing supporting evidence that the system can be modeled, to a first approximation, thermodynamically.

To date, those with a thermodynamic bias have dominated model experimentation on anthropogenic climate change. Dynamicists, however, are still significant to the evaluation and reception of such research because they form a more or less skeptical hinterland of climate modeling. This situation is likely to have influenced many political, media, and policy discussions about the science of climate change. Informally, dynamicists are sometimes critical of the existing models of the leading thermodynamicists, much in the vein of Scientist B.[7]

Dynamicists frequently have a background in numerical weather prediction (NWP) and university-based meteorological research. They are more concerned about the existence of model errors in the control run than are thermodynamicists. Provided they are not large, such errors are not thought by thermodynamicists to influence model response to CO_2 doubling—that is, they believe the difference between the control and perturbation simulations to be acceptably similar to the (unknown) difference between unperturbed and perturbed conditions in the real climate. This is a classic thermodynamic-type assumption, since it treats the climate system as one in which the quotient of the inputs and outputs remains (within limits) more or less constant. By contrast, the dynamicists are more inclined to regard errors in the control run as influencing the simulation's response to a perturbation (such as increasing greenhouse gases) in unrealistic ways. Hence they see the reduction of errors in the control simulation as critical, while thermodynamicists see diminishing returns for such effort.

A good example of a typical dynamicist concern is that to be adequate for climate change experiments AOGCMs should be able to simulate ENSO. Some dynamicists have argued that anthropogenic climate change will have some of its most significant effects through natural modes of variability, such as ENSO; hence model simulation of such processes is critical to understanding the broad-brush pattern of climate change (Palmer 1993).

The categories of "thermodynamicist bias" and "dynamicist bias" should be seen as ideal types. The issue is not whether the climate system is *either* thermodynamic or dynamic, since a key rationale of GCMs is, after all, to represent the interaction of these. Instead, the issue is the relative weight accorded to thermodynamic or dynamic considerations in models for studying anthropogenic climate change.[8] E is closer to the thermodynamicist "ideal type" than most modelers, while F expresses a more common balance between thermodynamics and dynamics in the following interview quotation:

A lot of the short-term variability in the atmosphere is . . . going to be governed by the ability to represent the atmospheric dynamics, whereas some of the long-term changes like response to CO_2—the overall level of warming—is much more sensitive to the model physics. . . . One of the arguments goes that one of the largest aspects of interannual variability is El Niño; therefore, a change in frequency of El Niño would be a substantial climate change. I think there's some truth in that but I wouldn't want to push it too far. But I would certainly like to be using a model that makes some attempt to produce oscillations that are like El Niño. But I think one can learn a lot even though the current climate model doesn't really simulate El Niño. (Scientist F, interview, March 11, 1993)

This quotation illustrates how scientific perceptions of the appropriate balance of thermodynamics and dynamics depend on the specific research questions posed and the time and spatial scales thereby implicated. Dynamics become more important for climate seers when the research explores short-term variability and regional climate changes.

Most climate seers to date have adopted a balance of thermodynamics and dynamics similar to that of Scientist F. This partly reflects the dominance of research questions devoted to the long-term effects of perturbation by CO_2 at large spatial scales; a quantitative description of the climatic effects of increasing greenhouse gas emissions has (to date) required a thermodynamically biased approach. Thus the commitment to a thermodynamics bias *is partly a function of the objectives and key research priorities and questions of the research* (i.e., a function of being a climate seer). In turn, the adoption of a thermodynamically oriented approach will have some bearing on what research questions and objectives are thought worth posing and feasible to approach. The research objectives and questions and the adoption of a thermodynamic bias come to mutually reinforce one another (although not irrevocably).

Model constructors appear to be more eclectic in adopting both thermodynamic and dynamic approaches. Perhaps this reflects their desire to build the "perfect" climate model (which would incorporate both these dimensions). However, it does appear that the constructors may orient themselves relatively more toward dynamics, perhaps because so much work has already been devoted to thermodynamics, which is therefore better incorporated into existing climate models. Some of the "big" remaining problems in anthropogenic climate change research, such as developing credible regional climate change scenarios, seem to require much work on representing dynamics. On these issues, the climate seers have made less progress.

Epistemic Lifestyles in Three Climate Modeling Centers

So far I have tried to account for empirically defined ways of being a climate modeler in terms of differences in research objectives, methods, assumptions, and experiences, which (taken together) I have called *epistemic lifestyles*. The following discussion explores these distinctions in three major climate modeling centers that were among the principal groups studying anthropogenic climate change as of the early- to mid-1990s: NCAR, GFDL, and the Hadley Centre (HC) of the UK Meteorological Office (UKMO).

All three centers are large, resource-rich organizations, running complex coupled AOGCMs in transient forcing experiments (i.e., simulations in which CO_2 concentration is increased by an increment each model year to the point of doubling and beyond). However, each center is involved in such research to a different degree.

GFDL, where model constructors are least in evidence, has had a major group devoted to the anthropogenic climate change issue since the 1960s. Such simulations have also been the *raison d'être* of the Hadley Centre since its establishment in 1990 (from a group working at the UKMO since the mid-1970s). However, HC devotes considerable efforts to model development and has many model constructors, who coexist relatively harmoniously with climate seers. At NCAR only a small team (three or four scientists) works in the climate seer mode. NCAR's prime organizational mission is to serve the U.S. academic climate research community through provision, development, and testing of the Community Climate Model (CCM). A large group of scientists at NCAR is devoted to the CCM. Their work includes helping university academics to use the model for basic understanding of climate and to answer more "academic" questions, such as studies of paleoclimates, vegetation-climate interaction, and specific parameterizations. This group is much less concerned with immediate policy-relevant questions, such as CO_2 doubling (global warming) experiments. This gives the model constructors at NCAR a strong position within the organization, while they are least dominant at GFDL. Divisions between seers and constructors are more evident in the U.S. centers, where it is not uncommon for criticism and disagreement to bubble up in informal discussions.

U.S. science, it has been noted, tends to be more contentious and fragmented than European science, whose disputes are less evident and possibly hidden from view (Jasanoff 1986). Difference is also more likely to emerge in the United States because of the multiplicity of centers and funding sources. While NCAR is funded by the National Science Foundation, and by a series of "soft" funders, including the U.S. Department of Energy (USDOE), GFDL is part of the National Oceanic and Atmospheric Administration (NOAA) within the U.S. Department of Commerce. Two other major climate change modeling centers in the United States are the Goddard Institute for Space Studies (GISS), part of the National Aeronautics and Space Administration (NASA), and a group at the Lawrence Livermore National Laboratory (LLNL), a USDOE-funded national laboratory whose work was formerly dominated by nuclear weapons design. But there are also many centers and departments in universities that work on climate change using models (U.S. Global Change Research Program 1994, 1995). Hence there are four major climate modeling centers and a host of smaller university departments, all feeding climate modeling results and insights into different parts of the U.S. government and into Congressional hearings. It is perhaps not so surprising that differences in objectives and scientific styles are accentuated in such a system, a phenomenon also observed in the energy modeling field in the United States during the 1970s and 1980s (Baumgartner and Midttun 1987). The pluralistic funding and policy advisory process in the United States encourages competition between centers, reflecting competition between different offices within government. Research organizations are motivated by this diversity (and by peer pressure) to carve out their own "niche" in the climate change modeling field, rather than to compete "head on." Differences in research style are thereby accentuated and embodied in organizations.

GFDL, with its very long control and perturbation simulations, is especially strong in the climate seer role (see the many papers by Manabe and his collaborators, e.g., Manabe 1997; Manabe, Stouffer, and Spelman 1994). NCAR is especially strong in climate model construction and dynamics. The climate seers at NCAR are a distinct group within the organization and have tended to concentrate on more dynamical aspects of coupled models, such as ENSO and the heat-transport effects of a high ocean model resolution (see papers by Washington and Meehl, e.g., Washington 1992; Meehl 1990). GISS, meanwhile, is especially strong in producing new policy-relevant climate model formulations and simulations and in communicating new findings effectively to the policy community and beyond. GISS's rapidly produced prediction of the climatic effects of the eruption of Mount Pinatubo is a good example of such a role (Hansen et al. 1992), while its climate sensitivity studies have demonstrated the strong but indirect effect of particulate aerosols on climate, and the need for new long-term monitoring of climate forcings and feedbacks (Hansen, Rossow, and Fung 1993; Hansen et al. 1997). LLNL has concentrated on the development of new computational techniques, such as massively parallel processing, and on interactions between atmospheric chemistry and dynamics.

The HC is quite different in that it alone constitutes the focus of climate change modeling in the United Kingdom. Hence it is competing internationally but not intranationally. The HC is regarded as the main national source of scientific input on climate change issues not only to the wider scientific community, but also to all departments of government, to industry, and to NGOs. It enjoys a close, hands-on relationship with the Department of the Environment quite unlike what is found in the United States, or indeed in most European countries. Thus competition between centers results in differentiation and specialization in the United States, while the very *lack* of competition for funding and national recognition results in a more comprehensive approach to climate modeling in the United Kingdom. The close connection to government policymaking may also contribute to the cohabitation of epistemic lifestyles. Policy "needs," expressed through a single large government research contract placed at the HC, facilitate this "unified" approach to climate modeling. The contract requires the HC to provide state-of-the-art, authoritative climate projections for the IPCC and to feed into internal negotiations within government on the national position vis-à-vis the FCCC. Modeling to address policy-relevant questions thus becomes a priority for research.

In the mid-1990s the HC responded to policy-led questions about the ability of its AOGCM to reproduce the past century record of global temperature change. The policy mandate authorized research leaders at the HC to request help from climate model constructors in order to build better representations of relevant physical processes, especially of sulfate aerosols, which climate seers could then apply in model runs. The hierarchical culture of the UKMO and HC greatly facilitated the ability of research leaders to implement a cohesive strategy, given that such leaders at the HC have tended to be climate seers and have exerted considerable control over the deployment of computer and personnel resources. By contrast, GFDL and NCAR are less hierarchically organized. They are characterized instead by distinct research "baronies," competing for credibility and resources.

The HC shares the so-called unified model with the weather-forecasting scientists at the UKMO. The "same" model is adapted and used by scientists with different sets of concerns and interests, and may consequently work as a "boundary object" (Star and Griesemer 1989) usable in exchanges among several different epistemic lifestyles. This contrasts with the situation at NCAR, where one finds separate modeling groups, each with a different funding stream, and a different modeling approach or style, each in some degree of competition with the others. It also differs from GFDL, where a distinct climate seer GCM group coexists with a range of other process specialists and modeling efforts. The climate seers have established a strong, distinctive approach at GFDL and they tend to be unwilling to accommodate the agendas and priorities of model constructors.

The unified-model approach of the HC is perhaps only possible in a cultural context where more hierarchical control and a strongly "hands-on" policy direction of research planning are accepted. Under these conditions research is inevitably less driven by the concerns of distinct peer communities, whether they be seers, model constructors, dynamicists, process specialists, or oceanographers. Another consequence of the more centralized control of collective work is that individual creativity and initiative may be more inhibited and secondary to collective goals. However, the constructors and dynamicists at the HC have other important ways to pursue their interests and objectives. For example, wide-ranging collaborations exist with other centers and universities to develop the parameterizations or other aspects of the climate model. This work occurs simultaneously with, but rarely feeds directly into, the configuration of climate change perturbation experiments. At the HC, the constructors and dynamicists are also closely linked to weather forecasting (NWP) research. Indeed, parameterization specialists work on both the climate- and weather-forecasting versions of the "unified model"; they hold joint appointments in the HC and UKMO. In other words, constructors and seers are accommodated at the HC—despite the overriding influence of the latter—by networks of research and peer-group authority that extend beyond the HC into the high-status UKMO forecasting research (where model construction and dynamics are especially valued) as well as into other universities and research centers.

The differences between the HC and the two American centers lead me to propose a third epistemic lifestyle—the *hybrid climate modeling policy style*—in which the policy-influenced objectives and priorities of the research organization, as defined by its leadership, take precedence over other individual or organizational motivations and styles. Such objectives and work program are decided in negotiations between senior climate researchers and a coterie of policymakers in government. The available resources are then deployed by the organization as a whole in pursuit of those objectives, largely regardless of whether they are more thermodynamic, dynamic, model construction related, and so on. While the other two lifestyles also imply a degree of collective work, this third lifestyle is distinct in the strong boundaries around, and the differentiating roles and status within, the organization. The quid pro quo for consensus surrounding the style is a double identity for some researchers: part and parcel of the collective effort, but also individuals with independent interests, objectives, and research collaborations.

Intraorganizational negotiations are also important in the U.S. centers, of course. There, however, policymakers rarely have the direct influence over decision making about research priorities observed at the HC. NCAR's orientation toward dynamics and model development created a general context in which the climate seers were put somewhat on the defensive. And despite the polyarchy of research baronies at NCAR, internal peer-community pressure toward dynamically oriented model construction appears to have been rather strong. In addition (or as a consequence?), the climate seers at NCAR adopted a management style of reasonable tolerance, leading to negotiation and mediation with the other approaches (on decision styles, see Downey 1992). This intraorganizational context may help account for the rather more "conservative" approach of the climate seers at NCAR, compared with those at other centers, with respect to shortcuts, approximations, and correction factors.[9]

Concluding Comments

I have described three empirically identified differences in ways of doing climate modeling, or *epistemic lifestyles*: climate seers, climate model constructors, and a hybrid climate modeling policy style. The range of factors influencing which epistemic lifestyle has been adopted in different modeling centers included the following:

· Disciplinary/research experience background, especially the historical success or otherwise of thermodynamically or dynamically biased research approaches

· Organizational location, objectives, main funders, main users and customers

· The role of academic collaborators and users of models (e.g., in the case of NCAR, oriented toward model construction and dynamics)

· The role of policymakers in negotiations over research priorities and directions (e.g., at the HC)

· The role of organizational culture, especially along the continuum between centralized research direction and separate research "baronies"

· The opportunities for different epistemic styles to treat the climate model as a boundary object and to develop coexisting alternative axes of interaction and authority (e.g., to NWP research in the case of the HC)

· The role of different national cultures of research, especially the degree to which research becomes fragmented and specialized, or cohesive and comprehensive, because of funding arrangements and connections to policy institutions

These distinctions (among others) provide some insight into the range of different opinions, practices, and priorities within the climate-modeling community. But what of the wider relevance of the existence of different epistemic styles? Do the arguments here shed any light on the political debate over the role of climate change science? A key aspect of that debate, as discussed in this book at several places (see especially chapters 1, 7, and 8), is the crucial role of a high degree of scientific consensus in constituting and reinforcing a set of political and social beliefs concerning the desirability of reducing greenhouse gas emissions. Given that GCMs are, arguably, the major pillar upholding the powerful scientific consensus, any difference of opinion within the scientific community concerning the status and validity of GCMs will have a wider political significance.

Political factions opposed to, or simply wary of, calls for action on greenhouse gas emission reductions will find solace and intellectual support in the critical hinterland of GCM modeling populated by the climate model constructors. The appeal is simple enough to understand: the GCMs typically used by climate seers are perceived by climate model constructors as rather crude tools for simulating the climate system. Can those "crude tools" really be trusted to generate reliable scientific knowledge of climate change and the role of anthropogenic influences? The climate model constructors (knowingly or not) provide much critical ammunition from a powerful "insider" position of knowledge and authority, on which the contrarians draw in challenging the cognitive authority of GCMs.

One wonders, however, whether those "crude tools" will ever become sophisticated enough to win over the contrarians. Are the contrarians assuming that scientific certainty over the question of anthropogenic climate change can be settled once and for all by a statistical level of proof at the 95 or 99 percent confidence level? This chapter has not set out to answer those questions, but the existence of epistemic lifestyles suggests that we should not expect definitive levels of scientific certainty and consensus at such statistical levels to emerge. We may instead have to come to see scientific uncertainty as a multifaceted, multidimensional concept. Statistical levels of certainty may be appropriate under certain assumptions, beliefs, and institutions but not under others associated with a different epistemic lifestyle.

The epistemology of science and the sociological analysis of science in practice illuminates the interweaving of academic disciplines, organizational and institutional objectives and trajectories, policy influences, and the epistemic "spaces" that emerge in our intellectual understanding of the climate system. Epistemic lifestyles is one way of indicating the discrete packages of viable scientific endeavor that emerge from this interweaving. Scientific certainty and consensus cannot simply sit uncontested above epistemic lifestyles, but have to be produced by, and in a negotiation between, those epistemic practices.

The existence of epistemic lifestyles suggests that the climate model constructors' reservations about current GCMs used in studies of anthropogenic climate change are not simply grist for the mill of the contrarian cause. While questions about the GCMs used by climate seers are at the fore in the concerns and motivations of the climate model constructors, this is not the only appropriate level at which to explore the beliefs of climate model constructors vis-à-vis anthropogenic climate change. Rather, a "higher-level" negotiation between the climate model constructors and climate seers appears to find more agreement on the likely role of human-induced greenhouse gas forcing of the climate system than the apparent criticism of current GCMs would suggest.

More research is required to understand the character of negotiations between epistemic lifestyles. A practical message for the climate-modeling community is that mechanisms for promoting negotiation between scientific viewpoints, based on a fuller understanding of where those positions come from, would help to present a coherent position on key scientific and policy questions, which nevertheless acknowledges the vital and necessary role of diversity in the practice of climate science.

Notes

An essay like this is really the joint product of a sociologist and the many climate scientists and other commentators and colleagues who have generously provided assistance. In particular, Ron Stouffer first suggested the distinction between dynamics and thermodynamics, while Syukuro Manabe and Michael MacCracken shared their correspondence with me, gave me permission to use the material, and provided me with feedback on the first draft. James Risbey and Peter Stone also helped in developing some of the ideas here in the context of a detailed study of flux adjustments in coupled climate models (Shackley et al. 1999). Paul Edwards has provided very supportive editorial guidance and feedback. I am also very grateful to the following scientists: David Bennetts, Byron Boville, Kirk Bryan, Ulrich Cubasch, Klaus Hasselman, James Hurrell, Jerry

Mahlman, Gerald Meehl, John Mitchell, Tim Palmer, Philip Rasch, Michael Schlesinger, Joseph Smagorinsky, Warren Washington, and Richard Wetherald. It goes without saying that none of the above bear any responsibility for my interpretations of what they told me. Thanks go to the Centre for the Study of Environmental Change at Lancaster University, to the United Kingdom's Economic and Social Research Council, and to DG XII of the European Commission for providing me with the opportunity to conduct this research. Travel was assisted by grants from the Royal Society (London) and the National Center for Atmospheric Research, and these are much appreciated.

1. *Bucket hydrology* is a simple way of representing the movement of water from the land surface (McGuffie and Henderson-Sellers 1997, 168). *Convective adjustment* is a way of representing the vertical heating in the atmosphere due to the transfer of energy by moist and dry convection; it does not aim to represent realistically the physical process of convection (McGuffie and Henderson-Sellers 1997, 111–113).

2. The *ENSO* is a large-scale fluctuation in ocean and atmospheric dynamics in the southern Pacific, extending also to the Indian Ocean and the mid-Atlantic. An El Niño–Southern Oscillation event starts when the usually cold water off the Pacific coast of South America becomes warm, indicative of a massive change in the direction of trade winds and ocean currents. ENSO creates large changes in precipitation patterns, causing flooding in some places and droughts in others. The global scale and significance of ENSO has only been realized in the past fifteen years or so. For a good discussion, see Glantz 1996.

3. *Bottom-water formation* refers to the importance of the "downwelling" at a few ocean locations of cold and/or dense saline water. In the northeastern Atlantic, for example, the warm saline waters from the equatorial Atlantic cool rapidly and sink. Such downwelling is an important component of the *thermohaline circulation*, a massive ocean-circulation pattern essential to the temperate climate of northwestern Europe. Scientist B is referring to the possible sensitivity of this circulation to the precise position at which deep water forms and, in its extent, to the precise locational input of freshwater in the North Atlantic region (see, e.g., Rahmstorf 1997).

4. The idea of "styles" of scientific thought and practice is well established in the sociology of science. See for example Fleck [1935] 1979; Rudwick 1982; Maienschein 1991; Knorr-Cetina 1991; Hacking 1992.

5. I owe this distinction between the thermodynamic and dynamic style to discussions with Ron Stouffer.

6. The ice-albedo feedback occurs when warming melts ice deposits, rendering the surface darker and therefore less reflective to solar radiation, so promoting further heating. Warming also increases the water content of the air, so promoting further heating through the potent greenhouse gas properties of water vapor.

7. Examples of academic and semipopular literature from dynamicists that promotes a more circumspect view of climate models includes Pielke 1991, 1994; Palmer 1993; Stone 1992.

8. One distinguished climate modeler, commenting on a draft of this chapter, argued that it is impossible for a climate scientist to know whether dynamics or thermodynamics is more important because climate "is sustained through close interaction of these two types of processes." Hence, to distinguish climate modelers in this way is "highly misleading" (letter, September 23, 1997). In response, I have downplayed the distinction, but there still seems to me, and to other climate modelers (see note 6 and the *climate modeling pyramid* of McGuffie and Henderson-Sellers (1997, 44)), something of value here as a subsidiary, lesser distinction to the climate seer/model constructor category.

9. For instance, NCAR climate seers—unlike their compatriots at the HC, GFDL, and Max Planck Institut für Meteorologie in Hamburg (MPI)—have not used flux correction in coupled models of the early 1990s (Shackley et al. 1999). Nor have they used so-called accelerated spin-up methods for getting ocean models into equilibrium, use of which has allowed GFDL and MPI to run very long control simulations. In both cases, the reasons given by NCAR climate seers are similar (if less pointed) to the criticisms of these techniques by dynamicists and model constructors (interview, April 11, 1994).

5

The Rise and Fall of Weather Modification: Changes in American Attitudes toward Technology, Nature, and Society

Chunglin Kwa

Public understanding of the environment rests on a complex and shifting blend of science and culture that fundamentally shapes both scientific practice and social policy. Today, for example, human modification of the atmosphere is seen by many as among the greatest threats to the earth's environment. This conception underpins worldwide efforts to understand and avoid anthropogenic climate change.

Yet Americans have not always viewed modification of the atmosphere in such starkly negative terms. For much of the postwar era, in fact, Americans supported scientific research on ways of deliberately modifying weather in an effort to control the fury of nature. Indeed, the United States supported a vigorous research program into deliberate weather and climate modification for three decades, from the 1940s until well into the 1970s. By the late 1970s, in the wake of rising concerns about negative human impacts on the environment, much of this support had vanished. Nevertheless, remnants of the promotion of deliberate intervention in atmospheric systems can be seen as late as 1980, in a report by the U.S. Department of Energy promoting efforts to control the climate through selective additions of carbon dioxide to the atmosphere, and even more recently, in a 1992 U.S. National Academy of Sciences report suggesting that "geoengineering" technologies be developed to counter the effects of global warming (U.S. Department of Energy 1980; Committee on Science, Engineering, and Public Policy of the U.S. National Academy of Sciences 1992, 433–464). By exploring how and why American attitudes toward deliberate atmospheric modification changed, this chapter seeks to shed light on the intersections between science and politics that today underpin public understanding of global warming.

In January 1971, Robert M. White, the Acting Administrator of the newly formed National Oceanic and Atmospheric Administration (NOAA), expressed his belief that deliberate human modification of the weather should be regarded as a viable technology.[1] White said that "almost certainly" there were means to increase precipitation in a predictable way, that efforts to reduce the winds of Hurricane Debbie in 1969 had produced "encouraging results," and that "significant attempts" could be made to reduce lightning to aid in forest fire suppression. He announced that the World Meteorological Organization (WMO) had officially recognized the importance of weather modification, and he applauded the steps taken by the U.S. federal government to advance the field (White 1971). In retrospect, White's speech appears to have been the swan song of weather modification. Two years later, the secretary-general of the WMO noted that "weather modification operations related to precipitation are a waste of money and effort" (Hart and Victor 1993, 676). And, while it took White and other meteorologists several years to arrive at a similar conclusion, their views became increasingly irrelevant as the U.S. government progressively eliminated funding for weather modification research over the course of the next decade.

Weather modification's proponents made so many grand promises about its future efficacy that it is tempting to believe that weather modification research ended simply because experiments showed that it did not work. This chapter argues otherwise. The field existed for most of its history with little clear evidence that rainmaking and hurricane abatement worked, but there was equally little clear evidence that it did not. No convincing failures accompanied any of the seemingly distinct, random decisions that ultimately sealed the technology's fate. The field could conceivably have continued for many years. Meteorologists certainly argued that its potential had not been exhausted, and that it should continue to be funded. Why then did America abandon its efforts at controlling the weather and climate?

The short answer, for which I will argue in this chapter, is that weather modification fell prey to large-scale changes in American attitudes toward technology, risk, society, and nature. These changes, embedded in broad patterns of societal change in the United States (see Hays 1989), turned what appeared, early in its history, as an uncertain but promising technology for mediating nature's violence into what appeared increasingly as a risk to the earth's fragile atmospheric environment.

Weather Modification as a Research Field

One of the ironies of weather modification research is that meteorologists, who became its strongest proponents later, were initially its largest skeptics. Vincent J. Schaefer is usually credited with having performed the first experiment in weather modification, in 1946.

Although Schaefer's experiment was in fact by no means the first in American history (Spence 1980), this experiment set in motion a chain of events that led to scientific recognition of weather modification as a research field more than ten years later.

Langmuir and Schaefer

During World War II, Schaefer was the laboratory assistant of Irving Langmuir. Langmuir was working for General Electric on some chemical studies related to warfare, partly under Army contracts. Studying the conditions under which aircraft icing occurs, they noticed that, occasionally, clouds at temperatures below freezing did not produce snow. This phenomenon had been well known among meteorologists for more than twenty-five years, but Schaefer began experimenting with it with the help of a GE household freezer. Exhaling his breath, he readily produced supercooled clouds. When one day he put a bit of dry ice in the freezer, he observed an instantaneous and spectacular change in the supercooled cloud. The same day, he also discovered that the best results were obtained with a minute amount of dry ice. Langmuir then arranged for him to test this result on a real cloud in an experiment conducted near Schenectady, New York, in November 1946 (Schaefer 1968a). The test successfully reproduced the laboratory observations of cloud physics.

In early 1947, a contract was signed between GE and the U.S. Army Signal Corps for Project Cirrus. Until its termination in 1952, Cirrus involved 225 flights to test various cloud seeding techniques. These tests demonstrated that like dry ice, silver iodide acted as a freezing agent. By burning silver iodide in acetone, it is possible to produce silver iodide smoke, enabling "cloud seeding" from the ground. Moreover, silver iodide also proved effective at higher temperatures than dry ice. Langmuir, for his part, claimed to have discovered that applying water to warm clouds would create "a chain reaction mechanism" and produce rain (Schaefer 1968a). Project Cirrus also triggered a flurry of activities by small commercial firms. By the early 1950s, 10 percent of the land area of the United States was under commercial seeding operations (Fleagle et al. 1974, 10).

Langmuir was not a meteorologist and proud of it. He was perpetually at war with the meteorological profession (Byers 1974). One example of how his claims enraged meteorologists involved an attempt to modify a hurricane by seeding it in 1947. The hurricane in question initially traveled in a northeast direction. After Langmuir seeded it, it changed its course and, going westward, struck the coast of South Carolina and Georgia. GE's lawyers requested that Langmuir not claim that his seeding had caused the change of direction as a result of the seeding. The U.S. Weather Bureau denied that there could be a causal connection between the seeding and the hurricane's change of direction. To support their views, Bureau experts pointed to meteorological records of an example of an earlier hurricane with a similarly erratic path.[2] Nevertheless, Langmuir did claim that his seeding had effected the hurricane's change of course. Countering criticisms made by the chief of the Weather Bureau, Francis W. Reichelderfer, Langmuir argued:

I pointed out to him that the larger the storm, and the more energy that is stored in it, the easier it should be at the proper stage in its development to get widespread effects. To assume that a hurricane could not be successfully modified by even a single pellet of dry ice is like assuming that a large forest could not be set on fire by such a small thing as a single match. (Langmuir in *Final Report of Project Cirrus*, quoted in Byers 1974)

On another occasion, Langmuir claimed that a 1948 experiment in New Mexico produced 0.37 inches of rain over an area of 4,000 square miles. According to a Weather Bureau meteorologist, however, numerous natural causes for the rain existed. Langmuir countered by asking, if the Bureau knew so much about it, why they had not forecast the rainstorm (Elliott 1974, 70). Langmuir used the occasion to suggest that weather modification might be used to reduce the uncertainty created by unreliable weather forecasts by bringing the weather under human control. These kinds of problems with attributing causation have been endemic in disputes over the efficacy of weather modification ever since.

Langmuir used statistical methods to argue for the efficacy of cloud seeding, but again, his results were severely criticized by the meteorological community. A 1950 evaluation published by the American Meteorological Society concluded: "It is the considered opinion of this committee that the possibility of artificially producing any useful amounts of rain has not been demonstrated so far, if the available evidence is interpreted by any acceptable scientific standards" (Byers 1974, 34).

Yet Langmuir stepped up his claims, asserting in 1951 that he had been able to produce periodic large-scale weather modification effects over a large segment of the country (Elliott 1974, 71).

The professional community of meteorologists reacted with extreme skepticism. Primarily oriented toward forecasting, the idea of modifying the weather and the climate seemed far removed from meteorological concerns. A case in point is Jerome Namias who, unlike most of his colleagues at the Weather Bureau, remained on friendly terms with Langmuir. Despite their friendship, however, Namias sought to convince Langmuir that the periodic effects he claimed to have produced could also be explained by the evolution of the general circulation of the atmosphere (Namias 1986, 20).

The controversy around Langmuir raised important concerns in the U.S. Congress. During the 1950s, Congress was generally supportive of weather control efforts. However, Congress also wanted to halt excessive claims by several small private firms that dominated press coverage of weather modification activities. To do so, it sought to place weather modification research on sounder scientific footing (Byers 1974), and to this effect Congress adopted and, in 1953, President Eisenhower signed a law creating the U.S. Advisory Committee on Weather Control. Captain Howard T. Orville was appointed chair. Orville was a retired head of the Navy weather service, a former chair of the American Meteorological Society, and a former member of the steering committee of Project Cirrus, and was thus familiar to both meteorologists and weather control advocates. At first, Orville seemed to offer partisan support to the cause of weather modification. However, the *Final Report* was rather cautious. In the course of its preparation, University of Chicago statistician K. A. Brownlee sharply criticized the statistical analyses offered in support of weather modification, and the controversy this generated was widely reported in the professional press (McDonald 1956). This probably helped temper the conclusions of the Orville committee. The committee's most substantial conclusion was that under certain circumstances, cloud seeding demonstrably increased precipitation by 10 to 15 percent in mountainous areas of the Western United States.

By the time the Orville committee had served its term, Congress had already achieved some of its aims. By 1956, commercial activity had dropped to about one-fourth of its 1952 peak, and it remained near that level for many years (Huschke 1963). In 1957, Congressional passage of the Water, Cloud Modification Research Act designated the U.S. National Science Foundation (NSF) as the coordinating agency for weather modification research, bringing weather modification under the umbrella of science. NSF welcomed weather modification research, but it also changed it. Earl Droessler, at the time program director for atmospheric science at NSF, championed weather and climate modification throughout his career. However, his programs shifted the emphasis in weather modification research from the design and implementation of experimental tests to investigations into the physical processes controlling cloud precipitation. NSF also merged weather modification research with theoretical work on the behavior of the (global) atmosphere. Among the tools used in both these research fields were computer models; in the latter field these models were similar to those used for forecasting the weather, especially long-term forecasting (see next section). At NSF, weather modification grew to become a central program in meteorological research, with funding levels reaching close to those for the rest of the atmospheric sciences combined by the early 1970s.

Langmuir died in 1957, undaunted, at the age of seventy-six. Reichelderfer remained less than friendly toward weather modification until his retirement as chief of the Weather Bureau in 1963 (Langer 1963). Vincent Schaefer would later move to an academic position at the State University of New York.

Von Neumann and Charney

During this early period of weather modification, one important exception existed to meteorologists' general skepticism toward altering the weather. Shortly after Langmuir's beginnings, John von Neumann announced his own interest in weather modification. Von Neumann, like Langmuir, was not a meteorologist. But unlike Langmuir, who behaved defiantly toward the meteorological profession, von Neumann worked closely with its establishment, and he was the architect of the wedding of meteorology and the computer. While Langmuir took a hands-on approach to weather modification, for von Neumann it lay somewhat further away in the future, because it was linked in his mind to the yet to be developed long-range forecasting. In an interview with the *New York Times Magazine* in 1946, von Neumann announced his "intention of developing a very high speed electronic computing machine and of applying it to the prediction of natural weather and of calculating the effects of human intervention in the natural processes of the atmosphere."[3] For von Neumann, prediction and intervention were closely interlinked.

Von Neumann's most important collaborator was Jule Charney, who joined the Meteorology group at Princeton in 1948. They worked together in 1949 on the design of a numerical prediction experiment and continued to collaborate through the 1950s on computer models of the atmosphere (for details, see Edwards, chapter 2, this volume). In 1955, von Neumann acted as one of the advisers of the Advisory Committee on Weather Control. During the preparation of its *Final Report*, von Neumann participated in a panel on "possible effects of atomic and thermonuclear explosions in modifying weather." During this panel's deliberations, he mentioned an idea of Charney's, namely, that changing the albedo of the earth (a measure of the reflection of radiative energy) might be the most effective way to modify weather and climate. Although, in von Neumann's opinion, nuclear explosions did not have much effect on the weather, he thought that more theoretical and computer studies were needed (Heims 1980, 283, 494), and he refused to temper his enthusiasm. He said at the time:

Our knowledge of the dynamics in the atmosphere is rapidly approaching a level that will make possible, in a few decades, intervention in atmospheric and climatic matters. It will probably unfold on a scale difficult to imagine at present. There is little doubt one could intervene on any desired scale, and ultimately achieve rather fantastic effects. (John von Neumann in 1956, quoted by Sen. Clinton Anderson in U.S. Senate 1964, 15)

Von Neumann also articulated another reason to develop weather manipulation: he feared that the Soviet Union might develop it first, and he warned that this might ultimately be a more dangerous threat than ICBMs.

1966: Expectations Peak

If the history of weather and climate modification were indexed to the degree of expectations it engendered about potential future successes, 1966 would be its apogee. Weather modification received substantial support that year from major public and scientific institutions. The Weather Bureau issued report in which it announced its ambition to "undertake a major role in weather modification research." The Bureau also recommended significantly increased support for microphysical studies of cloud dynamics and the processes controlling precipitation and macrophysical studies of the dynamics and thermodynamics of climate, both of which were central to weather modification as it was then conceived (Gilman, Hibbs, and Laskin 1965). Two reports issued in 1966 by the National Academy of Sciences and NSF also weighed in heavily. Both pleaded for a strong increase of funding to fundamental research in weather and climate modification. Both expressed optimism with regard to the principle of its possibility (National Research Council 1983; National Science Foundation 1965).

The climate in Congress and the Executive Branch was also favorably disposed toward weather modification in the late 1960s. Congresspeople such as Senators George McGovern of South Dakota, Clinton Anderson of New Mexico, and Alan Bible of Nevada made clear not only that they supported weather modification but also that they had not forgotten about Irving Langmuir and John von Neumann, in whose scientific credentials they continued to put great trust (U.S. Senate 1964). Stewart Udall, then Secretary of the Interior, also supported artificial rainmaking. It may have complemented nicely other water projects he cherished, such as the San Luis Dam and the Central Arizona Project (Worster 1985). Udall declared, notably:

What does appeal to me, because of its potential for all mankind, is the idea of enormous rivers of water flowing over us in the atmosphere; of huge pools of moisture poised above our heads; of enormous reservoirs in vaporous state sailing majestically over mountains, or bumping into them and dropping their precious burdens too soon. (Udall 1966)

After 1966, Congress appropriated increased funding for both fundamental research at the Weather Bureau and NSF and also for operational research carried out under the supervision of the Interior Department's Bureau of Reclamation.

Two technologies in full development were responsible for this optimism: satellites and computers. Satellites, in particular, gave meteorologists hope that the central problem of insufficient amounts of data for prediction, forecasting, and weather and climate modification would be solved (Malone 1967). On April 1, 1960, the world's first weather satellite, TIROS I, was launched by NASA. In the next ten years, twenty-two satellites followed. The launching of the Applications Technology Satellite (ATS I) in 1966 enabled meteorologists to continuously monitor weather conditions over a particular area of the earth's surface. Francis Reichelderfer and Harry Wexler at the Weather Bureau had promoted the use of satellites for meteorological purposes at least from the 1957 Soviet launch of Sputnik. Wexler saw to it that the design of the meteorological satellites was tailored to numerical forecasting, and when data came pouring in, computers proved indispensable in managing it (Nebeker 1995, 175). Computers, of course, also served meteorology theoretically through the development of ever larger and more detailed data sets and numerical models of the global atmosphere (see Edwards, chapter 2, this volume).

Weather Modification Projects

Too many weather modification programs operated during this period to enumerate, but some stand out, either because of their size or because of their scientific or political importance (National Science Foundation 1965).[4]

Project Climax, to increase the snowpack in Colorado, was carried out at Colorado State University by Lewis Grant, from 1960 to 1970, supported by NSF. Early parts of the project were inconclusive. By developing increasingly complex models of the atmosphere, however, researchers were "finally able to identify a number of combinations of upper wind in the direction and temperature in the air stream that could be successfully exploited to produce additional snow," according to Edward Todd of the National Science Foundation (U.S. House of Representatives 1976b, 110; see also National Academy of Sciences 1973, 80–83, for more information on the Climax experiment).

Begun relatively late, in 1972, the National Hail Research Experiment (NHRE) was the biggest NSF sponsored effort in the period, funded through NSF's Research Applied to National Needs program. NSF hoped to bring it quickly to a point where a mission-oriented agency, such as the Department of Agriculture, could take it over and turn it into an operational program. NHRE was carried out in northeast Colorado by the National Center for Atmospheric Research (NCAR). The research concept of NHRE had the peculiar feature that it was designed after hail suppression projects in the Soviet Union, which reportedly had been very successful (U.S. House of Representatives 1976b, 144–159).

Numerical cloud models were the outstanding feature of the Florida project under the guidance of Joanne Simpson at NOAA's Experimental Meteorological Laboratory (EML) in Miami. This project's models of individual clouds and cloud systems appear to have been the first "operational" computer simulation models to be tested in weather modification experiments.[5] Simpson directed Project Stormfury (see below) between 1965 and 1967, and the approach she used, called *dynamic cloud seeding*, was developed in part during these years. It was her stated aim to introduce the results of her work on cloud systems into large-scale global circulation models, thus contributing to "man's eventual control of the atmosphere" (Simpson, Wiggert, and Mee 1968).

The dynamic cloud seeding approach was directed at the buoyancy forces and circulation that sustain the cloud, bringing about invigorated cloud dynamics through experimentally induced releases of fusion heat. In a publication in *Science*, the project's initial results were described as "very encouraging," while those of an older approach, labeled "static," were dismissed out of hand as inconclusive and controversial (Woodley 1970). EML based its evaluations of cloud seeding on a comparison between model simulation results and the observed behavior of clouds. Moreover, the experiments were conducted using both statistical controls and randomization procedures. EML thus argued that its weather modification was science based in a way that previous cloud modification was not. Over time, reports from the project continued to be positive, and this attracted scientific and political attention. In 1971, during a severe drought, the governor of Florida requested that EML seed clouds on an "operational" basis. EML performed the seeding and afterward claimed to have produced an additional 100,000 acre-ft of water (Simpson, Woodley, and White 1972).

The Beginnings of Project Stormfury

Project Stormfury was one of the more successful of the several dozens of weather modification projects funded during this period. It is particularly notable for the historian, however, for its efforts to combine operational hurricane seeding with theoretical research of a more advanced character than at most other weather modification projects. In 1955, the Weather Bureau established the National Hurricane Research Project, based in Miami. It was to study the formation and dynamics of hurricanes, to improve hurricane forecasts, and to seek means for hurricane modification. Then and later, huge damages from Atlantic hurricanes were frequently invoked to argue for research funding into hurricane dynamics and control. (Hurricane Betsy, for example, was reported to have caused more than $1.4 billion (U.S.) in damage in 1965.)

The first hypothesis for how to modify hurricanes was developed by R. H. Simpson and Joanne Malkus (later, Joanne Simpson). They based their idea on a relatively simple, qualitative understanding of the structure of a hurricane. Hurricanes owe their famous vortex structures to an equilibrium between the centrifugal force of air particles traveling at high speeds around the hurricane and the centripetal force created by low pressure at the storm's center. Simpson and Malkus suggested that seeding near a hurricane's central eye would set in motion a chain of events, eventually reducing differences in the pressure gradient and, hopefully, dissipating the storm's strength (Simpson and Malkus 1964). At this point, computer models of hurricanes were neither in existence nor was their need felt.

In 1961 a much-improved technique of producing silver iodide smoke was developed at the Naval Weapons Center in California, which made it possible to bring large quantities of the freezing agent into the clouds within minutes. In the same year, hurricane Esther was seeded with this new technique and the results were "encouraging if nonconclusive" (Gentry 1974, 507). Within a few months, the Navy and Weather Bureau took these uncertain results as a sign of future promise and together organized Project Stormfury.

In 1963, Project Stormfury seeded the eye wall of hurricane Beulah, with what appeared to observers to be significant effect. Robert M. White, the new Chief of the U.S. Weather Bureau, called the results of the experiment "encouraging," but noted that it was difficult to tell whether the observed consequences of seeding were the result of "man's action" or of natural fluctuations (White and Chandler 1965). In 1964, Jule Charney and A. Eliassen published a model description of a hurricane that became the basis for all subsequent hurricane models at Project Stormfury (Charney and Eliassen 1964). The model's (still qualitative) mechanism consisted of a feedback between cloud formation and atmospheric pressure. Cumulus clouds supplied the necessary heat to drive the pressure depression, while the depression vortex produced moisture, thereby maintaining the cloud system. According to Charney and Eliassen, this interaction leads to large-scale self-amplification. In the assessment of meteorologists at the time, models based on this mechanism provided "realistic" results (Rosenthal 1974). Charney and Eliassen's paper did not mention the possibility of hurricane modification, but other models were constructed on the basis of it, notably by Stanley Rosenthal in 1967 and 1969. Rosenthal simulated the effects of seeding and helped sort out strategies of where and where not to seed a hurricane (Rosenthal 1974). But, according to its developers, the Rosenthal model did "not sufficiently simulate nature" (Gentry 1974, 406).

Confidence in the models received a boost in Project Stormfury's year of glory, 1969, when a series of project experiments appeared to show consistent evidence of the ability to affect Hurricane Debbie's dynamics. Gentry described it thus:

If one considers the sequence of events in the Hurricane Debbie (1969) experiment, the results are quite impressive: seeded five times on August 18, and the wind speeds decreased 30 percent; no seeding August 19 and the storm intensified; seeding on the inner wind maximum of the first four seedings on August 20 and it decreased, and seeding of the outer wind maximum on the fifth seeding and the winds decreased 15 percent. (Gentry 1974, 406)

Computer simulation of the seeding of Debbie showed confirmation of wind speed reduction in the model of Stanley Rosenthal, but not in another model (Gentry 1974, 406).

In subsequent years, however, progress was not so easily achieved. In 1969, NSF commissioned the RAND Corporation to provide an overview of the field of weather modification. RAND's overview stated notably: "Three years have provided nothing substantial to the timorous optimism expressed in the National Sciences (1966) report on weather and climate modification. On the contrary, the trend of research seems to indicate a growing respect for uncertainties and desire to unravel them" (RAND Corporation 1969, 223).

The RAND Corporation put the problem of hurricane modification succinctly: since there were too few hurricanes for any statistical evaluation to be meaningful, there remained only one alternative. It was absolutely necessary to

be able to predict confidently what the behavior of the storm would have been had we not interfered. And since today we cannot make the prediction, we cannot make the evaluation. . . . The central problem in tropical storm modification lies in learning whether or not we have indeed effected the modification. (RAND Corporation 1969, 225)

The RAND report reads as a plea for greater coupling of weather modification experiments to the production of new simulation models of atmospheric processes on multiple scales.

Project Stormfury, however, experienced hard luck. No more suitable hurricanes migrated into the seeding area (which had been severely restricted as a result of Langmuir's first hurricane seeding experiment). There was one exception—hurricane Ginger in 1971—but it was such a poor hurricane that the failure of the seeding experiment could have been predicted beforehand.

1970–1973: Doubts of a New Kind Begin to Emerge

In 1972, a cloud seeding operation near Rapid City, South Dakota, and the nearby Black Hills ended in misfortune. On the morning of June 9,

1972, the arrival of a moist band of clouds in this perennially dry area proved too enticing for weather modification experimenters from the South Dakota School of Mines to ignore.[6] During two flights, the researchers seeded the clouds with 700 pounds of table salt (sodium chloride). What the cloud seeding team did not know, however, was that strong winds generated by a high-pressure area in the region of Hudson Bay had stopped the eastward march of the clouds just above the Black Hills. The clouds, rendered stationary, remained overhead for more than ten hours, releasing their water content and causing a devastating flood that killed over 200 people.

A remarkable coincidence made Rapid City host town of the Third National Conference on Weather Modification only seventeen days later. In the margins of the conference, the flood was hotly debated and it was heard many times that the risks posed by weather modification experiments were sufficient to justify restricting them to computer simulation experiments (Dennis 1972). But this was not the opinion that prevailed. During the conference, a team of investigators commissioned by the South Dakota Weather Control Commission reported to the governor of South Dakota. The conclusions, which the conference subsequently endorsed, stated: (1) that the June 9 flood was a natural disaster, caused by meteorological conditions beyond human control; (2) that weather modification activities had not contributed materially to the flood; and (3) that weather modification operations should be resumed.

A year later, a scientist in Albuquerque, New Mexico, who had been a conference participant in Rapid City, published a "dissenting view" in a letter to the *Bulletin of the American Meteorological Society* (Reed 1973). In it, he argued that cloud seeding had triggered the Black Hills storm, basing his conclusions on several possible scenarios presented at the conference. He also mentioned the possibility that the seeding had produced "as yet unknown and unexpected dynamical effects." Based on statistical considerations, he also argued that experimenters had taken too great a risk in Rapid City. In the same issue of the *Bulletin*, no less than six sharp rebuttals were published. The critic's use of statistics and "risk-benefit analysis" was strongly disparaged, as was his reference to "unknown effects," which was declared to be speculative and unscientific. Among the rebuttals was one by the Weather Control Commission's team of investigators. In it, the authors argued that "the experience of those who have used salt has been that rainfall has been slightly increased from non-raining or lightly raining clouds." They also gave a fairly precise estimate of the volume of rain that could have been deposited as a result of the salt seeding, basing their estimate on a linear ratio between the amount of salt and the volume of deposited water. This estimate amounted to only a tiny fraction of the total water volume deposited during the flood (St.-Amand, Davis, and Elliott 1973; see Reed 1974 and following discussion). Two NCAR meteorologists were likewise unconvinced that a conclusive cause-effect relationship could be established between the seeding and the flood (Kellogg and Schneider 1974).

Public opinion in South Dakota appeared, at least initially, to accept the official findings. No organized opposition arose, and a study of public opinion, undertaken shortly after the event of the flood, showed that a majority of respondents (55 percent) thought that seeding had not produced the flood, while only 25 percent thought that the seeding had been a determining factor (Farhar 1974). Two years later, however, the group Citizens Against Cloud Seeding began to oppose weather modification experiments in South Dakota, despite expert disavowal of any causal connection. After many public meetings, this group succeeded in preventing state funds from being used to support cloud seeding after June 30, 1976 (Dennis 1980, 4).[7]

Citizens Action against Weather Modification

Subsequent episodes also illustrate the contested view of weather modification activities that had arisen by the mid-1970s, with the Rapid City disaster quickly assuming an iconic status in public deliberations about weather control. The first such episode I describe here involved commercial weather modification; in the other two episodes I describe, public institutions carrying out weather modification research were implicated.

The San Luis Valley of Southern Colorado is populated mostly by ranchers and farmers. A high proportion of them are growers of Moravian barley, who under normal circumstances thrive well in the arid climate of the valley. But barley is extremely vulnerable to hail, and, unfortunately, precipitation in summer quite often comes in the form of extremely large hailstones. In 1966, at the instigation of Coors Company, a commercial weather modification campaign was started to suppress hail and divert rainfall. During the first years, the project did not elicit much response, but this changed when 1970 turned out to be unusually dry. That same year, some 400 farmers formed a group called San Luis Citizens Concerned about Weather Modification. One of the allegations this group leveled against weather modification was that cloud seeding actually suppressed rain. In their view, silver iodide seeding would lead to the formation of small ice crystals that would be blown away by the wind before they could precipitate to the ground. The group raised popular support, and their actions induced the Colorado legislature to grant the state greater regulatory authority. In early 1972, the Colorado legislature enacted a weather modification bill. Under this law, cloud seeding would require a permit and a public hearing. The permit hearing for the San Luis valley experiment was held in July. Approximately 600 people attended, and the meeting lasted until 2:00 a.m. A clear majority opposed the granting of a permit. In November 1972, an advisory referendum was held in the five counties of San Luis Valley. The vote was overwhelmingly negative (Carter 1973).

Shortly thereafter, W. K. Coors, president of the Coors Company, wrote to barley growers in the valley, attaching paramount importance to "weather management":

Unless we have good assurance of an adequate weather management program for the 1973 barley growing season by March 1, 1973, the 1973 Moravian barley allotment for the San Luis Valley must be reduced 20 percent. On each successive year, assuming no weather management program in that year, an additional 20 percent cut in allotment must be made until the total Moravian barley production from the San Luis Valley has been reduced to 10 percent of the Adolph Coors Company needs compared with the 60 percent that it supplies now. 10 percent is the maximum amount we are willing to place at the mercy of the natural elements which in our experience will give us a good quality crop once in twenty years. (Lansford 1973, 659)

After a new round of hearings, the permit was finally denied in 1973. The barley growers found another brewery to purchase their crop.

A second episode occurred in Montana. In 1972, a public meeting ended a five-year practice of cloud seeding to increase snowpack over the watershed above Hungry Horse dam. The cloud seeding ostensibly increased the water available for hydroelectric generation at the dam. The argument against the permit was grounded in ecological terms. Opponents argued that the Wilderness Act prohibited cloud seeding above a wilderness area in the immediate vicinity of the dam (Elliott 1974, 84).

In 1975, a citizens group in the Potomac basin similarly published a strong attack on weather modification activities in the region, implicating, in particular, Project Stormfury. The group claimed that, since the initiation of weather modification research in the Washington area in 1957, the local climate had undergone a dramatic change, with rainfall consistently below average. The report alleged that a conspiracy of diverse interests supported clear weather, including recreation, airports, road builders, and the construction industry. According to the group, cloud seeding was responsible for 90 percent of the air pollution problems in the United States, and it specifically linked the use of silver iodide to a higher occurrence of lung diseases. Project Stormfury was credited with having kept hurricanes from coming up the Eastern seaboard. Pointing out that hurricanes contribute 30 percent of the precipitation in the area, the report also argued that "if hurricanes are not allowed to go through their full life cycle, nature will find some other way to maintain heat balance, and this new method undoubtedly will be more disastrous than the hurricane." Hurricane Camille had been a foretaste: "The most destructive tropical storm ever to transverse the east coast, [Camille was] without question due to the blunderings of the inept science" (Tri-State Natural Weather Association 1975).

All of these events illustrate the growing opposition to weather modification, not because people believed it did not work, but because they believed uncertainties in the method posed undesirable risks. Underlying this belief lay a growing sense that human activities could substantially and destructively affect the natural environment. Messing with Mother Nature had come to be viewed as inherently problematic in American public consciousness. Stanley Changnon, a meteorologist in the service of the Illinois State Water Survey and a long-time supporter of weather modification research, noted that 1972, the year of the San Luis Valley controversy and the Rapid City disaster, can be regarded as the year when environmental concerns over weather modification began

(Changnon 1973). From the early 1970s on, a new area of uncertainty was added to an already contested field.

The Rise of Inadvertent Weather and Climate Modification

What is it that brought about the demise of weather and climate modification? It is tempting to think that at some point its practical and theoretical impossibility imposed itself so strongly on the meteorological community that the field was simply abandoned. This appears not to be so. Rather, the demise of deliberate weather and climate modification appears linked to the growth of environmental concerns. Within the field of weather modification, consistent with growing American concerns about human impacts on the environment, a subfield emerged during the late 1960s and early 1970s concerned with "inadvertent climate modification" or the modification of long-term weather patterns through human activities that affect the atmosphere. From 1970 on, the possibility of inadvertent climate modification appears to have increasingly alarmed politicians and the public. Between 1973 and 1975, budget cuts virtually put an end to research on deliberate weather and climate modification. Notably, meteorologists at the time strongly protested these cuts.

The 1965 NSF report *Weather and Climate Modification* did not address the possibility of anthropogenic climate change under any label. This need not imply that the issue was not recognized by scientists at NSF. However, it does suggest that it was not yet linked to deliberate modification of the atmosphere. The report did discuss, however, the possibility of adverse ecological consequences from planned weather and climate modification, for which a working group of the Ecological Society of America was invited (National Science Foundation 1965, 61–70). The 1966 National Academy of Sciences report of the same title did discuss inadvertent modification of the atmosphere (National Research Council 1983, 82–108). The report identified several potential causes: (1) carbon dioxide, (2) urbanization, (3) forestation, deforestation, and agriculture, (4) supersonic transport aircraft, and (5) contamination of the very high atmosphere. On the basis of "computations on sophisticated numerical models" at the Geophysical Fluid Dynamics

Laboratory, the report downplayed the potential effects of increased carbon dioxide. Air pollution did raise substantial concerns linked to possible changes in precipitation patterns. The report described the effect of cities on regional climates as highly complex, and it recommended that studies of this phenomenon be intensified. The effects of supersonic transport were constricted to their introduction of water vapor in the stratosphere, and again on the basis of the numerical models mentioned earlier, the report deemed them relatively benign. The general tenor of the National Research Council report *Weather and Climate Modification* was that inadvertent modification of the atmosphere was not a great cause for concern, though further research was heartily recommended (National Research Council 1983). Apparently, this was the received view at the time. Roger Revelle, oceanographer and influential policy adviser to the House Science and Technology Committee, gave as his opinion that global warming as the result of a rise in atmospheric CO_2, or global cooling caused by air pollution, was at best a reason for "curiosity (rather) than apprehension" (Hart and Victor 1993, 656).

Some change of emphasis occurred in 1968. In the "conference summary" of the First National Conference on Weather Modification, weather modification enthusiast Earl Droessler wrote:

Inadvertent weather modification is moving rapidly into the forefront of the truly big, urgent problems to be attacked by meteorology and allied science. There is an impending crisis regarding the volume of air pollution across the nation, and people are beginning to demand purer air to breathe. A variety of observations and ideas were presented to the Conference suggesting that man-made pollution may already be modifying precipitation patterns on local and regional scales at least, and may be affecting the global heat budget on even larger scales. (Droessler 1968)

Droessler was referring to a contribution by Vincent Schaefer (1968b) in which he announced his discovery that an atypical form of snowstorm was the result of badly polluted air, while ecologist Frederick Sargent (1968) was given the opportunity to issue a warning with respect to several "imponderable ecological risks" associated with weather modification. Not mentioned in the summary report, but discussed at the conference, were the possibilities of climate modification. Some years before, the President's Science Advisory Council, in a report about pollution abatement, had considered the use of reflective material in the atmosphere to counter a greenhouse effect (Hart and Victor 1993). The deployment of such a technique would have amounted to full-scale climate modification.

The 1969 RAND corporation report discussed earlier found that climate modification was technologically feasible. Various possibilities were considered. One of these sought to remedy inadvertent climatic change:

> The diversity of thermal processes that can be influenced in the atmosphere, and between the atmosphere and the ocean, offers promise that if global climate is adequately understood it can be influenced for the purpose of either maximizing climatic resources or avoiding unwanted changes. For example, to avoid undesired planetary warming, ways might be found to drain additional heat to space. (RAND Corporation 1969, 230)

But that was the last of it. In the years to follow, no similar considerations were reiterated. It was for instance entirely lacking in a report by the President's Council on Environmental Quality (PCEQ) (see President's Council on Environmental Quality 1970). The Council made some recommendations with regard to the direction of research on weather and climate modification. Noting that only about 1 percent of federal government research monies for weather modification went to programs investigating inadvertent human modification of the climate, it recommended in particular a recognition of the significance of human-made atmospheric alterations and their significance for a possible impending change of the climate. Not until 1991 would the National Academy resurrect its arguments that planetary-scale geoengineering could be used to counteract global warming (NAS 1991).

In 1970, concerns about environmental degradation reached a temporary peak. The PCEQ was established through the signing into law of the National Environmental Policy Act by President Nixon, as the government's "policy arm" in environmental matters. It was the year of Earth Day (on April 22), and the year during which organizations such as Friends of the Earth, founded the year before, developed a widely acknowledged political influence. At the time, two environmental dangers were considered to be of the highest importance: water pollution (mostly by pesticides) and air pollution by automobiles, smokestacks, and so on. The latter concerns us here.

The PCEQ report noted that various forms of air pollution decreased the transparency of the atmosphere and that this might have a cooling effect: "If pollution were significantly responsible, then the world would face an important problem of manmade global climate modification." The inverse action of an increased carbon dioxide content of the atmosphere was also noted, but the cooling problem received priority in the considerations of the report, the more so because it was linked with climate changes that are already occurring.[8] Concentrating on the annual average at the earth's surface, the report noted that while this average began a climb between 1890 and 1940 of 1.1°F, it fell again between 1940 and 1970 with 0.5°F. The report further noted that ice coverage in the North Atlantic had been among the most extensive in over sixty years.

The report reflected prevailing opinions of climate researchers (see Hart and Victor 1993), including Reid Bryson at the University of Wisconsin, Madison. Bryson's theory of an impending cooling of the earth would receive much attention in 1975 (see below). Other concerns of the Council were also in step with the research agenda of the meteorologists. These include the issue of the climatological effects of urban heating and the pollution caused by jet aircraft. Missing among the council's concerns yet were the "polluting" effects of weather modification activities, because these did not become publicly contested for another few years.

The Actual Experience of Climate Change?

The PCEQ assessment that an actual cooling was already occurring and that it was linked to air pollution did not stand alone. Air pollution as a cause for climate cooling was given in the meteorological literature by several authors, but different causes were put forward as well, such as volcanic activity and solar variations.

Against this variety of proposed external influences on the climate, Namias held that the autonomous dynamics of air-sea interactions alone could be held responsible of what had been indeed the abnormally cold decade of the 1960s (Namias 1970). Namias invoked the amplification of the "long wave pattern," or Rossby waves, which affected the Northern Hemisphere, as the cause of the observed climatic fluctuations. This regional phenomenon Namias saw in turn produced by the abnormally warm sea surface in the North Pacific.

Namias's analysis by no means stopped concerns about the perceived cooling of the climate. A 1974 report in Nature noted a "growing public awareness in North America that something is going wrong with the weather." Apparently, the "person in the street" knew that frost was coming too early, and the local TV station's weather reporter commented on "three early cold waves" (Gribbin 1974). Similar observations were also made in scientific articles by Reid Bryson, who was the real subject of the *Nature* report. In an article on "The Lessons of Climatic History," in which Bryson included a section on "the present arctic expansion," he cited as observations that "the growing season in England has diminished by two weeks," "midsummer frosts returned to the upper Midwestern United States," "the average temperature of Iceland has declined to its former [pre-1945] level," and many others (Bryson 1975). Several meteorologists noted in the same year a trend toward cooler temperatures, which could be the forerunner of a new ice age (Hammond 1974). Two years later, however, two atmospheric scientists reported to have found a warming trend on the Southern Hemisphere. They suggested that in the long run global warming as a result of the greenhouse effect might overtake cooling mechanisms in operation on the Northern Hemisphere (Damon and Kunen 1976).

Between 1972 and 1976, climatic change had become a matter of serious concern, among various scientists, parts of the public, and some politicians. It has been noted that, compared to the previous decade, not very many new scientific findings had been made (Hart and Victor 1993). Yet global warming by rising CO_2 and global cooling by air pollution were taken considerably more seriously than in 1966. Recall also that proposals for deliberate modification of the climate to counter these possibilities, such as those put forward by PSAC in 1966 and by RAND Corporation in 1969, were no longer being put forward. Even toying with the idea of climate modification was out of the question. Stephen Schneider, who in the hot summer of 1988 would become an important spokesperson in favor of the recognition of the greenhouse effect, expressed himself in no uncertain terms against weather and climate modification. While his main argument was that "irreversible climatic disasters" could result from attempts to stabilize the climate with technological means, he also noted that rainmaking had become a tool of war in Vietnam, forfeiting its innocence (Schneider 1976).

The Erosion of Weather Modification

The seeds had been planted for a development that eventually would be fatal for the prospect of weather and climate modification. After 1970, the issue of inadvertent climate modification substantially contributed to growing pressure from outside groups against weather and climate modification. The meteorologists sought to stem the damage from public opposition. "Inadvertent weather modification" was co-opted as a separate research activity, for which funds were allocated. Project METROMEX, a large-scale example of this research, examined the impact of St. Louis on precipitation patterns in its surrounding area (Berry and Beadle 1974). But this did not prevent an erosion of Congressional support for the entire subject of weather modification, which declined steeply after 1973. The meteorologists blamed the public and finally accepted the inevitable. In 1975, the president of the American Meteorological Society admitted that too many uncertainties surrounded weather and climate modification to sustain a claim for generous public support (Atlas 1975).

Federal funding for weather modification had continuously increased from 1967 to 1972—it had in fact doubled during that period. Total funding had risen from $12.0 million in fiscal year 1972 to $15.2 million in FY 1973, to an awarded $25.2 million in calendar year 1973. In the five-year period ending in FY 1973, the federal government had spent $77 million, not including the Department of Defense expenditures of $21.6 million in Vietnam to make it rain on the Ho Chi Minh trail (Shapley 1974). But in January 1973, halfway through FY 1973, expenditures were cut back by $65 million at the direction of the Office of Management and Budget (OMB) (Fleagle et al. 1974). In FY 1974, federal support dropped 21 percent, when other research and development supported by the government increased. The losses were unevenly spread. The Bureau of Reclamation's budget was nearly halved, NSF suffered a decrease of 30 percent, while research on inadvertent modification almost doubled to a total of $2.9 million (Fleagle et al. 1974; Changnon 1975).

Stanley Changnon, by 1975 serving as president of the Weather Modification Association, blamed the negative attitude of OMB to the "paradox" that the weather modifiers had created themselves: on the one hand, they claimed to have a working technology; on the other hand, they requested funds for more basic research into weather modification (Changnon 1975). Changnon made it clear that the field still included quackery and was known for overstated claims. In hearings before the House Subcommittee on the Environment and the Atmosphere, he declared:

Thirty years of experimentation have provided a variety of scientific outcomes ranging from success to total failure with the outcomes too often dictated by the typical uncertainties associated with the vagaries of nature coupled with poorly conducted projects that were often too brief in their duration.

...Weather modification, I believe, is still an immature basically uncertain technology unable yet to demonstrate sizable weather alterations and economic benefits in many areas of the United States. (U.S. House of Representatives 1976b, 348–350)

Changnon went on to make a plea for new and higher efforts in weather modification, and he expressed his expectation that since an actual "climate change in the sense of increased variability" had been underway since 1971, possibly causing food shortages, the latter would place weather modification back on the national agenda again (U.S. House of Representatives 1976b, 356). From 1975 on, public concern about an impending climate change has indeed grown considerably stronger. This has done little to help the cause of weather and climate modification, however.

Joanne Simpson, by the early 1970s professor of environmental sciences at the University of Virginia, testified before the U.S. House of Representatives during hearings on the use of weather modification in Vietnam: "A paramount hard fact about weather modification is that there are today only about a half dozen programs over the whole world that have conclusively demonstrated that the treatment works" (Simpson 1975). She went on to say that her own projects (in Florida, since 1968) fell in the category of the successful, and Changnon and other members of the scientific community were willing to grant her that (Changnon 1975).

Similar statements were made by David Atlas, who in 1975 served a term as president of the American Meteorological Society. A year before he had become director of the National Hail Research Experiment, but now he urged his fellow meteorologists to be very cautious. Warning that the credibility of the entire science of meteorology might be injured by promising too much too soon, he also concluded that "only a handful of experiments have demonstrated beneficial effects to the general satisfaction of the scientific community" (Atlas 1975).

Yet Robert M. White, Administrator of NOAA, still saw things from the other side when he testified to Congress in 1975:

Deliberate environmental modification is a growing issue. Man is developing a capability to modify atmospheric processes in a conscious and directed way. After three decades of controversial experimentation, this capability is still primitive. Nevertheless this is a field of research and development of great future promise. . . . Over the coming decades, this primitive capability will be transformed into a broader capability for affecting weather processes. (U.S. House of Representatives 1976a, 4)

And Earl Droessler tirelessly enumerated all the successes to date of weather modification. He called on Congress to establish a National Committee on Weather Modification, and he specifically addressed himself to the Department of Agriculture, in the hope of finding a new sponsor for weather modification research (Droessler 1975). As late as 1983, Droessler, by then president of the American Meteorological Association, continued to call for more weather modification research (Droessler 1983). As if to illustrate their futility, Project Stormfury was formally terminated in that same year, twelve years after the last hurricane had been seeded.

Between 1973 and 1976, the field of weather modification had managed to fix itself on the (lowered) 1973 level. But after 1976, publicly supported weather modification projects were terminated one by one. During Congressional hearings in 1976 it was openly stated by NSF officials that the National Hail Research Experiment had been a failure. The technology of hail suppression was assessed in 1975 by a research group that included social scientists. Their assessment led to recommendations such as that made to the State of Illinois not to proceed with hail suppression experiments (Changnon 1977).[9] Funding for weather modification continued to decrease steadily and, by 1986, the NSF spent $1.5 million, while in 1972 it had been $5.4 million (not corrected for inflation; Changnon 1986). One of the last surviving weather modification research experiments, the 1982 Florida Area Cumulus Experiment–phase 2 (FACE-2), failed to stand up to a statistical test that was fashioned after double-blind clinical drug trials (Kerr 1982).

It cannot be expected that the Florida test is the definitive end of attempts to make rain. To begin with, the test results were contested within the research team itself. Hence, they likely mark only an end to this round of funding. New projects can spring up anywhere in the world at any time. However, any new project proposal to make rain or in other ways affect the atmospheric environment will find a set of societal values very different from those in the 1960s, when faith in science and technology was naively innocent of the potential consequences of large-scale human activities. This is once more illustrated by the story of Project Stormfury.

The Demise of Project Stormfury

In 1972, three years after the seeding of hurricanes Beulah and Debbie, the Navy unilaterally decided to withdraw from Project Stormfury. The Navy gave no reasons, but it was widely assumed that criticisms raised against the military application of weather modification techniques in Indochina had led to the Navy's decision (U.S. House of Representatives 1976b). The immediate result was that Stormfury was left for some time without airplanes to carry out seeding operations. The NOAA part of the Stormfury budget ($2 million annually) was also cut. An internationally agreed plan to move Stormfury's activities to the Pacific (instigated in part by the lack of aircraft) never materialized because of political resistance of some western Pacific nations, Mexico and Australia in particular (National Advisory Committee on Oceans and the Atmosphere 1976, 1972).

By 1973 the project's priorities had changed to improving the accuracy of hurricane forecasts, and more than ever it was felt that successful modification depended on advances in the predictive capabilities of numerical models. Apparently, strategies to inhibit the very development of a hurricane, or to deflect it back to the ocean, were altogether ruled out, since it was now recognized that hurricanes were responsible for much of the rainfall in the Southeastern part of the United States. This left reduction of wind speeds as the only acceptable goal. Yet a review panel of the National Academy of Sciences under the chairmanship of Thomas Malone declared hurricane modification a national goal and recommended greatly expanded expenditures (Hammond 1973).

Beginning in 1977, Project Stormfury engaged in two new lines of inquiry, one on cloud physics, the other on "vortex evolution" (Willoughby et al. 1985). Both led to the conclusion that modification of hurricanes was theoretically impossible. In fact, Stormfury's internal advisory panel (on which Edward Lorenz served since 1965) had urged, every year between 1962 and 1973, for research in these areas, apparently because it was concerned that Stormfury's claims with regard to hurricanes Beulah and Debbie might be unjustified (White and Chandler 1965).

Project Stormfury was formally ended in 1983. Although that seems surprisingly late, one should remember that the last hurricane seeded was Ginger in 1971. The year 1983 stands foremost as a date marking theoretical conclusions indicating that hurricane modification was impossible. Looking back on its history, Stanley Rosenthal commented:

In retrospect . . . the changes in the hurricane [Debbie]'s intensity were the result of natural evolution rather than a consequence of seeding. . . . The goal of human control was captivating and seemed to be physically attainable in the beginning. . . . Stormfury . . . had two fatal flaws: it was neither microphysically nor statistically feasible. (Willoughby et al. 1985, 512–513)

In 1996, the Hurricane Research Division put the following text on the Internet:

Perhaps the best solution is not to try to alter or destroy tropical cyclones, but just learn to co-exist better with them. Since we know that coastal regions are vulnerable to the storms, enforce building codes that can have houses stand up to the force of tropical cyclones. Also the people that choose to live in these locations should be willing to shoulder a fair portion of the costs in terms of property insurance—not exorbitant rates, but ones which truly reflect the risk of living in a vulnerable region. (Landsea 1996)

Weather and Climate Modification: A Technology Contested in Multiple Ways

Changnon and Droessler, each in their own different way, regarded the failure of weather modification research as bad luck and caused by a series of missed opportunities. If only the wonderful experiments by Joanne Simpson had not found a sequel in the sloppy experimental design in Oklahoma and Texas; if only a leading agency had taken firm direction of the research; if only nature had been a bit more cooperative and sent more hurricanes suitable for seeding to the Atlantic coast; if only a new drought had come that would have convinced people of the necessity of weather modification; if only the environmental tide would flow away again . . .

But weather modification research had lost its credibility to make references of the form of "inconclusive but promising," and not simply because statistical evaluations remained difficult to carry out, even after thirty-five years. To uncertainties about effectiveness, but also to competing economic interests and to legal constraints, uncertainties of a new type had been added. Environmental concerns were clearly of prime importance in bringing down the field of weather and climate modification. The first important budget cut occurred in 1972, at the instigation of OMB. At the time, this decision hardly could have been informed by doubts about the effectiveness of weather modification. Mistrust of the technical potential of weather modification was not seriously voiced for several more years. It is more plausible to assume that environmental worries were behind OMB's decision. By 1972, the Nixon presidency already had developed a strong policy profile on environmental matters.

As my narrative suggests, moral considerations played a substantial role in the discontinuation of weather modification research and activities. It is probably more apt to state that around 1973, moral considerations stemming from the newly emergent environmental movement merged with a new sense of risk perception. Weather modification has always been approached through the dual concepts of benefit and risk. Prior to 1973, it was expected that the former outweighed the latter. One typical example is a suit by resort owners against the City of New York, reported in 1966 (Carter 1966). According to the New York court, the interest of the resort owners in good weather was outweighed by the public interest in ending a drought. After 1973, this reasoning no longer prevailed, and not simply as a result of a quantitative change in the balance of profits and losses. The very terms of the debate changed meaning. Potential benefits had always been uncertain, but in the course of the 1970s, the concept of uncertainty changed from indicating lack of knowledge to denoting the intrinsic indeterminacy of atmospheric processes.

A similar evolution befell the concept of "risk." One could say that, before 1971, "benefits and risks" were somehow equal to "gains and losses," and that there would be a net gain from weather modification. Those at risk of loss could be helped by a proper insurance system that would protect them from "weather" damage. After 1973, people began to imagine the risks involved not in terms of discrete weather events but rather in terms of a change of climate. It was increasingly feared that changing the weather could lead to an inadvertent change of the climate, rendering problematic the belief that insurance based on extreme weather events would adequately keep risk at bay. In the course of three decades, American attitudes toward technology, nature, and society shifted dramatically. From hopes in the 1950s for a technology triumphant that would control the fury of storms to fears in the 1970s for the environment in the face of the growing destructiveness of human society's technological excesses, Americans reevaluated their relationship to nature and science. This, more than anything else, tipped the scales against weather modification.

We may note here that around the same time several more technologies designed to steer large natural and social systems underwent similar drawbacks. The field of urban planning experienced a crisis of its rational planning methodology (Lee 1973; Harvey 1986), while at the same time a large modernist housing project, Pruitt-Igoe in St. Louis, was literally blown to pieces because of its persistent unmanageability (Jencks 1984). The buildings were not even twenty years old. Ecology's first attempt to build comprehensive models to manage and steer large ecosystems was considered a failure (Boffey 1976; Kwa 1993). The 1973 oil crisis was a major cause that made American economists cease to believe in the existence of automatic stabilizers of the national economic system, and they gave up the kind of modeling through which they had hoped to identify them (Epstein 1987). And technocratic warfare in Vietnam was heading for collapse (Gibson 1986; Edwards 1996a). The style of technoscientific reasoning that underlay all the examples enumerated here (cloud systems and hurricanes, ecosystems and cities, national economies and regional battlefields) conceived of large systems as closed systems, amenable to steering in a manner that dealt with them comprehensively (Edwards 1996a; Kwa 1994). With the benefit of hindsight, we could say that all these systems were at once too comprehensive and not large enough. They could never really be closed, because there is always at least one variable changing behind the back of the analyst. Every attempt to force all the variables to behave properly leads to unruliness within the system and unforeseen consequences outside it, blurring the very boundaries between system and environment that were supposed to remain closed. The move to globalism, advocated at the time by many environmentalists, was a way to acknowledge the latter aspect (Jasanoff, chapter 10, this volume). Weather modification's downfall occurred at the same time, and probably for the same reasons, as when the other fields experienced their crises. The difference is that the weather modifiers had to be told, instead of discovering it by themselves.

But the history of weather and climate modification did not end entirely in tragedy. Out of weather modification research emerged continuing concerns about anthropogenic climate change and an increasing demand for computer simulation models of the climate. In some respects, meteorology has now become the glamour discipline for which so many early protagonists of weather modification had hoped. The atmospheric sciences are no longer the same, however. Weather modification has become a marginal concern of the American Meteorological Society. The Society has given it up almost entirely, with the exception of rainmaking. The Bureau of Reclamation and several private firms still seed clouds in America's West. Several of the latter, however, have transferred their activities to other countries.

Notes

1. The NOAA was created July 9, 1970, at the same time as the Environmental Protection Agency. Like the EPA, the NOAA grouped several previously independent government agencies, among them the Weather Bureau (now called the

National Weather Service). White had been chief of the Weather Bureau since 1963.

2. In 1974, the director of the National Hurricane Research Laboratory of the NOAA judged that "from results of seeding experiments in recent years, and from increased knowledge of the behavior of hurricanes, it seems very unlikely that the 1947 seeding could have had much effect on the hurricane except for the seeded clouds" (Gentry 1974, 506).

3. Recollection of Philip D. Thompson (1983). The *New York Times* article had appeared on January 11, 1946 (Aspray 1990, 351).

4. Twelve precipitation enhancement projects, of which two are mentioned in this chapter, are discussed in some detail in National Academy of Sciences 1973.

5. The laboratory belonged to the Environmental Science Service Administration, which after 1970 became NOAA.

6. The seeding was part of Project Cloud Catcher, a $675,000 experiment sponsored by the Bureau of Reclamation (Carter 1973).

7. According to Dennis, it was primarily the two dry summers of 1973 and 1974 that had led to public discontent about the failure of weather modification to do something about it.

8. Interestingly, Helmut Landsberg, in a long review article on "Man-Made Climatic Changes" (Landsberg 1970), also concluded that human-made aerosols constitute a more acute problem than CO_2.

9. Apparently, the NSF RANN program paid for the costs of the technology assessment project.

6

Scientific Internationalism in American Foreign Policy: The Case of Meteorology, 1947–1958

Clark A. Miller

Prologue to the Future

In the years following World War II, significant changes in the practices and organization of meteorology around the world made it possible, for the first time, to observe the atmosphere in its entirety in real time. In studying the evolution of other domains of scientific research in the modern era, social scientists and historians such as Bruno Latour, Robert Kohler, and Peter Galison have begun to highlight the connections between the material culture of scientific research—the organization of physical infrastructure, technological instruments, and human resources—and scientists' access to and understandings of nature (Galison 1997; Kohler 1992; Latour 1987). Cognitive developments, these authors illustrate, depend in important ways on successful achievements in the mobilization of social and material resources. In a 1955 review of postwar meteorology reminiscent of these writings, J. van Mieghem, a prominent meteorologist and a principal architect of the 1957–58 International Geophysical Year (IGY), observed: "The history of meteorological progress is inseparable from the history of successful development in the observation network" (van Mieghem 1955b, 7). What is particularly interesting about van Mieghem's review, however, are the implicit connections it and other contemporary writings reveal between "successful development in the observation network" and manipulations not only of material culture but of political culture as well. Creating a worldwide science of meteorology in the immediate postwar era, I argue in this chapter, was a deeply political as well as technical exercise that ultimately linked progress in meteorology to the reorganization of international politics that followed World War II (cf. the concept of *heterogeneous engineering*; Law 1986).

Conceptualizing scientific progress as dependent on social achievements in both material and political terms is particularly important today as the social sciences grapple with the challenges posed by climate change and other transnational environmental issues to conventional social science theories of politics and society. The politics of climate, for example, tends to take for granted scientific depictions that portray the atmosphere as part of a global system linking atmospheric change to ocean dynamics, vegetation patterns, ice flows, and human economies. Van Mieghem's world, however, looked very different. Situated just a few years after the formation of the World Meteorological Organization (WMO), meteorologists were still in the process of creating a worldwide network of observing stations complete with instruments for measuring atmospheric dynamics and energetics. Not until they had encircled the globe with standardized, networked observers during the IGY did they, for the first time, have the means to glimpse the atmosphere as scientists see it today. The launching of the first satellites during the IGY similarly heralded a new era in meteorology, but even satellites would not provide a global picture of the atmosphere until well into the 1960s. It is no accident that the annual measurements of atmospheric carbon dioxide concentration in Antarctica (later continued at Mauna Loa in Hawaii) and the development of computer models of the atmosphere's general circulation—which together form the basis of current scientific projections of global warming—began during the same period (National Academy of Sciences 1966; Weart 1997; Edwards, chapter 2, this volume).

The central narrative of this chapter is a history of worldwide cooperation in meteorology from the signing of the World Meteorological Convention in 1947 to the IGY in 1958. Historians have long recognized the importance of the early postwar period for the growth and development of meteorology. They have also recognized, even if they have not always framed it in these terms, the degree to which postwar growth in meteorology depended on worldwide changes in the social institutions and practices of meteorologists (Daniel 1973). In the two decades following World War II, meteorologists constructed what weather scientists had long dreamt of, a *réseau mondiale*, a worldwide network of coordinated research stations providing simultaneous, real-time information on "the state of the atmosphere on every point on [the earth's] surface" (Daniel 1973, 8). Throughout the 1950s and 1960s, meteorologists added new stations in Asia, Africa, and Latin America to their existing networks in the United States and Europe. They created a global tele-communications network linking meteorological research stations and forecasting centers; they coordinated international standards for ground-based, airborne, and ship-born data-collection systems; they shared in the development and use of high-speed data-processing facilities and arti-ficial satellites; and they trained innumerable new recruits worldwide. Their efforts paid off handsomely in improved weather forecasts and new understandings of global-scale atmospheric dynamics that continue today to underpin research on the physics of climate change, ozone depletion, acid rain, and desertification (World Meteorological Organization 1990).

The account I present here supplements this previous historical work by linking postwar developments in meteorology to concurrent changes in the politics of international relations. Progress in worldwide meteo-rological cooperation depended on more than just scientific dreams and technological coordination; it also depended on mobilizing political support for meteorological activities. Meteorologists needed to convince skeptical governments and publics that scientific and technical coopera-tion of the form they offered would bring benefits beyond scientific knowledge for its own sake. Governments had to be persuaded essen-tially to nationalize international cooperation in meteorology through the signing of the World Meteorological Convention, transforming what had previously been carried out by private individuals into an inter-governmental affair with potential geopolitical ramifications. Building worldwide research networks took considerable resources. It relied on the willingness of countries to allow weather scientists to cooperate with each other, and it necessitated granting meteorologists the authority to commit their governments to not only the operation of weather stations but also their intercalibration against world standards and the regular exchange of meteorological data. In many former colonies, now newly independent states, it also took convincing governments of the benefits and obligations of participation in the international community of nations. All of this was a deeply political exercise (see other accounts of the political aspects of technological standardization, e.g., Porter 1995).

The stabilization of worldwide political support for meteorological cooperation, I argue in this chapter, took place as part of the broader stabilization of a political culture of international governance that explicitly linked science, technology, and politics in a liberal vision of postwar order. In his study of science and democracy, Yaron Ezrahi notes that the United States and other Western democracies have evolved historically and culturally specific discourses that "use scientific knowledge and skills ... to ideologically defend and legitimate uniquely liberal-democratic modes of public action, of presenting, defending, and criticizing the uses of political power" (Ezrahi 1990, 1). Although Ezrahi limits his analysis to the organization of domestic governance, the history of postwar meteorology makes clear that many of these cultural resources were also adapted for use in foreign policymaking. Particularly in the United States, faced with the task of reconstructing international order after World War II, foreign policymakers sought ways to use what they perceived to be the political neutrality of science and technology as an instrument in the construction of liberal international organizations.

To better understand how American foreign policy linked science and technology to postwar order, I explore in this chapter the specific clusters of discursive and institutional norms and practices, or *modes of interaction*, policymakers developed between scientific and political cooperation. I argue that U.S. foreign policy articulated three such modes of interaction in the early postwar era that were ultimately integrated into the activities of international institutions: (1) intergovernmental harmonization, (2) technical assistance, and (3) international coordination of scientific research. This chapter illustrates, historically, how these three modes of interaction shaped meteorology through the activities of the World Meteorological Organization. Following the introduction, in the chapter's second section, I briefly describe the ideological construction of scientific and technical cooperation in postwar American foreign policy. In the third section, I turn to the negotiation of the World Meteorological Convention in 1947 that created the WMO out of its predecessor, the privately organized and operated International Meteorological Organization (IMO). I describe how plans for the WMO fit into America's foreign policy vision of the postwar order and how this vision helped shape the organizational structure of the institution. Having created the WMO, meteorologists then proceeded to use it to expand their observation networks around the world, primarily through two programs. In the fourth section, I describe meteorological participation in the UN Expanded Programme of Technical Assistance to Lesser Developing Countries, and, in the fifth section, I describe meteorological participation in the IGY. In both, I elaborate the political reasoning that led the United States to promote scientific and technical cooperation, as well as the effects that reasoning had on how meteorological networks expanded.

The third, fourth, and fifth sections each highlight how one aspect of the WMO (respectively, its negotiation and standard-setting activities, technical assistance activities, and participation in the IGY) maps onto a unique mode of interaction between scientific and political cooperation as articulated in postwar American foreign policy. In each of these three areas, the central tension of this chapter—the construction of the global or universal out of the disjointed and local—emerges as a particular policy problem faced in the pursuit of meteorological cooperation. What is of particular interest to me in this chapter is, in each case, first, the way an ideological commitment to the political neutrality of science emerges as an important resource in the efforts of meteorologists and other policymakers to resolve those policy problems and, second, the challenges that arose to their efforts to realize that neutrality in pragmatic, institutional terms.

I conclude the chapter with several suggestions for how the history of postwar cooperation in meteorology alters our understanding of the historical evolution of patterns of international governance in the second half of the twentieth century. Traditional scholarship in international relations has largely ignored or taken for granted the place of science and technology in international politics. Even where science and technology have taken center stage in international relations scholarship, as in research on epistemic communities, studies of international regimes have tended to "black box" the internal workings of science and technology as lying outside the domain of political analysis (see Jasanoff 1996c; Haas 1990a, 1990b, 1992). By contrast, coupled with other recent scholarship on science and postwar American foreign policy, the history of postwar international cooperation in meteorology suggests that ideas about science and technology played essential roles in under-pinning the postwar organization of international politics (e.g., Burley 1993). Efforts to incorporate expertise into the conduct of international affairs shaped the development of international institutions, the formal languages (increasingly quantified) used to assess state behavior and compliance with international regimes, and the organization of the foreign policy apparatus of the national state.

The modes of interaction between scientific and political cooperation articulated by American foreign policymakers and institutionalized in organizations like the WMO provided a normative framework against which policymakers could define and articulate states' (and indeed the international community's) interests as well as an institutional model for how states could pursue those interests through specific kinds of activi-ties. In this, they helped establish the legitimacy of newly emerging forms of multilateral institutions in international politics (see Ruggie 1993). If international relations scholars are to pursue these connections more fully—especially as scientists and other experts grow in importance in international environmental and economic governance—they will have to take seriously questions of how particular epistemic communities and particular epistemes come into being and acquire authority in particular international political contexts. This chapter seeks to explain this process for the international meteorological regime centered around the WMO.

Science, Technology, and Postwar American Foreign Policy

History demonstrates the importance Americans placed on political and economic cooperation in postwar American foreign policy (Cohen 1991). In the years immediately following World War II, American foreign policymakers saw themselves as leaders in a worldwide effort to reconstruct a stable world order that would serve American interests alongside those of the rest of the world. The United States pushed for, and obtained, the creation of the Bretton-Woods institutions regulating international finance, the United Nations, and numerous other new organizations in an effort to lay the groundwork for future peace and prosperity through a network of transnational institutions. As the Cold War emerged after 1947, liberal internationalism gave way to national security as the primary concern of U.S. foreign policy. While adding a hefty dose of realism to American attitudes, this shift ultimately also reinforced American efforts at international cooperation, within the West as a counter to the threat of communist aggression (external and internal) and with the Soviets as a basis for peaceful (or at least stable) relations among the superpowers.

American scientists and other policymakers also actively promoted international cooperation in science and technology after World War II. In new international institutions like the International Monetary Fund, the World Meteorological Organization, the International Civil Aviation Organization, the Food and Agriculture Organization, the World Health Organization, and numerous others, experts from governments around the world came together, often for the first time, to find grounds of common interest (Haas 1992; Adler 1992; Ikenberry 1992). Through programs such as the Marshall Plan, the United Nations Expanded Program of Technical Assistance to Lesser Developed Countries, and other scientific and technical assistance projects, scientists, engineers, economists, and other experts traveled frequently from the United States to Europe, Japan, and numerous developing countries to help provide scientific and technical knowledge and skills where they were needed. And, beginning in the late 1950s, scientists worldwide increasingly began to coordinate international research programs such as the IGY, the International Biological Programme, worldwide campaigns to eradicate smallpox and polio, and the efforts of the Consultative Group on International Agricultural Research to end world hunger through increases in agricultural productivity.

American efforts to promote scientific and political cooperation in the postwar era were intimately related. American foreign policymakers perceived the production, validation, and use of scientific and technological knowledge and skills as intertwined with the pursuit of a free, stable, and prosperous world order. American perspectives on the use of science and technology in postwar international politics were based on a model that implicitly linked the pursuit of science to the success of a liberal world order. Summarizing American attitudes towards science and foreign policy, a committee of leading scientists and government officials led by Lloyd Berkner wrote in *Science and Foreign Relations*, a 1950 review of U.S. State Department policies:

The international science policy of the United States must be directed to the furtherance of understanding and cooperation among the nations of the world, to the promotion of scientific progress and the benefits to be derived therefrom, and to the maintenance of that measure of security of the free peoples of the world required for the continuance of their intellectual, material, and political freedom.

Underlying the[se] objectives are the basic premises on which this report is based; namely, that certain definite benefits which are highly essential to the security and welfare of the United States, both generally and with respect to the progress of science, stem from international cooperation and exchange with respect to scientific matters. These premises may be itemized as follows:

Since science is essentially international in character, it provides an effective medium by means of which men can meet and exchange views in an atmosphere of intellectual freedom and understanding. It is therefore an effective instrument of peace.

The healthy development of American science and technology that is essential to our national existence requires that American scientists have free access to and be fully aware of scientific thought everywhere, and that they join in its creation. American access to foreign scientific sources implies of necessity a two-way flow of scientific information if our access is to be anything but sketchy and difficult. We in America are dangerously prone to underestimate the importance of foreign scientific progress. American preeminence as demonstrated thus far is in the application of scientific discovery; hence it is to our practical advantage to promote the fullest scientific intercourse.

Scientific developments are increasingly recognized to be essential to economic welfare; and from its economic welfare stems the political security and stability of any nation. The United States recognizes this truth not only with respect to its own economy and welfare, but also, through its programs of technical assistance, with respect to the other nations of the world. By thus strengthening and bringing about self-supported economies in free nations through the world, this country seeks to counter the occurrence of economic depression and thus to offset the threat of Communist infiltration.

National policy with respect to national defense similarly demands that the scientists of this nation be kept currently aware of the latest advances of modern technology, in whatever nation these may occur. (Berkner 1950)

Accomplishing the goals of American foreign policy meant, in other words, significant changes in the organization of both science and international politics. The second and fourth of these premises spell out most clearly the ways political goals directed scientific change. These premises reflected America's perceived need to protect itself against the possibility of future discoveries abroad that might threaten the security of the United States and its allies. The report argued that scientific preeminence was now central to economic prosperity and national security and could only be maintained through policies that ensured American scientists' access to their colleagues' work throughout the world. "American access to foreign scientific sources," the report argued, "implies a two-way flow of scientific information if our access is to be anything but sketchy and difficult." Hence, the report concluded that the State Department ought to support American scientists in their efforts to cooperate with their counterparts abroad. Only by encouraging international scientific cooperation, including the exchange of data, journals, and personnel, could the United States hope to keep abreast of all foreign discoveries. In the future, the State Department should facilitate the visits of U.S. scientists to foreign countries, help organize international scientific conferences, and ease the difficulties faced by foreign scientists in obtaining visas to travel in the United States or other countries. The overall argument was simple. U.S. policies should promote a particular form of international scientific cooperation that stressed the free and open exchange of information and people. Only in such a way, the report suggested, could U.S. scientists be guaranteed access to new discoveries in other countries.

The report's first and third premises, on the other hand, spelled out a vision for how a reconstructed science and technology could contribute to worldwide political change by helping to constitute a new international order. Science offered two particular benefits in the model of coupled scientific and political cooperation laid out in the report. First, the report ascribed a particular moral character to scientists that would allow them to escape the influences of political contestation through free communication and the international character of science. Second, the report developed a particular account of the relationship between science, wealth, and order that linked scientific research to the economic growth and prosperity Americans believed necessary for political security and stability. These two premises described a model for international political and economic cooperation based around scientific and technical cooperation that, by the writing of the report in 1950, had already formed the basis for numerous American efforts to organize postwar international relations.

In connecting science and technology to international politics in this way, postwar American foreign policy drew heavily on America's domestic experiences with integrating scientific and technical expertise into public governance. Ezrahi has argued that political cultures, including liberal democracies, develop specific symbolic and institutional resources for linking science and technology to the pursuit of particular models of politics (Ezrahi 1990; see also Porter 1995). Perhaps it would be more appropriate to say that the incorporation of expertise into government policymaking in the United States and Europe has constituted an important site for the continual articulation and rearticulation of liberal democratic norms and practices. Where Ezrahi and historians such as Theodore Porter focus on the evolution of domestic governing institutions, however, it is clear that the process of articulating liberal democratic norms and practices also took place in the realm of foreign policy. Following World War II, American foreign policymakers adapted ideas about the contribution of science and technology to political well-being to American foreign policy goals, articulating and institutionalizing three modes of interaction between expert and political cooperation.[1]

One mode of interaction centered around *intergovernmental harmonization*. In the immediate postwar era, Americans established formal institutional settings for scientific and technical experts throughout the United Nation's specialized agencies, the Bretton-Woods institutions regulating international finance, and other multilateral organizations (see, e.g., Ikenberry 1992). Anne-Marie Burley has argued that these American efforts reflected the experiences of the New Deal American state (Burley 1993). To postwar American foreign policy experts, like their New Deal predecessors, technical cooperation carried a strong instrumental purpose. According to Burley, the United States promoted expert organizations during the immediate postwar years as places where international technical cooperation could help rationalize policy choices, much as their counterparts had in American domestic politics a decade earlier. From this perspective, scientific and technical cooperation could enhance the likelihood of efficiently and effectively coordinating government policies in such areas as international finance, telecommunications, and civil aviation, providing a basis for worldwide economic growth. In retrospect, it seems clear that the participation of experts in international organizations also helped legitimize centralized political institutions to democratic publics. Much as the perceived objectivity of expert professions helped lend authority to a host of new federal agencies under the New Deal (Burley 1993; see also Ezrahi 1990, esp. 217–236), so too did it help secure legitimacy for the postwar order's new multilateral institutions by presenting them as technical organizations whose activities would not abridge states' sovereign rights.

A second mode of interaction centered around *technical assistance.* Important early arenas of postwar technical assistance included American efforts to reconstruct Japanese and German societies along a liberal democratic model. Here, as in efforts to promote intergovernmental harmonization, American foreign policy drew heavily on New Deal and wartime experiences with science and technology. At war's end, the central goal of American policies vis-à-vis the former Axis nations of Japan and Germany focused on their demobilization and demilitarization. By 1947, however, the dismantling of the Japanese and German wartime economies had precipitated economic collapse, social instability, and the fear of communist exploitation. In response, the United States began to search for ways of revitalizing economic activity in the two countries (Smith 1994). At home, the lessons of the Great Depression had led Americans to connect social and political stability with economic development. Wartime experiences, in turn, had offered policymakers a dramatic display of the possibilities of connecting scientific research and technological innovation to economic growth (see, e.g., Bush 1945 for the classic statement of this position). American policies in Japan and Germany coupled these two lessons, promoting scientific and technological research as a motor for economic revitalization and, hopefully, political stability. At the same time, American policies also took steps to dissociate German and Japanese research from military and government influence. To help prevent the future recapture of science and technology by a militarist state, American policies in both countries explicitly encouraged the integration of scientists into the international scientific community and the development of strong bilateral scientific relations with the United States. In Japan, U.S. administrators went further, developing specific policies to overhaul national scientific institutions and establish their institutional independence from political elites (Yoshikawa and Kauffman 1994).

Nor were American technical assistance activities limited to Germany and Japan. The winter of 1946 struck Europe particularly hard, and with European economies already in collapse, the United States stepped in. By May 1947, Truman and his Secretary of State George Marshall were announcing plans for the largest peacetime aid program ever. While most discussions of the Marshall Plan focus on its financial components, American policymakers also mobilized considerable technical assistance to help revitalize economic growth, as they did in subsequent aid packages to Greece and Turkey designed to counter communist military threats in Southern Europe. In 1949, the Truman administration extended support for technical assistance in American foreign policy to countries undergoing the transition from colonial dependency to independent statehood. Foreshadowing the creation of the UN Technical Assistance Program, Truman announced in his inaugural address of January 20, 1949: "We must embark on a bold new program for making the benefits of our scientific advances and industrial progress available for the improvement and growth of underdeveloped areas" (cited in McDougall 1997, 181).

A third mode of interaction centered around *international cooperation in scientific research*. Beginning in the late 1950s, American scientists and policymakers promoted international cooperation in scientific research. American scientists and policymakers believed that science could provide universally valid descriptions of global phenomena that would form the basis for shared understandings of transnational problems and cooperative efforts to find their solution. The model for such research was the 1957–58 IGY. Indeed, before the eighteen months of the IGY had finished, similar research programs had already begun in the fields of public health and agriculture. As one observer put it, "These projects must surely be welcomed by all internationally-minded persons as being indicative of a growing tendency to solve the world's problems by cooperative action on a world scale" (Ashford 1959, 60). Subsequent programs included the International Biological Program, the Man and the Biosphere Program, the International Hydrological Decade, the World

Climate Program, and others, bringing a whole array of issues to the world's attention. Nor were the areas concerned merely on the fringes of international politics. In 1958, the Eisenhower administration opened a long and involved negotiation with the Soviet Union on the issue of nuclear testing with meetings of scientific experts. The State Department intended the meetings to forge shared understandings between U.S. and Soviet scientists on the technical capabilities of existing verification systems. While the scope of agreement achieved during the meetings proved less extensive than participants had hoped,[2] the meetings themselves help illustrate the beliefs of American (and Soviet) policymakers in the possibilities offered by science for enhancing international cooperation. A year later, to prevent economic or military exploitation of Antarctica, the United States, the Soviet Union, and ten other nations signed the Antarctic Treaty. Testifying to beliefs in the ability of science to transcend political conflict, the treaty designated the continent as a scientific reserve and established the Scientific Committee on Antarctic Research as one of two policymaking bodies for the continent, alongside the intergovernmental treaty organization.

In the fields of economic and environmental cooperation, too, the belief that scientific research could underpin the formation of shared, transnational understandings of policy problems led to new forms of international institutions. In 1960, Western Europe, the United States, and Japan formed the Organization for Economic Cooperation and Development (OECD). By and large, OECD science policy can be understood within the first mode of interaction described above—that is, intergovernmental harmonization, along the model of the early UN specialized agencies, to promote the exchange of experts and expert knowledge between countries and to harmonize national policies in scientific and technical arenas. As Ernst Haas notes, however, the OECD offered two novel institutional developments: programs for coordinating scientific research internationally and its Directorate for Scientific Affairs. This latter organization "prided itself on being a 'look-out institution,' an advance warning system which seeks to identify and explicate problems facing industrial societies but which are not yet on any public agenda" (Haas, Williams, and Babai 1978, 159). This kind of "catalytic" role for science in raising concerns in the international community also characterizes the UN Environment Programme (UNEP), organized in 1972 at the UN Conference on the Human Environment in Stockholm and one of the few international institutions created during the 1970s. UNEP's organization is much smaller than other UN agencies and its mandate specifically forbids it from implementing international programs to protect the environment. Instead, over the years, it has used its resources to promote scientific research on the environment as a means of raising the visibility of environmental degradation and to alter the priorities of the rest of the UN (Caldwell 1990).

Why did other countries not oppose the incorporation of essentially American models of the relationship between expertise and political order into the organization of international politics? This question is beyond the scope of this chapter, but a few speculations seem plausible. First, the United States had enormous prestige in the immediate aftermath of World War II, and American negotiators and experts actively participated in the construction of postwar international organizations. In fact, the blueprint for many of these organizations was drawn up by American foreign policymakers during the war, so as to avoid the lack of preparation that many viewed as contributing to the failures of the League of Nations following the previous war. In addition, U.S. money provided the backbone of financial support for many of the new institutions. Second, many European nations had similar goals to the United States and seemed willing to defer to U.S. approaches to carrying them out. This is particularly noticeable in Ikenberry's (1992) account of the British involvement in negotiating the Bretton-Woods institutions regulating international finance. The British government wanted to stabilize international financial exchanges, just as the Americans did, but it did not appear overly concerned about the institutional form such stabilization took. Soviet aims toward international institutions in the early postwar era centered, on the other hand, on preventing the United States from using international organizations to promote a liberal, bourgeois, capitalist world order. Why then did they not object to what Americans clearly viewed as scientific cooperation that would enhance the success of liberal internationalism? I can only speculate here based on the work of historians of Soviet science and technology. To Soviet leaders, science and technology were instrumental resources for furthering state policy through the rationalization of the means of production (see, especially, Graham 1997). In this (and only this) sense, then, we might talk about the Soviet view of scientific and technical cooperation as outside of the realm of their ideological competition with capitalism. Soviet leaders may have believed, therefore, that limiting international institutions to scientific and technical cooperation would serve as a way of preventing those institutions from furthering the interests of Western political and economic agendas, a view that, if it correctly captures Soviet attitudes, appears to have been profoundly mistaken.

New Standards for Meteorological Cooperation

Postwar American foreign policy regarding science and technology had important implications for meteorologists. To be sure, Americans did not simply impose their vision of world order on international cooperation in meteorology. American scientists, State Department officials, and other foreign policymakers did have very clear ideas, however, about how cooperation in meteorology could help the world. They articulated these ideas during the negotiation, in Washington, D.C., in 1947, of the World Meteorological Convention and subsequently during the reign of the chief of the U.S. Weather Bureau, F. W. Reichelderfer, as the WMO's first president, from 1951 to 1955. As I demonstrate in the following three sections, these ideas helped to shape the organization's practices into conformance with the three broad modes of interaction described above.

One of the most important ways American foreign policy for science and technology affected international cooperation in meteorology was in supporting its transformation from a private to an intergovernmental affair. Through the creation of the WMO in 1951, meteorologists gained new resources and a new affiliation with the United Nations. They also gained the authority to commit their governments to world standards for meteorological practice. After 1951, meteorologists used this authority to substantially increase the frequency and content of international data exchanges and to integrate this plethora of new data into new computational approaches to weather forecasting aimed at improving agricultural production and creating safer transportation by air and sea. In the eyes of American foreign policymakers, however, the harmonization of meteorological practices served the international community in an even greater role as a model for other areas of international cooperation.

International cooperation in meteorology began in the mid-nineteenth century. In 1873, meteorologists formalized this cooperation in the IMO. Except for its first few years, however, the IMO was not an intergovernmental body. Although IMO members were, by organizational requirements, directors of national meteorological services, they did not act as official governmental representatives. Rather, they attended as private individuals concerned with the progress of meteorology. Consequently, their efforts to develop standards for measuring and communicating meteorological data relied on the voluntary cooperation of their home governments. Despite this lack of official status, however, meteorologists did manage to transmit limited information between their national networks—helping to advance fledgling weather prediction efforts—and to establish and maintain a small transatlantic meteorological community into the early decades of the twentieth century in the midst of a boom in cultural and intellectual exchanges in Europe and the United States (Daniel 1973; Iriye 1997).

But during the 1930s and 1940s this situation began to change. By the 1930s, many of the directors of the world's meteorological services believed that the IMO would receive substantial benefits by becoming an intergovernmental organization. The increased status would allow members to devote more resources to the IMO, would make their actions binding on their home governments, and would allow them to strengthen international standards and improve the coordination of international data exchanges. Most important, it would assure governments that they could entrust the IMO with the future of meteorological support for aviation networks. In the early 1930s, international aviation was expanding rapidly, necessitating the development of new networks to provide meteorological data at high altitudes and in new geographic areas. By acquiring governmental status, the IMO would position itself to take charge of these networks. Without it, governments might well turn elsewhere. During the mid-1930s, therefore, the IMO entertained discussions of negotiating a new convention.[3] Like many other international activities, their discussions were postponed by World War II.

Immediately after the war, however, the assignment of meteorological network building in support of civilian air navigation to the International Civil Aviation Organization (an intergovernmental body) seemed to confirm their worst fears. Resources that could have benefited meteorological cooperation went to another organization because of the IMO's private status (International Meteorological Organization 1947, 404–405, especially the comments of Chilver and Gold). By the time the IMO resumed its activities with a special meeting of its Conference of Directors in 1946, thoughts of an international convention that would bring official, intergovernmental status to worldwide meteorological cooperation were high on the organization's agenda (International Meteorological Organization 1947).

The IMO's desires to negotiate a new convention and establish an intergovernmental organization for meteorological cooperation fit well into American foreign policy planning. Throughout the 1930s and 1940s, American foreign policy planners in the State Department had begun to assert greater control over America's cultural and intellectual exchanges with other countries. Indeed, one of the hallmarks of the early postwar period is the extent to which international cooperation was increasingly nationalized under governmental control, making intergovernmental cooperation the norm across a wide range of formerly private activities (Iriye 1997). This conjunction of interests led the U.S. State Department and the chief of the U.S. Weather Bureau, F. W. Reichelderfer, with the agreement of the leadership of the IMO, to invite IMO members to attend the first regularly scheduled meeting of the IMO Congress, held in Washington, D.C., in September 1947, with the expectation that they would participate in negotiations to draft a new world meteorological convention (see IMO Circular no. 340, July 16, 1947, in International Meteorological Organization 1947, 396).

In the eyes of the U.S. State Department, however, intergovernmental cooperation in meteorology was not just any form of intellectual exchange. As expressed by Garrison Norton, U.S. Assistant Secretary of State, the negotiation of the WMO convention held important meaning for "a broader future than the science of meteorology alone." Norton was less an active participant in than an observer of and a host to the negotiations. Nevertheless, his opening welcome to negotiators in 1947

made clear his and the U.S. State Department's hopes for the new organization and its place in world affairs. "The Western hemisphere," he
said, had long been "particularly favorable for growth and development
of the ideal of international co-operation." America had for most of its
history disdained to interfere in the affairs of Europe, but the republics
of North and South America had seen the benefits of international cooperation among themselves and formed, in 1890, the International Bureau
of the American Republics and, in 1907, the Pan-American Union. The
headquarters of this organization, Norton noted, was the very building
in which the meteorologists would meet during their negotiations. Now,
in the postwar era, he intimated, American leadership would promote
international cooperation, such as that exemplified by the meteorological community's "long and successful history" of working together to
collect data on the atmosphere, as the central goal of world affairs. For
Norton, the symbolism of a transition to a new world order was perfect.
In its seventy-three-year history, the IMO's first meeting outside Europe
was taking place under U.S. auspices and would result in the negotiation
of a new example of international cooperation (Norton 1947; see also
McDougall 1997 for an account of American attitudes toward postwar
international cooperation).

Indeed, Norton's address indicated, world meteorological cooperation
offered more than just symbolism to the search for international cooperation; it offered a model for how to accomplish it. Much as the Berkner
Committee's report would a few years later, Norton's account of the
long-term success of meteorology stressed the moral character of scientists as one key element in securing international cooperation. Cooperation on technical issues helped rationalize public action, coordinating
governmental affairs without involving political wrangling. Norton's
account emphasized meteorologists' "global outlook," their "apprecia-
tion of international cooperation," and their ability "in the search for
scientific truths" to avoid "the more uncertain and selfish motives that
complicate and hinder co-operation in some fields of international interest." He made the following plea: "I hope you can always keep foremost
in your mind the technical and scientific nature of your work so that
your relationships may be as free as possible from the obstacles and problems of political science" (Norton 1947, 373). In the future, Norton

hoped, scientific cooperation in meteorology would provide an example for other areas of international cooperation to follow.

In drawing out his model of international cooperation based on meteorology, Norton also pointed to the "world-wide nature of weather science." Like an increasing number of other areas of human interest, the atmosphere was a global phenomenon and required worldwide knowledge in order to be understood in its entirety. This was merely the impetus to cooperation, however. There were plenty of global problems around for which no international cooperation had ensued comparable to that in meteorology. Meteorologists' success, Norton argued, derived from their possession of "a universal language ... in the form of an international weather reporting code" for exchanging weather reports on a daily basis. Over the course of the previous seventy-five years, meteorologists had developed standard symbols and measures for reporting the timing and amount of precipitation, temperature, winds, cloud cover, and other weather phenomena. "Language differences and international boundaries present no barrier to your exchange of information," Norton noted. Through their meteorological codes, Norton believed, meteorologists had rendered differences among nations irrelevant and had provided a standard basis for harmonizing government practices around the world. From Norton's perspective, anyone could communicate without loss of meaning or possibility of confusion using the symbols of science and mathematics.

And there could be no question for Norton that successfully standardizing international practices and effecting successful international cooperation entailed significant material benefits. In the case of meteorological cooperation, those benefits flowed from the better meteorological forecasts produced by worldwide harmonization of data collection, reporting formats, and forecasting practices. Improvements in these forecasts improved knowledge of the weather for "agriculture, commerce, industry, and transportation," increasing the safety of travel by sea and air and improving agricultural production.

I am informed that your Technical Commissions whose meetings have just been concluded in Toronto have proposed further improvements for exchange of weather information between countries and for uniform practices in charting the weather and sending forecasts and warnings of storms which affect travel by sea and air and in one way or another exert an influence on many phases of agriculture, commerce, industry and transportation. *As your new standards for observing and measuring the weather are studied and adopted in your forthcoming meetings here, you will strengthen the foundation upon which to build an era of progress that will bring you the gratitude of men and women throughout the world who use your weather services.* (Norton 1947, 373; emphasis added)

Best of all, such benefits were not limited to individual countries but were available to all: "The world needs your help in solving some of its problems and your efforts surely will bring greater comfort and safety and will contribute to a higher standard of living for all peoples." And in September 1947, Norton intimated, those problems were great, indeed, for Europe was in crisis. For the success of the Marshall Plan, European cooperation was considered essential. European leaders, therefore, met during late summer 1947 to prepare their request for American aid. It is not inconsequential, I suspect, that the final day of that conference in September was the day on which Norton spoke to the opening session of the meteorologists' meetings. After all, agricultural production and air and sea travel, two of the areas expected to benefit most extensively from better weather forecasting, could hardly have been far from anyone's minds in the State Department as they sought to figure out how to provide food and supplies for the upcoming winter in Europe.[4]

The vision for the IMO's future was clear in the eyes of America. Signing a new convention meant binding international standards for meteorological cooperation. Binding standards meant better-quality and higher-frequency data exchange. Better data exchanges meant better weather forecasts and better climatological data. These in turn meant better agricultural production and safer air and sea travel. The result would be greater economic prosperity in the West, more opportunities for cross-cultural exchange, and more social stability in the face of the Cold War:

In our welcome to you, we therefore have in mind a broader future than the science of meteorology alone. Certainly, we are interested in your technical success and in the new constitution for the World Meteorological Organization which you will consider here. But we also believe that your achievements will contribute in some measure to the aims of permanent world peace and prosperity towards which the nations of the world are working. (Norton 1947, 374)

Despite support for a new convention from the U.S. State Department and Reichelderfer's leadership of the drive for intergovernmental status, the negotiations were not easy. When IMO members convened in September 1947 in Washington to negotiate the treaty, a number of IMO members remained skeptical about the benefits of dramatically reorganizing meteorological cooperation. Many questioned the implications of change for the role of professional meteorologists in the organization. Would meteorologists retain control over the new organization, or would control be ceded to diplomats or international bureaucrats who had little notion of or interest in the requirements of successful cooperation in international meteorology? Would the focus of the organization remain technical, or would the new organization simply become another site for intergovernmental political battles that had by 1947 already become commonplace in the new United Nations? Would the new organization differentiate among members on the basis of political power, eliminating the long-standing equality of directors in the organization? Others raised important concerns about the activities of a new intergovernmental organization. What would the organization be authorized to do? Would it take on new activities that depleted resources available for the activities it had long supported? Would new activities, such as the development of an international institute for meteorological research, reduce the resources available for national research institutes and steer meteorological cooperation away from efforts to benefit local weather forecasting? Still others simply pointed to the seventy-year record of the IMO in promoting what they saw as successful meteorological cooperation and asked: Why risk change (International Meteorological Organization 1947, 402–406, especially the comments of Ferreira, Nagle, and Lugeon)?

To overcome these objections, early drafters of possible conventions from 1939 on had established several principles that would have to guide any new organization. It would have to: (1) ensure the independence of the organization from political control, (2) retain professional requirements for members and power for the directors of all national meteorological services, (3) continue to treat all meteorological services as equal under the organization's bylaws, and (4) maintain, as much as possible, the traditional priorities of international cooperation in meteorology, particularly in its efforts to achieve worldwide coverage. Participants in the 1947 negotiations in Washington, D.C., reaffirmed these principles during the opening discussions. Proponents of the new intergovernmental status for meteorological cooperation, quick to assuage any doubts, assured negotiators that these principles would continue to be served in any new organization. Reichelderfer, who would later become the first president of the new WMO, was prominent among those who supported the treaty and made these assurances:

Mr. Reichelderfer was of the opinion that the general situation existing at the time the Statutes of the IMO were established had completely changed during the twentieth century which has witnessed the development of technology and the modification of social, economic, and political conditions. Difficulties will doubtless be encountered in the application of a constitution, but they will be compensated for by great advantages. He was convinced that a solution could be found for any difficulties, that the rights of the smaller nations would be respected, that each Meteorological Service would have a voice within the framework of a Convention or constitution and that the influence of diplomats could be successfully deflected. (International Meteorological Organization 1947, 402)[5]

At the beginning of the second substantive discussion of the negotiations, therefore, four key instructions were submitted to the drafting committee: (1) retention of the worldwide character of the organization, (2) retention of its independence from governmental control, (3) retention of professional representation, and (4) retention of the equality of countries in the organization's decisions (International Meteorological Organization 1947, 407). As the negotiations progressed, however, a new vocabulary emerged among the participants, with profound consequences for the ability of the new organization to live up to these principles. Whereas early discussions had made reference to "nations," "countries," and "meteorological services," the opening of the eleventh meeting saw the concept of "Sovereign States" introduced into the negotiations. Subsequently, discussions focused almost exclusively on "members" and "States." The source of this new vocabulary, and its rapid adoption by the negotiators, was the insistence of American legal experts from the State Department, acting as advisors to the drafting committee, that formal participation in international institutions be limited to states whose governments retained the necessary authority under international law to accede to international agreements. Tacit in

American views was the assumption that only sovereign states held the legal authority to conduct their own foreign affairs.

The new vocabulary of "States" instead of "countries" superimposed a geopolitical imagination of the world over top of the geographic imagination that had previously organized meteorological activities. Whereas under the IMO, countries had served as areas of territorial coverage for purposes of membership, the designation of States in the WMO Convention now added a new set of nonterritorial considerations. These considerations put new constraints on any governmental organization that might emerge from the negotiations and acted to prevent several of the agreed-on principles from being achieved. Convention language divided countries into sovereign states and others, such as colonial territories, with full membership being accorded only to sovereign states. Diplomatic and legal principles came to dominate certain discussions within the organization, particularly as they affected membership, and some countries—such as Spain and China—were initially excluded from the organization on the grounds that their governments were not recognized as sovereign under international law. Not surprisingly, opposition to these developments was considerable. T. H. Hesselberg, the director of the influential Norwegian meteorological service, argued:

A few days ago the Conference was voting as technicians, but now the Draft which had come out from the [drafting committee] was based on considerations other than technical ones. What we had regarded as the fundamental brick in the structure of the WMO had been abolished on non-technical or purely legal grounds. Any World Meteorological Organization would probably eventually have to be linked to the UN, to conform with the general policy considerations of certain States. In such circumstances, and if there were some political changes at the UN level at any future time which would affect the position of any State, then the Meteorological Service of that State would probably have to be eliminated from the organization. This meant that in the future, the worldwide representation of the WMO, if it were ever attained, would be in constant jeopardy from non-meteorological causes, which was just what we were all anxious to avoid. (International Meteorological Organization 1947, 466)

The only principle not substantively changed from earlier drafts involved professional representation. The new WMO insisted that all members, whether sovereign states or otherwise, continue to be represented by the directors of their national meteorological services.

These changes in who could participate in meteorological cooperation mirrored broader debates within international society and in the early discussions of the United Nations over countries such as the Soviet republics, the People's Republic of China (PRC), and Spain. Long and bitter disputes over the rights of various representatives to attend and hold membership in the WMO surely must have led some participants to conclude that the diplomats had indeed come to power within the organization. These disputes make clear the extent to which meteorologists participating in the WMO carried their own normative agenda into these negotiations. Particularly noteworthy is the dispute over membership by the PRC. At the First Congress of the new WMO in 1951, the PRC requested membership in the organization. By geographic standards, China is not a small country, and many members of the WMO wondered how the organization could pretend to worldwide coverage if it failed to include China (World Meteorological Organization 1951b). Not only would the organization fail to achieve what they viewed as an essential ideal of global data collection, but one-fifth of the world's population would fail to benefit from the promise of better weather forecasting through international cooperation. Furthermore, the removal of an area equal to that of the United States would make it more difficult to provide accurate weather forecasts for neighboring countries as well. But the legal experts were insistent. Technical reasoning alone could not counter the fact that the Chinese government was not universally recognized as sovereign and so could not accede to the WMO Convention.

In the end, the directors of the world's meteorological services swallowed their concerns and negotiated and signed the new treaty. By 1951, a sufficient number of countries had ratified the treaty to formally establish the WMO. These events marked a crucial point in the history of postwar meteorology and the political positioning of meteorologists in postwar international affairs. The WMO's new intergovernmental status enabled its members to commit to it new resources to support meteorological research and forecasting. These resources brought access to new technologies—computers, satellites, telecommunications, and remote sensing—that transformed meteorologists' networks for collecting meteorological data and exchanging forecasts of the world's weather.

The signing and ratification of the World Meteorological Convention also gave the WMO the authority to commit world governments to the standardization of ground-based, airborne, and shipborne measurements of meteorological parameters. Prior to the Convention, the IMO had given high priority to the adoption of international standards of meteorological practice. Such standards helped enable easy transfer and integration of data into regional weather forecasts, helped guarantee that data were taken at regular intervals by calibrated instruments, and helped ensure international distribution in a sufficiently timely fashion for forecasters' needs. Because directors attended the IMO as private individuals, however, their recommendations were never binding on their governments but were limited by their ability to encourage their countries to voluntarily comply with IMO standards. In 1951, the First World Meteorological Congress reaffirmed that such standards would remain of highest priority for the WMO, with the added benefit that such standards would now be binding on national governments. The Congress established new Technical Committees to carry out several related tasks: to rewrite the technical regulations that governed the standards of international cooperation, to develop new international codes for the transmission of meteorological data, and to update the networks sending regular weather reports and forecasts around the world. During the postwar era, these activities and their acceptance by national weather services went a long way toward harmonizing the practices of meteorologists in the United States and Europe and providing the technical and organizational infrastructure for the data network that currently provides the world with the basis of its weather forecasts and for its understanding of the dynamics of the global atmosphere.

Last, but certainly not least, the designation of the WMO as a UN specialized agency granted powerful legitimacy to meteorology in world affairs. Drawing on this legitimacy, meteorologists extended their networks around the world.

Building States, Apolitically

The second area of international cooperation in which the WMO benefited considerably from American foreign policy initiatives was technical assistance. Between 1945 and 1960, the WMO made extensive progress in extending its observation networks into new geographic areas with little or no prior meteorological coverage. The majority of this activity took place under the auspices of the UN Expanded Program of Technical Assistance for the Economic Development of Under-Developed Countries, with which the WMO acted in partnership from its inception. WMO-facilitated projects helped countries organize their meteorological services, train beginning and advanced meteorological personnel, and upgrade their technological capabilities (although, in general, these programs did not support the purchase of new equipment). By 1960, the WMO was spending over $300,000 per year in support of technical assistance activities in more than thirty countries and was receiving requests for projects whose full support would have entailed twice that figure.[6] Perhaps the most important of these projects took place in countries such as Libya, Israel, Peru, Afghanistan, Jordan, Iran, and Taiwan, where WMO experts helped create meteorological networks that had not previously existed.[7] Needless to say, no comparable activity could have taken place under the auspices of the privately organized and financed IMO.

The WMO was one of the earliest recipients of money from the UN Technical Assistance Board (UNTAB), the organization responsible for overseeing the UN Expanded Program of Technical Assistance. At the meeting of the First World Meteorological Congress in 1951 to inaugurate the WMO, the delegates agreed that technical assistance to underdeveloped countries constituted a worthwhile task for the new organization.[8] On the table at that meeting was a request from the UN General Assembly to all UN specialized agencies to provide technical assistance to the new Libyan government in support of economic development and social progress. The Congress resolved unanimously to approve this action and offered its support in formulating and implementing a plan for the organization of a national meteorological service.[9] Over the next eight years, WMO experts consulted with the Libyan government, drew up a plan for the new service, established it, recruited and trained Libyan nationals in carrying out meteorological observation and forecasting, and acted for many years as the organization's acting director.[10] If one is looking for an example of the social and political work that goes into the construction of transnational epistemic communities, I can imagine few better than the activities of meteorologists in postwar Libya.

As Tony Smith and Walter McDougall have argued in their histories of American foreign policy, technical assistance projects like WMO's work in Libya were not simply random acts of kindness (Smith 1994; McDougall 1997). During the 1950s, a UN Trusteeship governed Libya, a former Italian colony, for the purposes of guiding the processes of decolonization and transition to republican democracy and independent, sovereign statehood. The United Nations, prompted largely by the United States, designated technical assistance programs to Libya and to other countries undergoing decolonization to help bring about a stable and secure transition. Through technical assistance, these new states would receive the material basis, technical training, and institutional infrastructure believed necessary for society and economy to function. In short, technical assistance was a form of state building.

Where Smith and McDougall fall short is in explaining why such assistance was technical. Indeed, Smith notes that, by the 1950s, American support for state building in countries such as Vietnam largely ignored advice on party building, the mechanics of elections, or the rule of law in favor of expert support for development projects. The answer, I believe, lies in viewing the incorporation of expertise into public policy as a site where the norms and practices of liberal democratic governance are continually contested and articulated. Throughout the nineteenth and twentieth centuries, American political institutions have struggled to find publicly acceptable ways of legitimating policy choices through the appropriation of politically neutral expertise (Ezrahi 1990). During the Progressive Era and New Deal, in particular, strong professional norms and a belief in the universality of science helped secure public acceptance of the growth of expert agencies in the federal government by appearing to render such agencies consistent with American ideals of individual freedom, democracy, and the decentralization of power (see also Porter 1995). Americans viewed expertise, in other words, as offering a buffer against the corruption of power by political interest.

By highlighting the scientific character of technical assistance, U.S. foreign policy leaders similarly sought to persuade a skeptical American public to accept American intervention in the domestic affairs of other peoples. Limiting assistance to scientific and technical expertise would prevent the use of such assistance by the American government to abridge or distort the sovereign rights of free peoples abroad. Such aid was represented as merely technical, and therefore politically neutral. Hence, American foreign policymakers argued, it did not involve taking sides in domestic politics. At the same time, Progressive ideologies, still powerful in America's foreign relations, continued to portray science and technology as instrumental in the advancement of peoples abroad. Both arguments were reinforced in the minds of the American public and political and technical elites by the lessons they drew from America's abortive experiences in colonial administration in the Philippines and Cuba and from other past efforts to supply technical assistance abroad (Curti and Birr 1954).

The argument of political neutrality was more difficult to sustain, however, in the eyes of policymakers abroad, particularly in recipient countries (Ferguson 1990). State building, from their point of view, often carried with it obvious political consequences that were difficult to disguise. The WMO technical assistance program in Libya, for example, presented an image to the world of merely providing technical training and education for meteorologists. The WMO program, however, also helped organize and train a federal civil service out of the three former colonial provincial bureaucracies that occupied the region, helping to shift power and authority within Libya to the new federal government (World Meteorological Organization 1954).

Only when technical assistance successfully integrated the political consequences that flowed from expert activities into domestic political norms and institutions were the results generally considered successful. The WMO technical assistance program in Israel provides a useful example where such integration seemed to work, for all that it was much smaller in scope than the program in Libya. In all, the WMO's technical assistance to the new Israeli state during the early 1950s consisted of two lecture series and advisory missions of a few months duration each, as well as fellowships for several Israeli meteorologists to study abroad. The Israeli requests for assistance, however, grew directly out of state-building concerns within the new polity. At the time, settling its large immigrant population in its arid regions was one of Israel's highest priorities. Yet the Israeli government believed it lacked the information necessary to make such a program work and so requested visits from specialists in indoor environment and agricultural micrometeorology in arid regions to provide the requisite expertise. Technical assistance could therefore be portrayed as instrumental to achieving Israel's own goals.

An Israeli meteorological service review of technical assistance in meteorology for the *WMO Bulletin* provides additional clues as to how Israel integrated ideas of foreign expertise and technical assistance into positive articulations of Israeli politics. The review cast WMO technical experts as non-Israeli arbiters to which the government could turn for "impartial assessment" of the country's development policies and capabilities for carrying them out. The review also suggested that the WMO's technical assistance program had helped alter the Israeli public's demands for meteorological information. According to the report, after the lecture series by WMO experts, the Israeli meteorological service received a substantial increase in requests for weather and climate-related information from farmers, homeowners, and industrialists in desert regions (Gilead 1954). Whether these requests for information reflected the creation of new expectations that the government should provide expert advice to its citizens, helped stabilize already existing tendencies in the Israeli polity, or merely raised the expectations of citizens about the quality of advice their government ought to be able to provide is not clear from the report. Regardless, the report indicates that the public's new demands for information provided the Israeli meteorological service with the necessary incentives and rationale to centralize its operations using older U.S. and European services as models and helped encourage Israeli politicians to support that centralization to provide better weather forecasting. When the WMO technical assistance program created its "Special Fund" in the late 1950s to fund more significant development programs, further centralization and modernization of the Israeli meteorological service was one of its first projects (World Meteorological Organization 1960).

Thus when, in cases such as Israel's, technical assistance not only provided reliable expert knowledge but also successfully integrated the political consequences of expert activities (and particularly foreign expert activities) into domestic norms and institutions of political governance, the results were often considered successful. In many other cases, however, the political changes wrought by technical assistance were not seen as legitimate. Over the course of the 1950s and 1960s, technical assistance projects came to be seen in many developing countries as serving the goals of international institutions and experts instead of the goals of national governments. States began to object to assumptions built into technical advice about which directions of development were most appropriate. This became particularly significant where technical assistance activities were clearly supporting international institutions' goals of integrating new states into international society in addition to or even opposition to supporting local development priorities. At this point conflicts began to arise over which goals took precedence when they pointed in different directions. Meteorology, with its links both to an international network of meteorological observers and also to the international civil aviation system, was particularly vulnerable to such criticisms. Who could say whether the principal concerns guiding the WMO technical assistance programs emphasized the benefits of new meteorological facilities to international scientific networks or to local weather forecasting?[11]

In institutions like the UN Technical Assistance Board, these kinds of concerns became flashpoints in determining whose priorities should control technical assistance. From the WMO's first agreement with the UNTAB in 1952 through the mid-1950s, meteorology received a certain priority in technical assistance activities. Meteorology was, as Western experts associated with the WMO and other international agencies such as the UN Food and Agriculture Organization and the International Civil Aviation Organization described it, a "key mission." Western experts argued that, through the development of a national meteorological service, a country could gain significant advantages in a number of different areas including air transportation, hydropower development, and agriculture (Treussart 1957). From the perspective of many developing countries, however, meteorology held relatively little attraction since it provided little immediate or direct economic stimulus and did not involve significant technological development. When control over development assistance was transferred largely to recipients in the late 1980s, projects organized through the WMO began to decrease precipitously.

Over the same time period, technological developments in the United States and Europe, combined with cost-cutting attitudes towards foreign assistance, also helped undermine the tacit assumptions on which support for far-flung meteorological networks was based. Beginning with the first weather satellites in the early 1960s, the impetus for expanding surface observation networks began to decline. Today, American and European meteorological services often find it easier to obtain worldwide meteorological data by organizing a few hundred thousand workers to build and fly artificial satellites than by maintaining long-distance, international data networks in remote areas. The financial resources stay at home. There is no need to negotiate the intricacies of multiple political environments nor to worry about the ability to "discipline" observers in disparate locations over long periods of time. Establishing calibration requires only a few ground stations scattered in U.S. territories, on islands, and on military bases around the world. Overall, for most U.S. meteorologists, satellites provide a considerably more global—and more reliable—set of meteorological data.

Meteorological cooperation is not merely technical, however. Part of the WMO's success in creating a worldwide meteorological data-collection and reporting system lay in the network of trust it had built to allow countries to be relatively secure in the belief that they would continue to benefit from worldwide meteorological cooperation. The benefits gained in local forecasting by receiving data on the global state of the atmosphere depended on everyone collecting and transmitting their data to each other. Now, however, countries fear that their data flows may be cut off by the United States or Europe in times of crisis or that, more mundanely, they may be asked to pay for meteorological data as the United States and Europe contemplate privatizing their meteorological services. These fears illustrate the extent to which changes in either technological practices or political norms and assumptions can imperil the social foundation on which scientific cooperation between countries is based and can thus threaten both scientists' access to nature and the long-term continuation of transnational epistemic communities. For the moment, most meteorologists continue to support the necessity of maintaining the ground network and the desirability of providing for the free exchange of data on the grounds that it provides a global public service and promotes international goodwill towards the United States. Whether these arguments will remain sufficient to secure the future of current forms of international meteorological cooperation in an era of tightening economic constraints and declining public support for foreign aid is an important question (World Meteorological Organization 1996).[12]

Symbolic Instrumentalism

The third area of international cooperation that substantially benefited meteorology during the early postwar period involved new initiatives in scientific research. In much the same way that international cooperation in standard setting and technical assistance received support from their contributions to American visions of political order, so too did new efforts to develop international scientific research programs. Since the seventeenth century, scientific research and travel have been international in the sense that scientists in Europe and the United States drew on each others' work through elaborate networks of correspondence, the publication of scientific journals, and the training of young scholars in foreign universities. Beginning with the IGY, however, governments added an important new component to international science by establishing and coordinating joint scientific research agendas. Along with this effort came the assumption that scientific research would also contribute to broader political goals (Kwa 1987; Hart and Victor 1993).

During the early years after World War II, international cooperation in scientific research—as opposed to technical assistance and intergovernmental harmonization—was not high on the American agenda. In 1947, for example, during the negotiation of the World Meteorological Convention, the French meteorological service proposed the formation of an international institute for meteorological research under the auspices of the new WMO. This proposal, seconded by South Africa, was also raised in the UN Economic and Social Council and at the First Congress of the newly formed WMO in 1951. The concept of an international research institute, however, received little support from many other countries, who feared it would duplicate the research capabilities of existing laboratories, have negative consequences for their own national research enterprises, or entail excessive financial obligations.[13] At both the First and Second World Meteorological Congresses, the majority of delegates, led by U.S. opposition, consigned the international meteorological institute to further study (see Annexes AF-IV and AF-V in World Meteorological Organization 1951b).

By the early 1950s, however, many U.S. scientists were clamoring for international cooperation in scientific research. Geophysicists, in particular, especially in the United States and Europe where local geophysical research had reached high degrees of sophistication, felt that fundamental questions about global-scale environmental processes would remain unsolved unless opportunities could be created to collect data on a worldwide basis. Such opportunities, they argued, required international collaboration (Berkner 1959).

The laboratory of the geophysicist is the earth itself and the experiments are performed by nature; his task must be to observe these natural phenomena on a global basis if he is to secure solutions and to develop adequate theoretical explanations. This is one of the compelling reasons for the worldwide scope of the IGY in 1957–58: to observe geophysical phenomena and to secure data from all parts of the world; to conduct this effort on a coordinated basis by all fields and in space and time so that the results secured not only by American observers, but by participants of other nations can be collated in a meaningful manner. Only through such an enterprise as the IGY can synoptic data be satisfactorily and economically acquired. (U.S. National Committee for the International Geophysical Year 1956, viii)

In 1950, these concerns led Lloyd Berkner and several prominent colleagues in geophysics to prompt the Joint Commission on the Ionosphere of the International Union of Geodesy and Geophysics (IUGG) and the International Council of Scientific Unions (ICSU) to propose a third international polar year. Over the next seven years, their efforts finally resulted in the 1957–58 International Geophysical Year, the first large-scale example of intergovernmental cooperation in scientific research and a model for numerous subsequent efforts to address global issues, including environmental degradation, agricultural production and world hunger, and even peace and security. Through the IGY, scientists worldwide participated in efforts to prioritize their research activities, to collect data according to internationally agreed-on protocols, and to submit their data for publication in centralized databases over a period, ultimately, of eighteen months.

Because of its prominence in international affairs and its long history of international cooperation, the WMO quickly assumed leadership in preparing and coordinating the meteorological research program of the IGY. The WMO's participation in the IGY began in 1951. In that year, the WMO Executive Committee accepted the invitation of the IUGG to appoint a representative to its newly formed commission to organize a third polar year. At the same time, the WMO also established a special committee to coordinate its participation (van Mieghem 1955a).[14] One of the WMO's first concrete acts was to resolve "that a Geophysical Year for the whole world would be of greater interest to the World Meteorological Organization than a Third Polar Year" (Resolutions 14 and 41, World Meteorological Organization 1951d), communicating to the geophysics community a message that received resounding support in other areas of geophysical research. Soon after, proposals for a third polar year metamorphosed into proposals for the IGY. During 1952 and 1953, the WMO coordinated the development of the meteorological program for the IGY. The Special Committee for the International Geophysical Year (CSAGI, Le Comité Spécial de l'Année Géophysique Internationale) adopted this program in 1954 (van Mieghem 1955a). Ultimately, worldwide cooperation in meteorological research under the leadership of the WMO was seen by scientists as one of the most successful dimensions of the IGY in generating scientific data on global scales and providing new insights into global-scale natural phenomena (Ashford 1959). The prior existence of international data collection networks and protocols, authoritative bodies capable of setting priorities, and institutional support for the publication of data substantially eased the burdens of building a globally coordinated research enterprise. At the same time, the WMO also benefited handsomely from the IGY in terms of physical infrastructure. During the IGY, meteorologists used new funds to extend their networks globally for the first time and to introduce new capabilities for research on upper atmospheric dynamics and the global energy budget.

Early and influential participation by the WMO also affected many aspects of the IGY research networks and scientific programs in other fields as well, bringing these other programs into line with important traditions in international meteorological cooperation: a strong emphasis on the intercomparison of instruments; worldwide extension of measurement networks; adoption of standard forms for submitting measurements; creation of international centers to coordinate the collection and dissemination of observational data, bibliographic materials, and documentation; free exchange of scientific information; and the central role of national agencies in formulating, funding, and implementing research under an international umbrella. International research agendas, however, continued to reflect national priorities since all research was to be organized, funded, and implemented by national scientific agencies.

The question remains, however, why American policymakers changed their minds regarding international research. The answer lies, I believe, in the conjunction of pressure for international research from scientists, arguments such as those presented in the Berkner report—*Science and Foreign Relations*—that scientific cooperation promotes an atmosphere of free intellectual exchange, and the emergence of a new class of transnational or global problems to which science presented itself as a solution. Based on arguments similar to those presented in the Berkner committee report, the Roosevelt and Truman administrations had supported international technical cooperation for its ability to harmonize government practices through standard setting and technical assistance. In making the case for such actions, however, their arguments had presumed the prior existence of shared interests among countries. Such interests were necessary preconditions to the technical activities of experts who should not presume to determine interests but rather should work to optimize the ability of states to achieve their collective interests by coordinating governmental practices. In contrast, by the mid-1950s, the Eisenhower administration appeared to believe that scientific research, by fostering an atmosphere of free intellectual exchange, could actually help forge shared interests on problems of global concern—an acute problem under the growing threat of the Cold War. By establishing objective, universally valid descriptions of global processes, scientists could create the basis for shared understandings of how those processes intersected with human activities. At the same time, the scientific community itself could act as a channel by which those views could influence the views of publics and political leaders in other countries (Adler 1992). Drawing on this model, the Eisenhower administration supported not only the IGY but also the formation of scientific organizations to support U.S.-Soviet arms control efforts (Zoppo 1962) and the creation of Antarctica as a scientific reserve to prevent its exploitation by the demands of the Cold War on international security (Peterson 1988).

U.S. emphasis on global-scale issues carried substantial consequences for meteorological cooperation during the IGY. In particular, the IGY ultimately weighted data collection priorities toward questions related to global-scale processes. In 1952, the WMO prepared the first draft of the meteorological program for the IGY. It quickly realized, however, that it could not hope to carry out all of its intended research during the eighteen months of the IGY. To sort out its priorities, the WMO turned to its committee responsible for promoting meteorological science, the Commission on Aerology. Arguing that opportunities for worldwide coordination of data collection would be as rare in the future as they had been in the past, the Commission returned a mandate similar to that later adopted by CSAGI for the IGY as a whole: "Absolute priority should be given to problems which [can] only be solved by effective worldwide cooperation" (van Mieghem 1955a, 9). Hence, the IGY gave priority to data-collection efforts that involved all meteorologists worldwide in coordinated exercises. Measured against the priorities of the American and European scientists who sat on these committees, this choice makes perfect sense. Utilizing scarce IGY resources for activities that did not promote such research would result in wasted opportunities. Activities that did not require effective coordination among all meteorologists could be carried out at any time. Measured against the priorities of those who hoped to use the IGY to gain government support for local research, however, the WMO's and IGY's choices supplied little comfort. Cooperation in meteorology under the IGY would result in little improvement for local forecasting in many areas of the world.

The effects of this prioritization can be seen on the expansion of meteorological observation networks. During the IGY, meteorologists significantly extended their observation networks, just as they had throughout the 1950s through the UN Technical Assistance Program. There were important differences, however, between the extension of meteorological networks accomplished through the IGY and those accomplished through technical assistance. Whereas, under the UN Technical Assistance Program, countries placed new research stations so as to facilitate local meteorological forecasting, during the IGY they placed new research stations to fill gaps in a global grid to better cover the atmosphere as a whole. Primary emphasis was placed on extending longitudinal lines of stations on the 10°E, 140°E, and 75°W meridians and latitudinal lines of stations along the equator, 15°N, and 30°N latitudes. Emphasis was also placed on providing research stations in unpopulated areas and on oceanic islands.[15]

Instrument selection, too, was influenced by the global criteria. One of the principal scientific obstacles to a better understanding of the general circulation of the atmosphere was a dearth of high-altitude data. Hence, the IGY placed priority on expanding the capability of research stations to increase the maximum altitude of their balloon flights. Similarly, instruments used for research activities aimed at determining the global energy budget, including radiation, ozone, and carbon dioxide measurements, also received priority in acquisition (van Mieghem 1955b; see also van Mieghem 1964). By contrast, providing funding for research stations in poor countries to acquire basic instruments of use to forecasting received lower priority.

The shift in U.S. perspectives toward support for international cooperation in scientific research had other consequences as well that are evident in the IGY and its aftermath. In particular, the impact of scientific cooperation on official foreign policy had to be carefully renegotiated, creating new definitions of national sovereignty. Prior to this shift, American foreign policy treated scientific cooperation as it did all other forms of intergovernmental activity. Such cooperation represented official American policy and so was subjected to American interpretations of international law. As a result, as I observed above for the rules governing participation in the WMO, Americans insisted that only states they recognized as sovereign could partake in international scientific cooperation with the United States.

However, during debates over the participation of the People's Republic of China (PRC) in the IGY this position was reversed. The

United States, having recognized Chiang Kai-shek's Nationalist Chinese government in Taiwan, refused to recognize the government of the PRC. U.S. policy, under conditions of nonrecognition, forbade U.S. delegations from negotiating with PRC representatives or being present in any meeting in which that country was represented. Consequently, when the IGY accepted the participation of the PRC at the third meeting of the CSAGI in September 1955, the U.S. delegation faced a significant dilemma. Should it leave the meeting, foregoing the possibility of continued participation in the IGY? Or, alternatively, could an exception be made for scientific meetings in which official U.S. participation did not constitute a political recognition of any other state whose scientists also happened to be present? The Eisenhower administration ultimately allowed U.S. participation in the IGY to continue, despite the presence of PRC representatives (Berkner 1959). Put simply, while the Chinese were not allowed to enjoy the fruits of sovereignty in any other area of international cooperation, after 1957 they participated regularly in international scientific cooperation.

This change in policy provides an important measure of the changing view of science and international politics in the 1950s. In subsequent years, the view that scientific cooperation should not count as official recognition (and hence should be exempted from diplomatic considerations in cases of disputed sovereignty) became the official policy of the International Council on Scientific Unions, the international coordinating body for scientific professional societies and the parent organization of the IGY. In 1958, at the conclusion of the IGY, ICSU adopted a resolution enshrining the U.S. State Department's concession on the PRC as a norm of international practice:

The General Assembly, in keeping with the purely scientific character of ICSU, approves the following statement: 1. To ensure the uniform observance of its basic policy of political non-discrimination, the ICSU affirms the right of the scientists of any country or territory to adhere to or to associate with international scientific activity without regard to race, religion, or political philosophy. 2. Such adherence or association has no implications with respect to recognition of the government of the country or territory concerned. (Greenaway 1996)

Scientists, in other words, had special rights, according to ICSU, to act outside the boundaries of states' foreign policies. The creation of these rights depended, however, on dissociating scientific activities from the

imposition of obligations on states' foreign policies. Scientific cooperation, therefore, according to ICSU's principles, does not carry with it any
necessary recognition of the sovereignty of the nations whose scientists
are cooperating. In support of this policy, ICSU has often asserted these
rights on the behalf of scientists worldwide, generally with the support
of the U.S. government. Since 1958, ICSU has fought a wide array of
battles to establish the legitimacy of its position vis-à-vis all governments,
seeking to secure the rights of scientists in all countries to travel and
associate freely with their colleagues abroad (Greenaway 1996). These
actions have had the dual consequence of encouraging acceptance of the
view that science is apolitical while simultaneously conducting a very
political exercise to encourage scientific freedom in all countries, freedom
that carries with it norms of association that are themselves also deeply
political.

In adopting ICSU's views, however, the United States has continued to
insist that it retain control over all research carried out using U.S.
support. To date, international research programs continue to follow the
same pattern of organization that guided IGY research: international
coordination of nationally organized, funded, and implemented research
projects, with the degree of coordination varying greatly from one issue
to another. Consequently, tacit assumptions embedded in the organization of many programs have implicitly supported one country's or group
of countries' priorities over others', much as the IGY supported the
global priorities of U.S. and European scientists to the detriment of other
scientists' efforts to improve their own local research capacity. Science,
today, continues to carry an internationally independent image similar
to that envisioned by Berkner during American scientists' dispute with
the State Department over participation in the IGY:

Now this is a very interesting point, because there are two points of view that
can be taken here. One is the point of view of the State Department at that time,
that everything in the international field had to be subordinated to US objectives
of the time. The other point of view was that the world is a kind of organism,
in which politics is one part of the organism, science is another part, religion is
a third part, etc., and each element of the organism should be permitted to function, with the hope that if there's a malfunction somewhere in the system, the
general health of the other elements will help to correct the malfunction. (Berkner
1959)

The reality of science's connections to the politics of states is far more complex, however. The challenge for emerging global institutions in areas such as climate change, I believe, is to maintain widespread support for the view that science can contribute to the solution of international problems by transcending national peculiarities while simultaneously exposing for international discussion and negotiation the tacit connections between scientific research and state policies that shape its organization and conduct (for an explicit working out of this problem, see my thoughts in chapter 8, this volume).

A New World, a New World Order

Scientists are frequently described today by scholars in science studies as resource accumulators, entrepreneurs enrolling heterogeneous allies in the construction of ever-widening networks of material flows and social influence (Latour 1987). In many ways, that description closely fits meteorologists in the immediate postwar era. Through the WMO, atmospheric scientists built worldwide networks of observing stations, all making standard observations at standard times and reporting their measurements to a small number of central institutions who then collated, processed, and re-presented the data back to national meteorological services. This network might even be fruitfully understood as a network of centers of calculation (Latour 1990). I have tried to argue in this chapter, however, that scientists' ability to create and shape the material cultures through which they produce knowledge also depends on their ability to articulate connections between their work and the evolving clusters of norms and practices that govern the deployment of science in broader sociopolitical order. As Yaron Ezrahi points out, for example, theories of liberal democracy have relied heavily since the seventeenth century on attestive norms and practices also found in science. Only after World War II, however, did support for scientific research in the United States and Europe become a principal responsibility of public institutions as ideas about science and technology became linked to broad ideas about American wealth and security (Dennis 1994; Ezrahi 1990). These clusters of norms and practices, or *modes of interaction*— similar to what James Pocock has described as modes of civic consciousness (Pocock 1975)—delimit both science and politics. They provide not only the instrumental and symbolic basis for legitimating particular kinds of scientific expertise, knowledge, and activity but also the authority to use that science to support the maintenance of public order and social trust within particular political systems. They describe, in other words, what comes to be understood as both legitimate scientific practice and legitimate political authority. They define science and politics.

In this chapter, I have identified three modes of interaction that emerged to characterize the deployment of meteorological science in postwar American foreign policy specifically and international politics more generally: (1) the coordination and standardization of government practices through technical cooperation, (2) the building of state capacity to promote economic development and its attendant securing of social stability; and (3) the raising of concern about problems of a transnational and often global character.

These observations help explain why postwar meteorological cooperation was able to expand in the directions that it did, while not expanding in others. The three modes of interaction I have identified significantly influenced the development of meteorology and its policy influence in the years after World War II. As Jasanoff (1997a) has argued, probing the processes by which expert communities expand their networks and institutions is essential to understanding how they acquire power and authority in international politics. By offering policymakers a discursive framework for how science could contribute to their foreign policy goals, these modes of interaction secured for meteorology a formal institutional role in the UN system and enabled meteorologists to gain access to resources that they had not possessed before: resources for constructing new observation stations and recruiting new observers, resources for developing and enforcing standard practices and instrumentation, and resources for gaining accreditation in political contexts where otherwise they would have received little notice. The latter, in combination with the perceived economic benefits of meteorology, played a prominent role in persuading former colonial states and developing countries to allow meteorologists to extend their observational networks into these areas.

The discursive framework put in place around these modes of inter-action also delimited the directions world meteorological cooperation could take. Participation in international meteorological cooperation was structured in part by the evolution of international law as applied to multilateral institutions. Throughout this period, the PRC faced enormous difficulties in participating in the WMO's activities. Territorial or colonial meteorological services in the WMO also saw their membership and voting rights reduced as a consequence of a new insistence on the priorities of sovereign states within international institutions. In addition, the limited scope of international agreement on the goals of international meteorological cooperation shaped the kinds of observational networks and institutional capabilities the WMO could build. Policy restrictions prevented both the construction of national meteorological observing networks through funds provided by the IGY and, vice versa, the construction of networks designed to facilitate global atmospheric science through UN Technical Assistance programs. In much the same fashion, the WMO was largely prevented from funding an international meteorological institute or conducting its own meteorological research, although it could and did act to facilitate and coordinate national research agendas.

These observations also help explain why postwar international cooperation more broadly took some of the institutional forms that it did. Ruggie (1993) has recently observed that multilateralism emerged in the years following World War II as the dominant institutional form for international organizations. This chapter helps extend that argument, exploring how multilateral organizations came to adopt certain kinds of expert-based functions in the early postwar era—many of which remain key functions of international institutions today.

While the broader consequences of this development have not been the central focus of this chapter, several examples are suggestive. First, it seems apparent that international scientific and technical cooperation helped to underpin the legitimacy of international institutions following World War II. In the postwar era, the number and scope of international regimes has expanded dramatically. U.S. policies toward meteorological cooperation in the 1940s and 1950s reaffirm the conclusions of Anne-Marie Burley that ideas about how science and technology could contribute to international peace and prosperity played key roles in convincing foreign policymakers to create such regimes in the immediate postwar years and in shaping their organization and activities (Burley 1993).

Even in the wake of subsequent reevaluations of the utility and success of many international regimes, it is clear that many of these same ideas about science and technology continue to influence political scientists' and policymakers' views of international institutions. In their works examining the performance of international environmental institutions and environmental aid programs, for example, Haas, Keohane, and Levy have argued that international institutions are accorded the greatest legitimacy by policymakers and serve most effectively in achieving successful international cooperation when they engage in three activities: (1) raising concern about transnational issues, (2) coordinating intergovernmental interaction, and (3) building state capacity to implement policy (Haas, Keohane, and Levy 1993).[16] The similarity between these descriptions of effective institutional practice and the modes of interaction described in this chapter suggests the extent to which international institutions and international relations scholars have internalized norms and practices first expressed in postwar American foreign policy.

At the same time, the history of meteorological cooperation also suggests that we adopt two critical perspectives on Haas, Keohane, and Levy's recommendations. First, Haas, Keohane, and Levy suggest that increasing the role for science in international institutions can help achieve the goals of international governance. The history of meteorological cooperation suggests, however, that how science is organized and integrated into political frameworks (including, in turn, shaping those frameworks) may play a critical role in determining its persuasiveness and effectiveness. In addition, the modes of interaction developed in meteorology and other early postwar regimes were established to fit particular kinds of policy problems in a particular international context. In recent years, however, the international system has witnessed a transition from American and Soviet dominance to a much more multipolar world and the emergence of widespread beliefs that global environmental and economic concerns may require deep commitments to global governing institutions. Both of these transitions may sufficiently alter the kinds of policy problems and political contexts faced by international cooperation to render existing forms of international organization obsolete. Hence, simply increasing the use of science in international regimes, without paying close attention to the tacit political assumptions laden in particular arrangements of expertise, may do little to improve the effectiveness of international institutions.

Second, the dramatic expansion of technical experts in the WMO and other specialized agencies of the United Nations also appears to have contributed to the growing belief among international policymakers that quantitative information can serve as the basis for international regulatory action and compliance (Chayes and Chayes 1995). One of the principal activities of many international organizations, such as the UN Food and Agriculture Organization, has been the production of international databases describing national activities. Similarly, in many international economic regimes, data on financial and trade practices have played growing roles in the postwar era in bilateral and multilateral economic arrangements and disputes. Today, the creation of an international system of accounting for greenhouse gases is one of the central activities of the climate regime. Where prewar foreign policy operated almost exclusively on the assumption that diplomatic negotiations should operate secretly, reserving knowledge to elite foreign policymakers, postwar institutions have increasingly come to depend on the ability of a wide variety of experts and publics to have open access to knowledge about state behavior in order to establish international accountability. (This argument parallels that of Porter 1995 regarding the role of quantification in securing publicly authoritative knowledge of state behavior in American domestic politics.) Consequently, methods for transparently producing and validating that knowledge have become increasingly important (and increasingly subject to dispute) in a variety of international regimes (see Miller, chapter 8, this volume).

Finally, the mobilization of experts in international politics has contributed to altering the organization of the state for foreign affairs. Over the course of the past half-century, the U.S. State Department has lost considerable power and authority over the conduct of U.S. foreign policy. Political commentators as far apart politically and professionally as George Kennan and Theodore Lowi have severely criticized this development. Kennan, concerned about the failure of the State Department to develop a professional diplomatic service, has argued that the weakness of the State Department has left foreign policy subject to the whims of domestic political interests (Kennan 1951, 1997). Likewise, for Lowi, the decentralization of power and authority in foreign policy reflects the same decline in the ability of the American government to use state power to further social goals that has hamstrung it domestically (Lowi 1969).

In early accounts of the rise of experts in American political affairs, Price (1965) and Skolnikoff (1967) note that this shift in power from Departments of State to other government agencies resulted from the need to mobilize experts to pursue new foreign policy goals after World War II. The examination of the early history of the WMO in this chapter extends their analyses by exploring in detail why that mobilization took the particular institutional form that it did. Embedded in the World Meteorological Convention is the insistence that states' representatives to the WMO be the professional meteorologists who direct their national meteorological services. Other international organizations such as the Food and Agriculture Organization have adopted similar practices. The result has been a significant shift of foreign policy responsibilities from Departments of State to other government agencies as the participation of experts in international institutions has become central to international affairs. Today, large numbers of federal agencies are active daily in the conduct of foreign policy and, in the United States, for example, important international negotiations are overseen by specially constituted interagency task forces.

There were alternatives to this development, among them Kennan's suggestion that the State Department increasingly turn to experts for the conduct of foreign policy. The Berkner report in 1950, for example, encouraged the State Department to make a number of internal reforms, including the hiring of large numbers of scientists for its embassies and operations throughout the world. By 1951, however, the battle to incorporate large numbers of experts in the foreign service had effectively been lost. Despite the recommendations of the Berkner committee, the Department was never able to mobilize the kind of expertise required to participate in the host of technical institutions constructed in the years following World War II, and it quickly gave up trying. Few scientists were ever hired, not as a result of a decrease in the importance of science and technology in foreign affairs but because the necessary expertise already existed in other government departments. In its participation in international scientific institutions, technical assistance projects, and joint research programs, the United States has relied almost exclusively on experts from other federal agencies and nongovernmental institutions. In turn, it has relied extensively on interagency task forces to coordinate U.S. positions in international negotiations and meetings.

Ultimately, several factors contributed to the inability of the State Department to incorporate greater numbers of experts in its activities. Replicating other federal agencies' existing expertise within the State Department proved unacceptable within the American polity for reasons of cost, political control over foreign policy, and concern among scientists that expert issues not be subjected to the ideological concerns of the State Department. Coordinating domestic and international concerns on substantive issues was easier in a single agency, and involving existing government institutions (instead of creating new ones) helped speed the creation of effective international institutions.

In terms of the conduct of international politics, this decentralization has had significant consequences that have not yet been subject to sufficient analysis. First, international politics has become a domain in which a wide range of experts from numerous government agencies now participate, introducing a host of new concerns into the calculation of national interest on any particular issue. Second, these new participants have created opportunities for the development of transnational communities, much like the community of meteorologists who organized and operated the WMO. On occasion, as Peter Haas and others have pointed out, these communities are able to wield a certain power in international politics by forging shared values and shaping the solution of international problems (Haas 1992).

This decentralization also speaks to the future evolution of liberal models of global governance. Anne-Marie Slaughter points out, for example, that over the past decade, judges, parliamentarians, and others involved in the processes of civil governance have begun to establish transnational networks (Slaughter 1997). She argues that these networks constitute a new model of liberal international governance—

independent, transnational organizations. While these developments may be new to judges, they are not new to meteorologists, economists, physicists, marine scientists, and other experts involved in numerous areas of state policymaking. These groups have been building transnational communities for over forty years (see, in addition to the account of meteorology developed here, e.g., Adler 1992; Ikenberry 1992; Haas 1990a, 1990b). One possible explanation for the difference between these two periods of transnational community building merely amalgamates my own conclusions with Slaughter's. Namely, the events of the early postwar era constituted the emergence of a new model of liberal international governance in which experts came to play important roles in shaping certain areas of state behavior; similarly, the events of the past decade constitute yet another transformation of models of international governance. This solution relies, however, on an a priori demarcation between expert and civic domains of governance and legitimation that is increasingly difficult to sustain in the light of contemporary scholarship in science studies.

On the other hand, a potentially more interesting explanation of the relationship between these two eras of changing patterns of governance may emerge if we ask whether or not the same kinds of activities that characterized meteorological cooperation in the 1950s now characterize judicial cooperation in the 1990s. Are judges, in other words, able to cooperate precisely because we now perceive judicial competence in more expert-like ways, as potentially universalizable and professionalizable? Or, conversely, is it that we now see expertise itself as more political and hence draw inferences for judicial cooperation from success in other areas of (expert-dominated) politics? In either case, the historical experiences of organizations like the WMO and the Bretton-Woods institutions may provide us with important insights into what makes for effective transnational cooperation as well as the conditions under which such cooperation can break down. That intergovernmental cooperation among judicial authorities should look different to political analysts than intergovernmental cooperation among meteorological authorities testifies to the success of the international system at separating out distinct areas of nominally scientific and nominally political policymaking. At least until recently, when carried out in accordance with expectations derived from one of the three modes of interaction I have identified, expert cooperation has been widely viewed as exempt from the normal concerns of states for national sovereignty. How was this separation initially accomplished and maintained over time? What is contributing to its breaking down today? Examples like the construction of a meteorological community in Libya under UN trusteeship also prompt us to ask questions about the source of these community-building efforts. Who is joining these transnational communities of judges and parliamentarians? Why? What efforts are being made to extend these communities to developing countries? What implications do these efforts have for concepts of justice, equity, and democracy?

Last, but certainly not least, international scientific and technical cooperation has altered the symbolic meaning of important ideas within global society. Meteorologists built a worldwide observational network during the early postwar years. That network opened the possibility of representing and investigating atmospheric dynamics on global scales. Prior to the formation of the WMO, no tools existed to allow scientists to investigate the atmosphere as a global system. "Seeing" the atmosphere as a global commons, in other words, depended on creating the necessary material and political cultures to make global measurements possible. Ironically, the very success of meteorologists in creating these cultures and changing public representations of the atmosphere has helped undermine the political settlements that enabled intergovernmental cooperation in meteorology in the first place. The public construction of global environmental problems such as climate change and ozone depletion, to which the WMO's networks have contributed considerably, has raised the specter of problems for which existing models of international governance are already proving inadequate. How science fits into global governance and what that science will look like—the modes of interaction that will characterize the future of international science and politics—are under discussion today in ways they have not been since 1947.

Notes

1. The argument presented here should not be read to suggest that U.S. leaders instituted a coordinated, hegemonic policy agenda across the wide range of international institutions created in the immediate postwar era. Nevertheless, by

looking across a variety of regimes, it is possible to detect consistent patterns in the political mobilization of expertise during the postwar period that correspond broadly to ideas being articulated by American foreign policymakers.

2. The participants' failure to agree on numerous issues, many of which centered around the disagreements over the trustworthiness of various strategies for organizing verification efforts, illustrates recent social science claims about trust in systems with interdependent technical and human systems (see, e.g., MacKenzie, 1990).

3. In 1939, T. Hesselberg, director of the Norwegian meteorological service, led the drafting of a proposed world meteorological convention in Berlin. Subsequent drafts were prepared and submitted to the Paris meeting of the International Meteorological Organization Executive Council in 1946 (see, e.g., International Meteorological Organization 1947, 384–388).

4. "For two successive years unusually severe droughts have cut down food production. And during the past winter storms and floods and excessive cold unprecedented in recent years have swept northern Europe and England with enormous damage to agricultural and fuel production" (Acheson [1947] 1969).

5. Other key supporters of the treaty, such as T. Hesselberg, director of the highly influential Norwegian meteorological service, concurred: "In order to avoid drawbacks affecting the future structure of the IMO, the previously existing privileges of the IMO were maintained to the fullest extent possible in the Berlin Draft of 1939 and it was emphasized at the London Conference in 1946 that the IMO must retain its worldwide character and its independence. Furthermore, these principles govern the provisions incorporated in the Paris Draft of 1946. However, some of the drafts recently presented seem to affect these principles. Mr. Hesselberg did not object in any way to a Convention and he even saw that it would be advantageous provided the two principles in question, that is to say the worldwide character and the independence of the Organization, are ensured" (International Meteorological Organization 1947, 403).

6. WMO's support from the UN Technical Assistance Program amounted to anywhere from 1% to 3% of the total funds disbursed for economic development in any given year.

7. Details of many of the WMO's technical assistance programs can be found in the pages of the *World Meteorological Bulletin*. From its inception in 1952, the *Bulletin* dedicated several pages in each issue to updates of ongoing technical assistance activities.

8. WMO participation in the UN Expanded Program of Technical Assistance for Economic Development of Under-Developed Countries was initiated in Resolution 10 (I) of the First World Meteorological Congress (World Meteorological Organization 1951a). The organization of the WMO program was modeled on background documents prepared by Sir Nelson Johnson, the last president of the IMO, and by the U.S. government. This documentation can be found in Annex AF-VIII, "Technical Meteorological Assistance to Underdeveloped Countries," submitted on the instructions of the president of the Executive

Council of the IMO, and Annex AF-IX, "Participation of the World Meteorological Organization in the United Nations Technical Assistance Program," presented by the United States (World Meteorological Organization 1951c).

9. Background documentation for the delegates was provided in the form of UN resolutions 387 (V), "Report of the United Nations Commissioner in Libya, Reports of the Administering Powers in Libya," and A/1727, "Technical and Financial Assistance to Libya," in Annex AF-X, "Technical Assistance to Libya," in World Meteorological Organization 1951c.

10. Descriptions of the WMO technical assistance program to Libya can be found under the heading "Technical Assistance Program" in numerous issues of the *WMO Bulletin* during the 1950s (e.g., World Meteorological Organization 1954).

11. The WMO technical assistance project in Jordan provides a good example of the difficulty of determining where development priorities lay and who set them. Nominally advising the Jordanian government on agricultural meteorology, the British expert assigned by WMO to Jordan was located at the Jerusalem airport, where the meteorological service primarily served the need for aviation. He also served as the secretary for the National Climate Committee, a body made up of "the Ministry of Agriculture, the Directorates of Civil Aviation, Forestry, Irrigation and Water Power, the United States Foreign Operations Administration and the United Kingdom Air Ministry" (World Meteorological Organization 1954, 130; emphasis added).

12. Meteorologists have also turned to growing public concern for climate change as a source of support for their networks. In recent meetings of the institutions of Framework Convention on Climate Change, WMO leaders have made strong appeals for new international support for the meteorological network so that it can continue to provide countries with climatological information relevant to public policy (United Nations Framework Convention on Climate Change Subsidiary Body for Scientific and Technological Advice 1997).

13. In fact, much of the early support for an international meteorological institute can probably be viewed less in terms of support for such an institute per se then as support for coordinated international cooperation in meteorological research. In his opening presidential address to the first meeting of WMO's Commission on Aerology (responsible for atmospheric sciences) in 1953, J. van Mieghem argued in support of the development of several international meteorological institutes on the grounds that solving fundamental problems in meteorology can only come through international collaborations of observers. Within two years, however, van Mieghem supported a different "model" of international research based on coordinating national research programs (cf. van Mieghem 1953 with van Mieghem 1955a).

14. In addition to his role as president of the WMO Commission on Aerology, van Mieghem assumed the tasks of the reporter for the IGY program in meteorology and assumed the role of the WMO's representative to CSAGI.

15. Van Mieghem (1955a) notes: "Now that observing stations exist in most inhabited regions, it is becoming ever more obvious that the large gaps in the various geophysical observing networks over the oceans, polar regions, and deserts constitute barriers to progress [in geophysical science], and the need everywhere for synoptic [sic] observations is daily becoming more pressing. Synoptic studies constitute the only method of investigating phenomena on a planetary scale." This philosophy clearly drove the IGY extensions of the meteorological network. Within a decade, however, satellite observations (made at only one point at any given point in time, but aggregated globally over periods of weeks and months) and computer models would offer very different sets of practices for studying planetary phenomena.

16. My argument here augments their finding in two important ways. First, we can now see that these particular institutional configurations are not products of necessity but of particular norms and practices of international politics and the assumptions that they embed about the nature of states and of scientific knowledge. Thus, second, as we begin to see new assumptions about science and new sets of international practices in the wake of the end of the Cold War and the rise of global environmental issues, we may see new institutional configurations arise (Keohane, Haas, and Levy 1993; Keohane 1996).

7

Self-Governance and Peer Review in Science-for-Policy: The Case of the IPCC Second Assessment Report

Paul N. Edwards and Stephen H. Schneider

In the spring of 1996, the Intergovernmental Panel on Climate Change (IPCC) released its long-awaited Second Assessment Report (SAR) on possible human impacts on the global climate system. The report's eighth chapter concluded that "the balance of evidence suggests that there is a discernible human influence on global climate" (Houghton et al. 1996, 5)—a phrase that has since become probably the single most-cited sentence in the IPCC's history. The Global Climate Coalition (an energy industry lobby group) and a number of "contrarian" scientists immediately launched a major, organized attack designed to discredit the report's conclusions, especially those relating to the crucial question of whether human activities are responsible for changes in the world's climate.

Led by the eminent physicist Frederick Seitz, these critics claimed that the IPCC had inappropriately altered a key chapter for political reasons. They alleged that the IPCC had "corrupted the peer review process" and violated its own procedural rules. These charges ignited a major debate, widely reported in the press, lasting several months.

The accusations of corruption reach a fundamental issue in the emerging global climate regime, namely, how the IPCC as a self-governing institution can maintain scientific integrity in the face of intense political pressures (both internal and external) and tightly constrained deadlines. In this chapter we consider these charges on three levels. First, we evaluate their accuracy as specific challenges to the IPCC peer review process, and note how the IPCC rules of procedure might be clarified to avoid them in the future. Second, we explore their meaning against the larger background of the IPCC's role in the politics of climate change. Finally, we use this episode to examine more fundamental questions about the role of formal review mechanisms in certifying scientific knowledge produced for policy contexts, and about the relative importance of those mechanisms in different national and cultural contexts.

The IPCC Second Assessment Report

The IPCC is an office of the United Nations Environment Programme and the World Meteorological Organization. Its purpose is to evaluate and synthesize the scientific understanding of global climate change for national governments and UN agencies, as expert advice for use in the ongoing negotiations under the Framework Convention on Climate Change (FCCC). The agency's nominal goal is to represent fairly the full range of credible scientific opinion. Where possible, it attempts to identify a consensus view on the most likely scenario(s). When consensus cannot be reached, the agency's charge is to summarize the major viewpoints and the reasons for disagreement. IPCC reports are intensively peer reviewed. They are regarded by most scientists and political leaders as the single most authoritative source of information on climate change and its potential impacts on environment and society.

Like all IPCC assessments, the SAR contained three "Summaries for Policymakers" (SPMs), one for each of the IPCC's three Working Groups: climate science (Working Group I), impacts of climate change (Working Group II), and economic and social dimensions (Working Group III) (Bruce, Lee, and Haites 1996; Houghton et al. 1996; Watson, Zinyowera, and Moss 1996). Since the full SAR stretches to well over 2,000 pages—most of it dense technical prose—few outside the scientific community are likely either to read it in its entirety or to understand most of its details. Therefore, these summaries tend to become the basis for press reports and public debate. For this reason, the Working Groups consider their exact wording with extreme care before they are published. At the end of the IPCC report process, they are approved word for word by national government representatives at a plenary meeting attended by only a fraction of the lead scientific authors.

The SPM for Working Group I, which assesses the state of the art in the physical-science understanding of climate change, contained the following paragraph:

Our ability to quantify the human influence on global climate is currently limited because the expected signal is still emerging from the noise of natural variability, and because there are uncertainties in key factors. These include the magnitude and patterns of long-term natural variability and the time-evolving pattern of forcing by, and response to, changes in concentrations of greenhouse gases and aerosols, and land surface changes. *Nevertheless, the balance of evidence suggests that there is a discernible human influence on global climate.* (Houghton et al. 1996, 5; emphasis added)

Three-quarters of this paragraph consists of caveats about uncertainties and limitations of current understanding. Nonetheless, its now-famous closing sentence marked the first time the IPCC had reached a consensus on two key points: first, that global warming is probably occurring ("detection"), and second, that human activity is more likely than not a significant cause ("attribution"). Like this summary paragraph, the body of the report discussed—frequently and at length—the large scientific uncertainties about attribution. The Working Group carefully crafted the SPM's "balance of evidence" sentence to communicate the strong majority opinion that despite these uncertainties, studies were beginning to converge on a definitive answer to the attribution question.

The SAR was fraught with political significance. Official publication of the full report occurred in early June, 1996. At that point the Second Conference of Parties to the FCCC (COP-2) was about to meet in Geneva; the session would determine some of the starting points for the Kyoto meeting in 1997, where binding greenhouse-gas emissions targets and timetables were to be negotiated. A sea change in American climate policy was widely rumored. Since the Reagan administration, official U.S. policy had sanctioned only voluntary, nonbinding emissions targets and further scientific research. If the United States were to abandon its resistance to binding emissions targets and timetables, a strong international greenhouse policy would become much more likely. Since the more-research, no-binding-targets position was officially based on assertions that scientific uncertainty remained too high to justify regulatory action, the SAR's expressions of increased scientific confidence were viewed as critical.

The rumors proved correct. On July 17, 1996, then–U.S. Under-Secretary of State for Global Affairs Tim Wirth formally announced to COP-2 that the United States would now support "the adoption of a

realistic but binding target" for emissions. The exact degree to which the
IPCC SAR influenced this policy change cannot be known. But Wirth
certainly gave the impression that the report was its proximate cause.
He noted in his address that "the United States takes very seriously the
IPCC's recently issued Second Assessment Report." He then proceeded
to quote the SAR at length, proclaiming that "the science is convincing;
concern about global warming is real" (Wirth 1996; emphasis added).

"A Major Deception on Global Warming"

On June 12, 1996, just days after formal release of the IPCC SAR and
scant weeks before the COP-2 meeting in Geneva, the *Wall Street Journal*
(WSJ) published an op-ed piece entitled "A Major Deception on Global
Warming." The article was written by Frederick Seitz, president emeritus
of Rockefeller University. Seitz is not a climate scientist, but a physicist.
Nevertheless, his scientific credentials are formidable. He is a recipient
of the National Medal of Science and a past president of both the
National Academy of Sciences and the American Physical Society.

In his article, Seitz accused some IPCC scientists of the most "dis-
turbing corruption of the peer-review process" he had ever witnessed
(Seitz 1996, A16).

Seitz's Accusations

Seitz's distress stemmed from the fact that the lead authors of the SAR's
Chapter 8—on detection and attribution—had altered some of its text
after the November 1995 plenary meeting of Working Group I (WGI),
in Madrid, at which time the chapter was formally "accepted" by the
Working Group. According to Seitz, since the scientists and national governments who accepted Chapter 8 were never given the chance to review
the truly final version, these changes amounted to deliberate fraud and
"corruption of the peer-review process." Not only did this violate normal
peer review procedure, Seitz charged; it also violated the IPCC's own
procedural rules.

Quoting several sentences deleted from the final version of the chapter,
Seitz argued that the changes and deletions "remove[d] hints of the skepticism with which many scientists regard claims that human activities are

having a major impact on climate in general and on global warming in particular." Without directly attributing motives, Seitz implied that the changes had been made in the interests of promoting a particular political agenda. Seitz said that Benjamin D. Santer, lead author of Chapter 8, would have to shoulder the responsibility for the "unauthorized" changes. Seitz was not present at the IPCC meetings. He did not contact Santer or anyone else at the IPCC to verify that the changes were indeed "unauthorized" before publishing his op-ed piece.

Responses from Santer and the IPCC

Santer responded immediately, in a letter co-signed by some forty other IPCC officials and scientists (myself among them—SHS). They said that Seitz had misinterpreted the IPCC rules of procedure. Rather than being "unauthorized," they wrote, the post-Madrid changes were in fact *required* by IPCC rules, under which authors must respond to comments submitted during peer review or arising from discussions at the meetings (Santer 1996a).

Commentators at the Madrid meeting had advised making changes to Chapter 8 for two reasons. First, they urged clarification of the meaning and scientific content of some passages in accordance with the recommendations of reviewers (including some criticisms introduced at the Madrid meeting itself). Second, they thought the structure of the chapter should be brought into conformity with that of other SAR chapters. In particular, a "Concluding Summary" was removed from the final version, since no other chapter contained a similar section. (Chapter 8, like all the rest, already had an "Executive Summary.") Sir John Houghton, in his capacity as co-chair of WGI, specifically authorized that these changes be made, though he did not review their wording.

Santer, in consultation with other Chapter 8 authors, made the suggested changes in early December. The entire SAR, including the newly revised Chapter 8, was "accepted" by the full IPCC Plenary at Rome later that month.

Santer made the changes himself, and the final version of the chapter was not reviewed again by others. However, as he and his colleagues continually stressed, this procedure was the normal and agreed IPCC process. Santer et al. pointed out that no one within the IPCC objected (or had ever objected) to this way of handling things. Replying separately in support of Santer and his colleagues, IPCC chair Bert Bolin and WGI co-chairs John Houghton and L. Gylvan Meira Filho quoted the official U.S. government review of Chapter 8, which stated explicitly that "it is essential that . . . the chapter authors be prevailed upon to modify their text in an appropriate manner following discussion in Madrid" (Bolin, Houghton, and Meira-Filho 1996).

Further Exchanges

The *Wall Street Journal* op-ed was not the first time charges of suppression of scientific uncertainty in Chapter 8 had been aired. On May 22, a few days before the Seitz op-ed appeared, the small journal *Energy Daily* reported the same allegations in considerably greater detail (Wamsted 1996). The *Energy Daily* article also reported their source: a widely circulated press release of the Global Climate Coalition (GCC, an energy industry lobby group).

In its June 13 issue, the prestigious scientific journal *Nature* also reported on the GCC allegations (Masood 1996). The *Nature* report, unlike the Seitz and *Energy Daily* articles, included explanations of the revision and review process from Santer and the IPCC. Under the hot-button headline "Climate Report 'Subject to Scientific Cleansing,'" an accompanying editorial argued that the GCC analysis was politically motivated and generally false. But the editorial also noted that the Chapter 8 changes might have resulted "in a subtle shift . . . that . . . tended to favour arguments that aligned with [the SAR's] broad conclusions" (*Nature* editors 1996, 539).

The *Wall Street Journal* op-ed set off a lengthy chain of exchanges lasting several months. The main participants in the public controversy were Seitz, Santer, other Chapter 8 authors, the chairs of the IPCC (Sir John Houghton and Bert Bolin), and climate change skeptics S. Fred Singer and Hugh Ellsaesser. Singer, in particular, made the charges of political motivation explicit. In a letter to the *Wall Street Journal*, he wrote that Chapter 8 had been "tampered with for political purposes." The IPCC, he claimed, was engaged in a "crusade to provide a scientific cover for political action" (Singer 1996, A15).

Semiprivately, in e-mail exchanges involving many additional participants (and widely copied to others), the debate became intense and sometimes quite bitter. Santer, who felt forced to defend himself, spent the majority of his summer time responding to the charges. Previously a quiet, private man known to scientists primarily as a proponent of the rigorous use of statistical methods, Santer rapidly became a public figure, submitting to dozens of interviews. The drain on his time and energy during this period kept him from his scientific work, he said (personal communication, 1996).

Both the public and the private exchanges themselves became objects of further press reports, widely disseminated by the news wire services. As they went on, the debate spread from the initial issues about peer review and IPCC procedure to include questions about the validity of Chapter 8's scientific conclusions. Even before the report was formally published, climate-change skeptics had claimed that Chapter 8 dismissed or ignored important scientific results that disconfirmed the global warming hypothesis. They argued that the allegedly illegitimate changes to Chapter 8 made this problem even more acute (Brown 1996).

The Chapter 8 Revisions and IPCC Self-Governance

As a hybrid science-policy body, the IPCC must maintain credibility and trust vis-à-vis two rather different communities: the scientists who make up its primary membership, and the global climate policy community to which it provides input. Independent self-governance is one of the primary mechanisms by which it achieves this goal. The IPCC's rules of procedure spell out a variety of methods designed to ensure that its reports include the best available scientific knowledge and that they represent this knowledge fairly and accurately. Chief among these is the principle of peer review, traditionally one of the most important safeguards against bias and error in science.

Seitz, the GCC, and others accused the authors of Chapter 8 of fraud on two counts. First, they alleged that the changes made to Chapter 8 after the final IPCC plenary violated the IPCC's own rules of procedure. Second, and more seriously, they charged them with violating the fundamental standards of scientific peer review. In this section, we argue that IPCC rules were not violated in the case of Chapter 8. In addition, we argue that in practice the IPCC process correctly reflects the essential tenets of peer review. However, we also show that the IPCC rules do not specify adequate closure mechanisms for the report-drafting process. We demonstrate that the two-level certification process ("acceptance" and "approval" of IPCC documents) is poorly specified as well, and can invite misinterpretation by determined critics.

In their responses to the Seitz/GCC charges, the Chapter 8 authors claimed that IPCC governance rules required them to make the changes advised immediately before and during the Madrid WGI Plenary. Analysis of the IPCC rules suggests that the real situation is more ambiguous. Yet they had three very good reasons for believing this to be the case.

First, the rules require authors to respond to commentary, to the best of their ability and as fully as possible (Intergovernmental Panel on Climate Change 1993). Working Group co-chairs have broad discretion to define this process and set time limits for it. Nowhere do IPCC rules explicitly address the question of when a report chapter becomes final (i.e., when all changes must cease). Therefore, Santer et al. correctly understood that the Working Group Chairs and the Plenary meeting itself would define the endpoint of the revision process.

Second, report chapters are "accepted" rather than "approved." Acceptance constitutes IPCC certification that the drafting and review process has been successfully completed. It is an expression of trust in the authors and the process, and is explicitly distinguished from "approval," or detailed review on a line-by-line basis. Operating under these definitions, the IPCC Plenary "approved" the WGI Summary for Policymakers (SPM), but "accepted" Chapter 8. In other words, Plenary acceptance did not imply word-for-word review of the chapter. Instead, it indicated trust that the authors had responded appropriately and sufficiently to the review process. Therefore, the Chapter 8 authors believed that the rules permitted them to make changes when explicitly requested to do so by the IPCC Plenary, or in response to peer comments received at or immediately prior to the Plenary.

Third, no IPCC member nation ever seconded the Seitz/GCC objections (Bolin 1996). (Ninety-six countries were represented at the Madrid plenary.) From this, above all, we can safely infer that Santer et al. proceeded exactly as expected. They believed that they were following IPCC rules, and this made perfect sense within the established informal culture of the IPCC.

However, a careful reading of the IPCC's formal rules reveals that in fact the rules neither allow nor prohibit changes to a report after its formal acceptance. The legalistic Seitz/GCC reading of the rules is not, therefore, completely implausible—even if it was, as we believe, primarily a smokescreen to divert attention from the clear consensus that attribution could no longer be considered unlikely.

Our analysis suggests a significant flaw in the rules as currently written. While "approved" documents (the SPMs) clearly must not be altered once approved, there is no precisely defined closure mechanism for "accepted" documents (full-length Working Group reports and their constituent chapters) (Intergovernmental Panel on Climate Change 1993). The Seitz/GCC attack has effectively demonstrated that a hybrid science/policy organization like the IPCC needs better, more explicit rules of procedure. This minor virtue aside, however, the Seitz/GCC reading violates the spirit and intent of the IPCC process.

The IPCC is run by scientists. Its participants think of it primarily as a scientific body. By the standards of many political organizations, its formal rules of governance are not very extensive. They are also not very specific. The rules purposely leave undefined the meaning of key terms such as "expert" and important processes such as "taking into account" comments. Under the rules, Lead Authors carry full responsibility for report chapters, and the IPCC leadership retains very broad discretion, subject to Plenary "acceptance" and "approval" by national governments.

There are good reasons for this arrangement. Formal governance is relatively unimportant in scientific culture. This is true because scientists generally belong to small social groups endowed with strong and deeply entrenched (informal) norms. In addition, since scientific methods and results are constantly changing, too much focus on formal rules would inhibit progress. Likewise, formal rules are not very important in the day-to-day functioning of the IPCC. Instead, informal rules based on the everyday practices of scientific communities guide the bulk of the work

(Collins and Pinch 1993; Gilbert and Mulkay 1984; Latour and Woolgar 1979; Merton 1973).

Maintaining this informality is quite important for effective scientific work. Yet it is not without dangers, especially in a situation where almost any scientific finding can have political implications (Jasanoff 1990; Jasanoff and Wynne 1998). Just as in any other politicized realm, without clear procedures to ensure openness and full rights of participation, dissenters may find—or believe they have found—their voices ignored. One of the IPCC's most important features is its openness and inclusivity; balancing this against scientific informality will require constant vigilance, and perhaps a reconsideration of the formal review process.

From the point of view of political legitimacy, then, acceptance of reports before final revision is clearly a risky proposition (Jasanoff 1991). But from the viewpoint of scientific legitimacy, ongoing revision is a normal feature of the research cycle. Even after a multistage review process, minor flaws can be found and improvements added. This is not unlike the common situation in which an author makes minor changes to galley proofs—changes not subject to peer review. Thus, in the case of the IPCC, adding a final approval stage to the already long and cumbersome review process would be unlikely to add significantly to the scientific credibility of the final result. While it needs to revise its rules to better protect itself from accusations of political capture, the IPCC must also, at all costs, avoid becoming a science-stifling, inflexible bureaucracy.

In fact, in late 1999 the IPCC finalized a major revision to its rules of procedure, in response to considerations that included the Chapter 8 controversy. According to David Griggs of the IPCC Working Group I Technical Support Unit, one of the major changes is the introduction of "Review Editors." These editors

will assist the Working Group Bureaux in identifying reviewers for the expert review process, ensure that all substantive expert and government review comments are afforded appropriate consideration, advise Lead Authors on how to handle contentious/controversial issues and ensure genuine controversies are reflected adequately in the text. (Griggs, personal communication, July 1999)

The new rules should make disputes such as the Chapter 8 controversy less frequent and, perhaps, provide new mechanisms for resolving them without resort to salvos in the popular press—although the already-adequate existing mechanisms did not prevent Seitz and his colleagues from sidestepping them in order to attack the IPCC.[1]

The Chapter 8 Revisions and the Peer Review Process

As we noted above, one of the most important standards of scientific accountability holds that publications must be reviewed by expert peers before results are released. Seitz and the GCC accused the IPCC of violating this standard, too. Were they right?

The Peer Review Process

Peer review is among the oldest certification practices in science, established with the *Philosophical Transactions of the Royal Society* in 1655 (Chubin and Hackett 1990, 19).

In a typical peer review procedure, scientists write articles and submit them to a journal. The journal editor sends the article to several referees, all of them experts in the authors' field ("peers"). Peer review at journals is usually "blind": authors are not informed of the referees' identity, though the author's name may be known to the referees. Blind review operates on the principle that free expression of criticism is more likely when referees, who often know authors personally and want to maintain good relations with them, can say what they think without having to consider authors' reactions, especially to negative evaluations. Many journals use a more stringent "double-blind" procedure, in which neither referees nor author(s) are informed of each other's identity. Double-blind review is based on the principle that criticism is more impartial when authors' identities are unknown to referees, who might be swayed in either positive or negative directions by authors' reputations, personality traits, and so on. A similar process is normally applied to grant proposals (Kassirer and Campion 1994). Standard peer review procedure varies by field and by journal or grant agency. Few journals in the atmospheric sciences, for example, use double-blind review.

However, the fundamental purpose of peer review is to strengthen the quality of work by subjecting it to criticism and evaluation by those best qualified to judge it. Many paths can lead to this goal; it does not depend on blind procedures, and not all journals employ them. Some journal editors (like myself, as editor of *Climatic Change*—SHS) go so far as to encourage referees to reveal themselves. In any case, many scientific communities are small enough that even double-blind referees and authors can often guess each other's identity.

Referees can typically choose one of three recommendations: acceptance, rejection, or acceptance after certain specified changes are made ("revise and resubmit"). The last of these responses is by far the most common. The authors then rewrite their article in response to the reviewers, and the editor serves as referee on the issue of whether the revisions have satisfactorily answered reviewers' criticisms. The process often goes back and forth several times, with several rounds of revision, until a suitable compromise is achieved among reviewers, authors, and the editor. (At grant agencies, grants officers fill a role analogous to the journal editor. The process usually requires resubmission of the grant application in subsequent rounds of the funding cycle.)

Does Peer Review Work—and for What?

Certainly peer review is imperfect. Not all referees do their job well, and personal, political, and social factors can all enter into the process in unseemly ways. Empirical evaluation of peer review's effectiveness is decidedly mixed.

But as we will show, the question of whether peer review "works" depends largely on what one thinks peer review is for. A science-studies approach to peer review gives a new perspective on its purpose, one more consonant with the high esteem in which scientists hold the process and with its role in IPCC assessments. First, however, let us briefly review the major criticisms of the process.

Numerous studies of peer review have judged that it fails as a dependable indicator of research quality. Several experiments have shown that agreement between referees on the same article is generally only slightly better than chance (for summaries of these see Cichetti 1991; Marsh and Ball 1989). A major, long-term study of grant proposal review at the National Science Foundation concluded that "funding of a specific proposal . . . is to a significant extent dependent on the applicant's 'luck' in the program director's choice of reviewers" (Cole 1992, 99). Other studies indicate that peer review suffers from systematic "confirmatory bias"—that is, the tendency to rate more highly studies that confirm existing beliefs, regardless of their quality (Bornstein 1991; Cole 1992; Mahoney 1977; Ross 1980). A related critique views peer review as a form of censorship which effectively blocks expression of innovative ideas that challenge dominant scientific paradigms (Moran 1998). Finally, peer review cannot reliably detect fraudulent science (Chubin and Hackett 1990, chap. 5).

Several scholars have claimed that far from assuring impartiality, "blind" peer review in fact encourages two kinds of counterproductive, unethical referee behavior. First, anonymous referees—whose rewards for their efforts are minor to nonexistent—may tend to minimize the time they devote to review, even to the point of approving work they have not actually read. Second, anonymity allows referees to engage more easily in personally or politically motivated attacks on others' work; as we pointed out above, even double-blind review may not prevent authors and referees from guessing each other's identities in a small field (and wrong guesses can turn out to be even more harmful than right ones). One major study of peer review recommended eliminating the "blind" system by requiring referees to sign their reviews: "This would hold reviewers publicly accountable for their decisions and would take a step toward acknowledging the value of reviewers' work. No longer would it be convenient for a reviewer to trash another's work. Nor would it be advisable to endorse unexamined work" (Chubin and Hackett 1990, 204).

Some commentators have recommended abolishing the system altogether (Roy 1985). Others see it as withering away of its own accord under the influence of reviewer fatigue, ever-expanding numbers of publications, and new electronic media that can circumvent the process, either deliberately or not (Judson 1994).

A Revised Conception of Peer Review

Most of these criticisms of peer review depend on a particular (and often tacit) view of its purpose, namely that peer review acts as a kind of "truth machine," automatically separating "good" science from "bad." This view—which may be more common among those *studying* peer review than among participants in the process—implicitly assumes that scientists (peers) agree very closely about most things, so that the opinion of one scientist about an article ought to be similar to that of most others. That this should turn out, empirically, not to be the case is surprising only if one subscribes to what Stephen Cole, author of the NSF peer review study mentioned above, calls "the mythology that scientists do not or should not disagree." He cites the University of Chicago statistician William Kruskal, Jr.: "Careful objective studies of expert judgment typically find them disagreeing more than had been expected and more than the experts themselves find comfortable. . . . Variability is usually considerable, the more so in close cases" (Kruskal, cited in Cole 1992, 100).

Yet disagreement among peers is undesirable only if it is interpreted (wrongly, we would argue) as arbitrary. Taking a science-studies view of the process, Cole goes on to observe that disagreement among experts is basic to scientific practice:

> The great majority of reviewer disagreement observed [in our empirical studies] is probably a result of real and legitimate differences of opinion among experts about what good science is or should be. . . . Contrary to a widely held belief that science is characterized by agreement about what is good work, who is doing good work, and what are promising lines of inquiry, this research indicates that concerning work at the research frontier there is substantial disagreement in all scientific fields. (Cole 1992, 100)

In fact, disagreement is vital to science, since it drives further investigation. But if expert judgment varies too widely to provide a quasi-mechanical means of winnowing out bad science from good, why is peer review important? The answer depends on one's conception of its role and purpose.

We maintain that peer review ought to be regarded as a human process whose primary functions are to improve the quality of scientific work, to maintain accountability both inside and outside the scientific community, and to build a scientific community that shares core principles and beliefs even when it does not agree in detail (Haas 1990a, 1990b). For example, peer review helps to minimize errors; reviewers frequently catch mathematical and methodological mistakes. Reviewers frequently also suggest better methods, recalculate numbers, and offer solutions to unresolved problems. Peer review also helps to distribute new research results, and helps ensure fair distribution of credit for work done. It acts as a certification mechanism, a barrier to entry, and a disciplinary device (in many senses) (Foucault 1977; Kuhn 1962; Merton 1973). Despite the severe sound of these latter functions, they are in fact vital to building any coherent knowledge community.

Peer review can be also described as an institutionalized form of the "virtual witnessing" process by which science establishes factual knowledge (Shapin and Shaffer 1985). It ensures that at least a few relatively disinterested parties have carefully scrutinized the experimental procedure and the reasoning and agreed with the conclusions drawn by the author(s).[2] It is a form of accountability, a way for the community to rehearse (and enforce) its fundamental norms and practices. For some or all of these reasons, nearly every scientist regards peer review as an extremely important mechanism, even though most are aware of its problems.

Several empirical studies have reached positive verdicts on the ability of the process to improve the quality of publications despite its acknowledged failings (Abelson 1980; Daniel 1993). Under this conception, as long as fundamental standards of scientific practice are met, the purpose of peer review is to minimize disagreements, but not necessarily to resolve them—since disagreement is viewed as a natural and unavoidable element of science as a human practice. Such a concept is, in fact, explicitly recognized in the IPCC rules of procedure, which specify that

lead authors should clearly identify disparities of view for which there is significant scientific or technical support, together with the relevant arguments. . . . It is important that reports describe different (possibly controversial) scientific or technical views on a subject, particularly if they are relevant to the political debate. (Intergovernmental Panel on Climate Change 1993)

Finally, and most significantly for our purposes here, peer review plays a major role in establishing the credibility of expert knowledge for policy purposes (see Miller and Edwards, chapter 1, this volume). The power of "virtual witnessing" stems from its basic (and basically democratic) tenet that any suitably qualified person could (at least in principle) play the role of witness. Symbolically, if not literally, it establishes the openness of science to the whole human community. Echoing this point, Chubin and Hackett call peer review the "flywheel of science, if for no other reason than that it symbolizes the professional autonomy and the accountability of science to the society that sustains it. Peer review communicates and enforces the terms of a social contract" (Chubin and Hackett 1990, 216). As a "flywheel" of accountability, peer review dampens the influences of personal, social, and political interests that might otherwise affect science-for-policy. It also renders publication of both "junk science" *and* true paradigm-challenging innovation considerably more difficult.

Thus we believe that while its problems should not be ignored, its virtues must be recognized. Peer review must stand as a basic norm of scientific practice. But its purpose should be clarified. It is not a truth machine, but a human technique for quality improvement, accountability, and community building. With this conception of peer review in mind, we can now return to the Chapter 8 controversy, where both sides accepted peer review as a fundamental standard. The only question was whether the IPCC authors had attempted illegitimately to circumvent it.

Did the Chapter 8 Revisions "Corrupt" the Peer Review Process?

The first thing to note is that IPCC reports are not primary science, but assessments of the state of the field. In other words, they do not constitute new research, but analysis and evaluation of existing, previously peer-reviewed research. IPCC authors sometimes incorporate research that has not yet been reviewed or published. In such cases, manuscripts must be made available to both IPCC authors and reviewers. This often means that peer review is already in process, but not yet completed. It also allows the IPCC to consider non-peer-reviewed, but potentially valuable sources such as "articles published in industry or trade journals; proceedings of workshops; reports and working papers of research institutions, private firms, government agencies, and nongovernmental organizations; contractor reports prepared for government agencies, firms, industry groups, and other nongovernmental organizations; and books that have not been peer reviewed" (Leary 1999). Such sources must be specifically flagged as "not peer reviewed," and IPCC authors are routinely directed to

critically assess [their] quality and validity. . . . Don't just cite results from non-peer reviewed sources without assessing their quality and validity. (Actually, you should be doing this for peer reviewed sources as well—our job is to assess the state of knowledge, not just report what's in the literature). Basically, your expertise substitutes for the peer review process for material that has not been peer reviewed. (Leary 1999)

Indeed, IPCC rules specify that assessment report authors must rely upon "the peer-reviewed and internationally available literature, including scientific and technical publications prepared by national governments and scientific bodies [and] the latest reports from researchers that can be made available in preprint form for IPCC review," in addition to IPCC-prepared supporting materials (Intergovernmental Panel on Climate Change 1993).

Nevertheless, as syntheses and evaluations IPCC reports must be (and are) subjected to their own peer review process, partially described above. The IPCC peer review procedure is far more extensive and inclusive than most. Most IPCC members (including nonspecialists, such as governments and lobby groups) receive draft IPCC documents and may submit "peer" comments. IPCC rules specify that draft chapters be circulated to

· Specialists who have significant publications in particular areas
· Lead Authors, contributors, and reviewers on the IPCC lists maintained by the Working Group and Subgroup co-chairs
· IPCC participating countries and organizations
· Specialist reviewers nominated by appropriate international scientific and technical organizations (e.g., WMO, UNEP, ICSU, Third World Academy of Sciences, FAO, IOC, World Bank, Regional Development Banks, OECD) (Intergovernmental Panel on Climate Change 1993)

Chapter authors are required to "take into account" all comments, although the meaning of this phrase is deliberately left vague. Given the volume of commentary and the many duplicate and irrelevant comments received, responses may be no more than a couple of words. Yet in aggregate, this extremely extensive peer review process typically leads to hundreds or even thousands of changes, since each document typically goes through several drafts.

Some of the most outspoken global warming skeptics in fact participated in the formal peer review of the SAR, including Chapter 8. Since

Seitz is not a climate scientist, and Singer is no longer active in research, they did not qualify as formal reviewers. However, Singer regularly attends IPCC meetings. In 1995, as the IPCC prepared the SAR, Singer was present at both the Madrid meeting and the IPCC plenary at Rome. Representatives from a number of NGOs which typically take a skeptical stance, including the Global Climate Coalition and several energy and automotive industry lobbies, also participated in the SAR peer review. Other skeptic referees included Patrick Michaels, Hugh Ellsaesser, and MIT meteorologist Richard Lindzen, an outspoken critic of some aspects of climate modeling. Lindzen was later appointed a Lead Author for the next IPCC Assessment Report. Thus the skeptical views of the Chapter 8 critics were already very well represented in the SAR peer review process, all the way through Chapter 8's formal acceptance at Rome.

Did the Chapter 8 authors "corrupt" the IPCC peer review process? Let's look at how it worked in IPCC Working Group I (WGI). In July of 1995, the third installment of the WGI drafting and review process for the SAR took place in Asheville, North Carolina. This meeting, like all other IPCC processes, was characterized by exceptional openness to critique, review, and revision. About six dozen climate scientists from dozens of countries took part. The meeting was designed to make explicit the points of agreement and difference among the scientists over exceedingly controversial and difficult issues, including Chapter 8—the most controversial.

New lines of evidence had been brought to bear by three climate modeling groups around the world, each suggesting a much stronger possibility that a climate change signal has been observed and that its pattern (or fingerprint) is matched to anthropogenic changes. Ben Santer, as a Convening Lead Author of Chapter 8, had assembled the results of a number of modeling groups. He presented the results of his group's effort not just to Chapter 8's Lead Authors and contributors, as is typical in IPCC meetings, but to the entire scientific group assembled at Asheville.

In this setting, Santer had to explain this work not only to his most knowledgeable peers, but also to scores of others from diverse scientific communities. These included stratospheric ozone experts such as Susan Solomon and Dan Albritton, satellite meteorologists such as John

Christy, and biosphere dynamics experts such as Jerry Melillo. Climatologists such as Tom Karl and I (SHS) were also present, along with the heads of national weather services and other officials from several countries who served on the IPCC's assessment team.

Not everybody present was equally knowledgeable on the technical details of the debate, of course. Perhaps only 25 percent of those assembled had truly in-depth knowledge of the full range of details being discussed. However, all understood the basic scientific issues and most knew how to recognize slipshod work—to say nothing of a fraud or a "scientific cleansing"—when they saw it. Even the less familiar participants thus served an essential role: they acted as technically skilled witnesses to the process of open debate.

This remarkable session lasted for hours. (In fact, it was continued less formally after dinner by roughly a dozen scientists, who spent nearly three hours discussing the final paragraph of the "Detection Section" of the Summary for Policymakers (Ben Santer, personal communication).) Though occasionally intense, it was always cordial, never polemical. As a result, the wording of Chapter 8 was changed. Ideas and concepts were somewhat altered, but basic conclusions by and large remained unchanged—because the vast bulk of those assembled were convinced that the carefully hedged statements the lead authors proposed were, in fact, an accurate reflection of the state of the science based upon all available knowledge, including the new results.

This was peer review at ten times the normal level of scrutiny! It would be almost inconceivable for the editor of a peer-reviewed journal to duplicate this process. A few referees and an editor can only hope to execute the reviewing role a fraction as well as the remarkable, open process at Asheville. Moreover, after the Asheville meeting, two more IPCC drafts were written and reviewed by hundreds of additional scientists from all over the globe.

It is true that the Asheville meeting was not a "blind" review, since everyone was in the same room. Under these circumstances, reputations, personalities, institutional politics, and the simple fatigue induced by long meetings probably played some role, one that might have been reduced through a more formal procedure where authors and respondents were more distanced from each other. Yet the Asheville meeting was only one part of a much more extensive process that did include the formal review described above. Furthermore, as we pointed out above, "blind" procedures are only one way to achieve the fundamental goals of peer review—and not necessarily the most effective one. Our claim here is that the quality improvements generated by the extensive and inclusive IPCC peer review process far outweigh the disadvantages of the open-meeting format for the final stages of peer review. If the real purposes of peer review are, as we have argued, quality improvement, accountability, and community building, then the IPCC process is as near to an ideal example as it may be possible to find.[3]

Furthermore, science-for-policy (such as the IPCC assessment reports) operates under severe time constraints not present in pure research. Reviewing and assessing the entirety of the rapidly growing climate-related scientific literature in two to four years is a vast project. Even full-time professional assessors would find this challenging, let alone the volunteer members of the IPCC. The IPCC's attempt to include a very wide range of interested parties in the review process greatly increases the work involved in responding to peer commentary. If the IPCC assessment reports are to serve their key function as input to FCCC climate negotiations, at some point the review-and-improvement cycle must stop.

An Open Process of Scientific Debate: Witnessing in Action

At Madrid, Santer presented Chapter 8's conclusions to the national delegates of ninety-six IPCC member nations. The conclusions were not presented alone, but followed a presentation to the plenary session of the scientific evidence contained in Chapter 8. Nevertheless, several countries objected to the Chapter 8 conclusions. Most of the objections came from OPEC or a few developing nations. One delegate, from Kenya, moved to have the chapter entirely dropped from the final report.

In response, the meeting's chair—following procedures often used at IPCC Plenary meetings to resolve disputes—called for a drafting group to revise the detection and attribution section of the Summary for Policymakers and to inform the Chapter 8 lead authors of various delegates' concerns. Nations complaining about the Chapter 8 draft were invited, indeed expected, to meet with Lead Authors, first to hear the scientists' point of view and then to fashion new, mutually acceptable language.

This breakout group worked for the better part of a day. Delegates from over half a dozen countries—including the Kenyan who had publicly advocated dropping the chapter—met with about half a dozen Chapter 8 authors, including Santer, co-Lead Author Tom Wigley, and scientists Kevin Trenberth, Michael MacCracken, John Mitchell, and me (SHS). The Kenyan sat next to me. Initially, he was confused by the discussion and somewhat hostile. We had many side conversations about what was being discussed: models, data, statistical tests, and various climate forcing scenarios. Although he was not a front-rank climate researcher, this delegate was a trained scientist. He began to grasp the nature of the Lead Authors' arguments, listening carefully to about half of the breakout meeting.

Ironically, the Saudi Arabian delegation sent no representative to this most controversial drafting group, even though Saudi Arabia had led the opposition in the plenary meeting. During the Chapter 8 debate, Saudi delegates often issued objections soon after receiving notes from the Global Climate Coalition representative. (Nongovernmental organizations were also represented at Madrid. For example, Singer—President of the Science & Environmental Policy Project and a self-proclaimed skeptic—raised a number of issues from the floor.)

Later in the plenary meeting, when Santer presented the drafting group's revised text, the Saudi delegates once again objected. Santer forcefully challenged them. Why, he asked, had no Saudi attended the breakout group—if their objections had some basis in science? The head Saudi delegate haughtily announced that he did not have to account to a lead author for his decisions about which drafting group to attend. Besides, he said, his was "only a small delegation" of a few people.

At this point the Kenyan delegate rose to speak. "I'm a member of a small delegation too," he said. (He was the only Kenyan representative.) "But somehow I managed to attend this most important drafting session. As a result, I am convinced that Chapter 8 is now well written and I have no objections to its inclusion in the report." (A paraphrase of his words from memory, by SHS.) The impact of his intervention was stunning, stopping with a few words what appeared to be a mounting movement of OPEC and LDC opposition to Chapter 8 before it could garner any further support.

Later on I (SHS) privately congratulated the Kenyan for having the courage to object publicly, observe privately, and then reevaluate his position before the entire plenary. He said he was not sure his country would approve of his stance, but having witnessed the debate process for several hours, he had become convinced it was honest and open. That was all he needed to change his opinion from preconceived skepticism to support of the Lead Authors' conclusions.

What this courageous delegate did was the essence of good science. He allowed his initial hypothesis to be subjected to new evidence, tested it, and found it wanting. He then listened to arguments for a different point of view, subjected them to the tests of evidence and debate, and reached a new conclusion.

Contrast this open IPCC process with that of the critics led by Seitz and the Global Climate Coalition. The latter first presented their technical counterarguments in the editorial pages of the *Wall Street Journal*. Some alleged—falsely, and without evidence—that Chapter 8's conclusions were based upon non-peer-reviewed articles (Santer 1996a). The Seitz/GCC group charged that the minor changes made to Chapter 8 during the post-Madrid revision process had somehow dramatically altered the report. Without a shred of evidence, Singer and others asserted that the changes constituted a politically motivated "scientific cleansing."

These irresponsible claims were not reviewed by a single independent, expert peer before being published—in the opinion pages of a business daily and a few news magazines. We leave it to readers to reflect on how the "flywheel" of peer review might have moderated the assertions of Seitz, Singer, and the GCC, had they been subjected to it.

The Scientific Results behind the Chapter 8 Conclusions

In a nutshell, the new evidence reported to IPCC and later published in *Nature* was based not on new empirical or theoretical results, but on new ways of asking climate models the right questions. In the past, critics such as the University of Virginia's Pat Michaels had correctly argued that direct observational evidence of global-warming effects (i.e., "signals") in the climate record were not very well matched to CO_2-only model results. For example, CO_2-only models suggested that the Earth's surface should have warmed up 1°C rather than the 0.5°C observed in the last century. Additionally, CO_2-only models suggested that the Northern Hemisphere would warm up more than the Southern Hemisphere. Such models also, however, suggested the stratosphere would cool as greenhouse gases increased. This clearly was happening, although at least part of that cooling can be attributed to stratospheric ozone depletion (Santer et al. 1996b).

The Earth's warming of 0.5°C during the twentieth century could be explained simply by asserting the trend to be a natural fluctuation in the climate. The IPCC scientists attempted to estimate the likelihood that natural events were responsible for the observed surface warming. They concluded that this was possible, but improbable. Critics, meanwhile, simply asserted that the warming was natural, *without characterizing the probability that this was the correct explanation.* Even if it did go unchallenged in a number of op-ed articles, this is a scientifically meaningless claim.

What is the probability that a half-degree warming trend in this century is a natural accident? This cannot be answered by looking at the thermometer record alone, since a globally averaged record is not reliable for much more than a century, if that. It is like trying to determine the probability of "heads" in a coin toss by flipping the coin once. Instead, climate scientists look at proxy records of climate change over long periods of time, such as fluctuating time series of tree ring widths, the deposits left from the comings and goings of glaciers, and the fluctuations of various chemical constituents in ice cores. These records, while not direct measurements of global temperatures, are nonetheless proportional to components of the climate in different parts of the world, and provide a rich record of natural variability.

This record (as summarized in Chapter 3 of the SAR) suggests that the warming of the last century is not unprecedented (Houghton et al. 1996). But it also is not common. Perhaps once in a millennium, such proxy records suggest, an 0.5°C global century-long trend could occur naturally (Schneider 1994). In my judgment (SHS), this circumstantial evidence implies that a global surface warming of half a degree has about an 80 to 90 percent likelihood of *not* being caused by the natural variability of the system. More recent evidence dramatically demonstrates that the last fifty years of the twentieth century saw a temperature rise distinctly larger than that of any period in the last 1,000 years (Mann, Bradley, and Hughes 1999).

Natural climatic forcing factors, such as energy output changes on the sun or peculiar patterns of volcanic eruptions, could cause the observed climate trend. However, each of these climate forcings has a peculiar signature or fingerprint. For example, energy increases from the sun would warm the surface, the lower atmosphere, and the stratosphere all at the same time. On the other hand, greenhouse gas forcing would cool the stratosphere while warming the lower troposphere. Aerosols from human activities, particularly the sulfates generated in coal- and oil-burning regions of the U.S. Northeast, Europe, and China, would cool the troposphere mostly during the day and not at night, and would largely cool the Northern Hemisphere, especially in the summertime when the sun is stronger.

This aerosol effect has turned out to be very important. Indeed, adding sulfate aerosols to greenhouse gas increases in the models led to a dramatic boost in the confidence that could be attached to the circumstantial evidence associated with climatic fingerprints. That is, when the models were driven by both greenhouse gases globally, and sulfate aerosols regionally, no longer did the Northern Hemisphere warm up more than the Southern Hemisphere, or all parts of the high latitudes warm substantially more than the low latitudes. Instead, a different fingerprint pattern emerged. Moreover, this pattern in the models showed an increasing correlation with observations over time—precisely what one would expect in a noisy system in which the human forcing increases with time. By itself, the pattern still has roughly a 10 percent chance of being a random event. However, when taken together with good physical theory and knowledge of ice age–interglacial cycles, seasonal cycles, volcanic eruptions, and now more consistent fingerprints, the vast bulk of the scientific community felt it was not irresponsible to assert that there was a higher likelihood that human climate signals had been detected. Taken together, all this circumstantial evidence was the basis for Chapter 8's now-famous claim that "the balance of evidence suggests a discernible human influence on climate."

At that point in the evolution of knowledge about the Earth's climate system, this was no longer a radical statement. It reflected a lowest-common-denominator consensus view of the vast majority of scientists. It did not say that a climate warming signal had been detected beyond any doubt. Neither we nor any other responsible scientists would make such a claim. But it did offer good reason to begin to plan, responsibly, for the possibility—which we now see as more likely than not—that the global climate will warm by at least one or two degrees centigrade during the next fifty years (further support for the likelihood of this outcome appears in Wigley, Smith, and Santer 1998).

The Meaning of *Consensus*: Responding to Climate Change Skeptics

To ignore contrarian critics would be inappropriate. Occasionally, non-conventional outlier opinions revolutionize scientific dogma (Galileo and Einstein are only the most oft-cited examples; see Kuhn 1962). However, we believe that news stories are grossly misleading and irresponsible if they present the unrefereed opinions of skeptics as if they were comparable in credibility to the hundred-scientists, thousand-reviewer documents released by the IPCC. The general public—or lay politicians—cannot be expected to determine for themselves how to weigh these conflicting opinions. And to publish character-assassinating charges of "scientific cleansing" without checking the facts is simply unethical, by the generally accepted standards of scientific practice.

The journalistic doctrine of "balance," while perhaps appropriate in two-party political systems where the "other side" must always get its equal coverage, is inappropriate if applied literally to multifaceted scientific debates, and it has nothing to do with peer review by experts, especially in the sense we have advocated here. In climate science, wide ranges of probabilities are attached to a whole array of possible outcomes (Morgan and Keith 1995; Nordhaus 1994). Scientific controversy simply cannot be trivialized into a false dichotomy between those who assert that human effects are likely to produce a catastrophic, "end of the world" crisis, "balanced" against those who assert that at worst nothing will happen and at best it will all be good for us. "The end of the world" and "no impact at all" are the two least probable cases (see Schneider 1997).

This is not just a problem for journalists. It also affects scientists. In communication with the public, we sometimes tend to focus our attention on controversies at the cutting edge of the art, rather than present clear perspectives on what is well understood—separating what is truly known from what is merely probable and both of these from what is highly speculative. This, combined with the propensity of the media to focus on "dueling scientists" and extreme, outlier opinions, leads to a miscommunication of the actual nature of the scientific consensus (see the chapter on "Mediarology" in Schneider 1989).

"Consensus," as we understand it, refers not to a single, exact prediction to which every scientist assents—an impossibility in this field— but to a generally agreed range of possible outcomes. This kind of consensus takes disagreement on details (and even, occasionally, on major points) for granted, both as an unavoidable element of a still-inexact science and as an important motor of scientific progress. Peer review, especially the inclusive and open form adopted by the IPCC, helps to build and maintain it. Consensus of this type is vital to the policy process. In essence, the policy question is to decide how much of current resources should be invested as a hedge against potential negative outcomes. This clearly is a value judgment. It is precisely the kind of judgment that the public and the policy establishment (not scientists) should make, but it can only be made if the decisionmakers—who are not, and are not going to become, experts—are aware of the best range-of-probability and range-of-consequences estimates of the responsible scientific community (see Moss and Schneider 1997).

Faxes sent by special interests to every major journalist on the planet or every significant elected and unelected official—what we like to call the "one fax, one vote" syndrome—are not very good sources of credible knowledge. Vastly better is the work of groups such as the IPCC and the National Research Council, which although slow, deliberative, sometimes elitist, and occasionally dominated by strong personalities, are nonetheless the best representation of the scientific community's current general opinion.

This kind of scientific consensus is not the same thing as "truth." Once in a while, the skeptics are right. Indeed, we are certain that some aspects of the current vision of climate change will turn out to be of minor impact, while others will prove to be more serious than currently thought. That is why assessment needs to be a continuous process, and why all policymaking requires "rolling reassessment." The IPCC, or its progeny, need to be reconvened every five years or so. Only with this input can the political process legitimately decide, and re-decide, to crank up its efforts at mitigation—or to crank them back down, depending on what is learned in each new assessment about the climate system, the impact of climate change on environment and society, and the effectiveness and distribution of mitigation costs. This ongoing and open process of refinement of knowledge is the only way that a complex system can become adaptive. Only an adaptive system can minimize the likelihood of making major mistakes, either by overinvesting in environmental protection or by allowing nasty experiments to be performed on "Laboratory Earth" without any attempt to anticipate or slow down the potential negative, irreversible consequences (Schneider 1997).

If the IPCC is to maintain its credibility as a hybrid scientific/political organization, peer review must remain a fundamental formal principle of its self-governance and a basic informal principle of its consensus-building process. Correctly conceived not as a truth machine, but as a technique for improving the quality of science and for moderating the influence of personal, social, and political factors on scientific results, peer review is a powerful technique for generating credible, trusted (and trustworthy) knowledge. The IPCC's widely inclusive, extremely intensive peer review process has opened the debate about climate change to a far wider range of actors than is usually consulted in science. By doing so, it has created a fairer, more thorough, and hence more powerful method for reaching consensus on the knowledge required for good public policy.

Notes

A previous version of this chapter was published as P. N. Edwards and S. H. Schneider (1997), "The IPCC 1995 Report: Broad Consensus or 'Scientific Cleansing'?," *Ecofables/Ecoscience* 1 (1): 3–9. The authors wish to thank Simon Shackley, Ben Santer, and Michael Oppenheimer for their helpful comments on an early draft.

1. The new rules are available at http://www.ipcc.ch/about/procd.htm.

2. "Virtual witnesses" "watch" a scientific experiment by reading its written description, "witnessing" it secondhand, and validating it by agreeing that it was correctly performed and that the reasoning used to reach conclusions was correct.

3. See Miller, chapter 8, this volume, for a more negative view of the IPCC's inclusiveness.

8

Challenges in the Application of Science to Global Affairs: Contingency, Trust, and Moral Order

Clark A. Miller

In this chapter, I explore three challenges to the application of science to global affairs: (1) the contingency and uncertainty inherent in knowledge about the global environment, (2) the need in global environmental policy contexts to secure credibility for scientific claims among far-flung, often highly diverse audiences, and (3) the often highly contested moral choices embedded in particular systems for producing and warranting policy-relevant science advice in international organizations. The efforts of experts and other policymakers to cope with these challenges are critical to the constitution of new governing institutions capable of equitably and effectively managing environmental change on planetary scales. To understand, evaluate, and contribute to such efforts, I argue, social scientists must adopt a more reflective approach to theorizing the relationship between, on the one hand, knowledge and ideas, and on the other, social and political institutions.

For many social scientists, global environmental issues are important sites in the construction of novel forms of social order on worldwide scales. The prominence of phrases like *global civil society*, *global polity*, and *global environmental governance* in recent scholarship reflects a growing sense that something new is happening in international environmental regimes that cannot be captured by traditional models of international relations that conceptualize states as independent rational actors (see, e.g., Litfin 1998; Young, Demko, and Ramakrishna 1996; Yearley 1996b; Lipschutz and Mayer 1996). Rather, global environmental regimes appear in these writings as institutions in which a wide array of societal actors, from states and NGOs to corporations and individuals, are busily working out, often in the face of serious opposition, new arrangements for living together in a worldwide community. In this chapter, I explore climate change as just such a site of societal reconstruction and resistance.

Implicitly or explicitly, most accounts of global civil society and global governance tie the emergence of new patterns of worldwide social and political interaction to cognitive convergence. New forms of social order emerge, according to this perspective, from the development of shared ideas among government officials, scientists, and citizens around the planet (Litfin 1998; Lipschutz and Mayer 1996; Haas 1990a, 1990b). Agreement on the risks of environmental change and the need for political cooperation among all the world's peoples, these accounts suggest, increasingly motivates people to band together in new, transnational communities. Consider, for example, the concept of epistemic communities. As described by Haas (1990b), these communities form around shared factual and causal understandings of global environmental change and proposals for policy change. Moreover, once established, they use their authority as "experts" to persuade other people to adopt the same ideas and, having done so, to agree to the creation of new environmental regimes (see also Haas 1992; Haas and Haas 1995).

In this chapter, I argue against this narrative. Theories of cognitive convergence generally posit the emergence of shared ideas as causal variables without exploring in detail the question of how particular ideas acquire credibility and authority among diverse audiences and therefore come to be shared in the first place (see, e.g., Jasanoff 1997a). Examining this fundamental question below, I suggest that what happens in international institutions is, in actuality, frequently the reverse of the causal story adopted by conventional accounts. Rarely do people adopt convergent ideas and then decide to band together in communities or form new institutions; rather, they come to share ideas as a result of social interactions that help constitute the community in the first place (e.g., Zehr 1994; Jasanoff 1993; Taylor and Buttel 1992). Shared understandings of nature and society, in this sense, may sometimes inspire social reorganization. But before they can do so, they must emerge from and attain widespread credence in detailed, day-to-day negotiations of meaning and practice (Lynch 1990). Frequently—as in the situations I describe in this chapter—these negotiations simultaneously constitute new social orders, so that shared understandings and new institutions arise together (Latour 1988). Institutions can thus play as important a role in the construction of new ideas as ideas do in stimulating institutional change. An adequate explanation for changes in global order must therefore take into account how cognitive understandings of global civil society and the global environment are *coproduced* with the social arrangements that connect up their meaning with the activities of individuals around the planet (Jasanoff, chapter 10, this volume).[1]

This alternative view of the emergence of global civil society carries potentially important implications for our basic understandings of international politics. To date, theories of international relations have tended to assume that the production of knowledge takes place outside of the domain of social analysis—and therefore that ideas can be treated as independent sources of power in society. By contrast, if the production of shared knowledge is itself a deeply political process, a complete understanding of changing patterns of global environmental governance must investigate not only what happens after ideas acquire consensual status but also how and why those ideas—and not others—acquired credibility and authority. Note that this is not an assertion that science is shaped by political forces. Nor is it an indictment of the role of science in global environmental policymaking. Rather, it is to say that the processes by which policy-relevant knowledge is produced, validated, and used to make global policy are part and parcel of the political foundations of global governance being built in emerging environmental regimes and must be analyzed as such.

From a more pragmatic perspective, the view of knowledge and politics presented here also carries important implications for the design and evaluation of international institutions. Recent scholarship in international relations has sought to revitalize the study of international institutions and to find universal standards for their proper design and evaluation (e.g., Keohane and Levy 1996; Chayes and Chayes 1995; Haas, Keohane, and Levy 1993). These approaches are grounded in the belief that the effectiveness of institutional designs can and should ideally be assessed on instrumental criteria. The recognition, however, that shared understandings of environmental risks are, in part, a product of social negotiation calls into question the ability of instrumental reasoning to provide a privileged standpoint from which to judge institutional performance. Consequently, the search for new criteria of institutional design and evaluation must take other forms. A more reflective and potentially more promising approach may be to take seriously the proposition that conceptual models of global nature and global society as well as social norms and practices for producing knowledge and managing social affairs are constantly being re-negotiated in specific institutional contexts. If we adopt this view, then two questions become important for social scientists. First, how do particular understandings come to be shared by people across the globe (Jasanoff and Wynne 1998)? And second, according to what criteria should the processes by which this happens be evaluated?

Linking Science and Politics in the Climate Regime

A clear articulation of these arguments can be made about efforts to relate science and politics in the context of the climate regime.[2] The need to link science to politics has become widely recognized in international policymaking over the past twenty years, particularly around environmental issues (Chayes and Chayes 1995; Haas and Haas 1995; Young 1994; Benedick 1991; Haas 1990a, 1990b). However, most scholars and policymakers have generally taken this relationship to be insufficiently problematic to warrant detailed attention (see Jasanoff and Wynne 1998; Jasanoff 1997a; Global Environmental Assessment Team 1997; Miller et al. 1997; Litfin 1994, arguing more recently for the importance of research into science-policy linkages in international environmental issues). Detailed examination of the negotiation of scientific advisory arrangements within the climate regime reveals, I will illustrate below, that how science should relate to international politics has been deeply contested among the participants in international institutions. As nations entered into negotiations on climate change in the late 1980s with the creation of the Intergovernmental Panel on Climate Change (IPCC), no shared understanding existed for how scientific advisory processes should be instituted in practical or normative terms. Questions such as who should be granted expert authority, what should count as evidence, who should be allowed to subject expert claims to critical inquiry, and who should have the authority to make these judgments generated intense debate in international negotiations, both among Western governments and, especially, between North and South. Virtually all of the chapters in this volume have identified challenges that underlay these disagreements: divergent national expectations about climate science and its role in public policymaking, heterogeneous distributions of scientific resources, normative assumptions embedded in climate science discourses, and contingency and uncertainty in climate modeling. Only over time, through complex negotiations, have participants in the institutions of the climate regime found ways to begin to overcome these challenges to create globally credible science advice and to link it to policy choices.

To explore how governments from around the world came to share at least temporarily settled models for producing, validating, and using expert knowledge within the climate regime, I analyze in this chapter the activities of the Subsidiary Body for Scientific and Technological Advice (SBSTA) of the UN Framework Convention on Climate Change. Since its creation in 1995, SBSTA has constituted one of the key sites where questions about the proper organization of science advice in international governing institutions are being addressed and, occasionally, settled. Although its formal function within the regime was initially uncertain, SBSTA has subsequently emerged as the principal forum in which regime participants have articulated and negotiated among competing models of institutional design for providing expert advice about climate change. It has served, in other words, as a space where governments (and to a lesser extent NGOs) can deliberate the ground rules by which scientific experts and knowledge claims receive accreditation within the institutions of the climate regime. Settlements arrived at in SBSTA have thus created an important part of the normative and institutional contexts that will mediate future interpretations of climate change and choices about human responses to it within the climate regime.

Under SBSTA's auspices, policymakers have grappled with questions about what it means, in practical terms, to organize science advice in international institutions. To date, however, they have received little help or guidance from scholars whose work might provide useful analytic perspectives on the relationships between global environmental science and politics (for an exception, see Jasanoff and Wynne 1998). International relations scholars, for example, have unquestioningly encouraged the expansion of scientific expertise in international regimes (Chayes and Chayes 1995; Haas, Keohane, and Levy 1993). Largely absent, however, have been detailed, empirical studies of scientific knowledge and public policymaking of the kind undertaken in recent years in domestic political contexts (Bimber 1996; Nelkin 1992), let alone more theoretically informed studies of the relationship between expertise and democratic governance in the fashioning of contemporary social order (Jasanoff 1996a, 1996c, 1990, 1986; Porter 1995; Yearley 1991; Ezrahi 1990; Brickman, Jasanoff, and Ilgen 1985; Wynne 1982). Those few scholars who have examined the workings of the climate regime closely have focused either on the IPCC (e.g., Shackley and Wynne 1995; Boehmer-Christiansen 1994) or the Intergovernmental Negotiating Committee and its successor, the Conference of the Parties (e.g., Bodansky 1994). In neither case, however, have they examined how the activities of such organizations help interactively define what will count as both good science and good governance in international institutions. Put differently, they have yet to come to grips with the mechanics of *coproduction*.

It is precisely the simultaneously linked production of scientific and political organization in international institutions that I want to explore in the activities that have taken place under SBSTA's auspices. In the following sections, I explore the challenges that participants in SBSTA have faced in organizing scientific advisory processes; I analyze how they have managed to overcome these challenges in several instances and to make progress toward globally credible science advice; and I offer suggestions for how SBSTA's experiences might be generalized to other international institutions.

In carrying out this analysis, I draw heavily on data from participant observations I made in early 1997. During SBSTA's fifth meeting, February 25–28, 1997, in Bonn, Germany, I had the opportunity to observe and interview numerous scientists, government officials, and NGO representatives participating in SBSTA. I attended meetings of the SBSTA plenary, special seminars held for government and NGO representatives, and one meeting of SBSTA's informal working group on methodologies. My interviews and observations focused on participants' expectations and understandings of SBSTA as an organization, as well as the history of its efforts to construct various institutional arrangements for science advice. In addition, I have also drawn on two documentary records: the official publications of SBSTA and summaries of SBSTA deliberations reported in the *Earth Negotiations Bulletin*. The latter is a publication of the International Institute for Sustainable Development, a Canadian NGO. I document these records in the appendix to this chapter.

Creating a Space for Deliberating about Science Advice

The creation of SBSTA and other international scientific advisory organizations such as the IPCC extends a long-term trend in the evolution of Western democratic forms of government. Over the course of the twentieth century, the world's liberal democracies drew increasingly heavily on science and other forms of expertise in the formulation and legitimation of public policy. Dramatic expansions of public support for scientific research and for the involvement of experts in public policymaking have come as Western governments have expanded their authority to regulate social welfare, environmental protection, and public health and safety. Scientific objectivity has come to represent, in Western democracy, an instrumentally effective force in the pursuit of public action, authority, and accountability, buttressing the authority of centralized regulatory institutions (Jasanoff 1996b, 1990; Porter 1995; Ezrahi 1990). Today, the creation of institutions like SBSTA and the IPCC reflects a growing effort to use science in a similar fashion in international politics, thus helping to legitimize a deepening and expansion of the role of international regimes in grappling with threats of environmental degradation around the world (see Miller, chapter 6, this volume, for a historical exploration of earlier postwar efforts to use science in the pursuit of particular models of world order).

Problems like climate change pose foundational questions about the future of such initiatives—how will countries learn to balance the high risks of action and inaction; to cope equitably with heterogeneous costs, risks, societies, and environments; to integrate national and international institutions; and to meld value commitments to environment, development, and human rights on planetary scales (Rayner and Malone 1998;

Litfin 1998; Young, Demko, and Ramakrishna 1996; Chayes and Chayes 1995)? For many policymakers in the climate regime, science seems to offer important, and possibly unique, resources for helping policymakers address these questions in ways that can secure worldwide public trust. Responding to these assumptions, governments have established a host of new scientific advisory processes to produce and validate knowledge related to the activities of the climate regime and to incorporate that knowledge into policy choices. Thousands of scientists, government officials, and representatives of nongovernmental organizations from numerous countries have participated in these processes under the auspices of the IPCC and SBSTA.

SBSTA was created by Article 9 of the UN Framework Convention on Climate Change as one of three institutions jointly responsible for treaty oversight. The text of Article 9 reads, in its entirety:

Article 9: Subsidiary Body for Scientific and Technological Advice

1. A subsidiary body for scientific and technological advice is hereby established to provide the Conference of the Parties and, as appropriate, its other subsidiary bodies with timely information and advice on scientific and technological matters relating to the Convention. *This body shall be open to participation by all Parties and shall be multidisciplinary. It shall comprise government representatives competent in the relevant field of expertise.* It shall report regularly to the Conference of the Parties on all aspects of its work.

2. Under the guidance of the Conference of the Parties, and drawing upon existing competent international bodies, this body shall:

(a) Provide assessments of the state of scientific knowledge relating to climate change and its effects;

(b) Prepare scientific assessments on the effects of measures taken in implementation of the Convention;

(c) Identify innovative, efficient and state-of-the-art technologies and know-how and advise on the ways and means of promoting development and/or transferring such technologies;

(d) Provide advice on scientific programmes, international cooperation in research and development related to climate change, as well as on ways and means of supporting endogenous capacity-building in developing countries; and

(e) Respond to scientific, technological and methodological questions that the Conference of the Parties and its subsidiary bodies may put to the body.

3. The functions and terms of reference of this body may be further elaborated by the Conference of the Parties. (Mintzer and Leonard 1994, 352; emphasis added)

While the creation of SBSTA might be taken as evidence of a shared understanding among negotiators of the role science should play in international politics, this conclusion is warranted by neither the history of SBSTA's creation nor a close reading of its authorizing text. SBSTA emerged as part of a wide-ranging reconfiguration of the climate regime between 1990 and 1995. From 1988 to 1990, intergovernmental discussions about climate change took place within the Intergovernmental Panel on Climate Change (IPCC). By 1990, however, many developing countries were dissatisfied with the IPCC. Whereas the IPCC represented climate change as a problem of global environmental limits (mirroring the view prevalent in most Western nations), most developing countries saw it as a problem of overconsumption in the North (e.g., Jasanoff 1993).

The IPCC's unresponsiveness to this alternate view of climate change led developing country governments to reject a request in 1990 by the UN Environment Programme that the IPCC open formal negotiations on the climate issue in early 1991. Instead, they voted overwhelmingly in the UN General Assembly to create a separate institution, the Intergovernmental Negotiating Committee, to house the negotiations. Later in 1991, scientists participating in the IPCC succeeded in internally reorganizing the IPCC into the form it takes today, focusing the panel's activities on providing assessments of the risks of climate change, establishing uniform rules of peer review and expert selection across the institution, and creating technical support units whose self-described function is to isolate the IPCC's participants from political interests who might seek to influence its findings. All of these changes were intended by the IPCC to strengthen its appearance as a scientifically objective body. However, the reforms had little effect on the IPCC's credibility with developing countries. Consequently, developing countries rejected efforts by the United States and the European Union to incorporate the IPCC into the institutional framework of the UN Framework Convention on Climate Change as it was being negotiated in 1991–92. Instead, a compromise led to the insertion of provisions for a novel Subsidiary Body for Scientific and Technological Advice (SBSTA) into the treaty's text to manage the regime's perceived need for expertise and its relations with "existing competent international bodies"—as close as the treaty text comes to mentioning the IPCC.

As can be seen in the treaty text authorizing SBSTA, however, the Framework Convention (as is typical for such documents) does little to specify exactly what SBSTA should do and how it should be organized to do it. The treaty establishes certain parameters of membership (italicized above, although even here countries have interpreted this phrase in a multiplicity of ways, sending delegates with a wide array of backgrounds to SBSTA; over time, most governments have tended to send the same delegates to SBSTA as they do to the Conference of Parties as a whole) as well as guidance on the tasks to be performed by SBSTA (paragraph 2). Other areas of potential interpretive flexibility, however, such as criteria for defining the competence of experts or the relevance of domains of expertise are left undefined. The phrase "drawing upon existing competent international bodies" has been interpreted by many countries as implying that SBSTA should develop some kind of relationship with the IPCC. Again, however, the details of what this relationship should look like are not specified. From the treaty's signature in 1992, then, until the formal constitution of SBSTA at the first meeting of the Conference of the Parties in Berlin in 1995, SBSTA remained a largely unknown factor in the climate regime. Evidence of whether and how shared understandings of the proper place of science in international politics have emerged in the climate regime can only come, then, from events subsequent to SBSTA's creation.

Deliberating about Science Advice within SBSTA

The sense of uncertainty surrounding SBSTA has largely continued unabated since its inception. Implicitly and explicitly, participants in SBSTA have struggled to find generally acceptable criteria and procedures for selecting experts, weighing evidence, establishing institutional mandates, and conducting reviews. Frequently, however, this has proven difficult. (*Pace* theories of international regimes, which have tended to emphasize the development of convergent norms and practices as a necessary condition for effective environmental protection, cultural anthropologists have argued that an explicit recognition of the heterogeneity of discourses that underlie these disagreements is essential to achieving sustainable development; see, e.g., Rayner and Malone 1998; Thompson, Rayner, and Ney 1998a, 1998b). The initial effort to create two new Technical Advisory Panels (TAPs) illustrates these difficulties. At the first meeting of SBSTA in Berlin in 1995, a proposal was introduced to supplement the IPCC (which supporters argued should continue to conduct risk assessments for the climate regime) by constituting two additional technical advisory panels to address questions related to standard methodologies and technology transfer. Over the course of several SBSTA meetings, however, deep-seated divisions emerged among participants over how to organize these panels. Issues on which participants differed included the following:

• Expert affiliation: Would experts on the panels be invited from governments, the private sector, nongovernmental organizations, international organizations, or universities?

• Method of appointment: Would experts be appointed by governments, nominated by governments and appointed by the Framework Convention Secretariat, or nominated by governments and appointed by SBSTA?

• Balance of experts: Would experts from any country be allowed to participate (in an open-ended structure) or would there be a regional/geographic balance, a balance between Annex I (industrialized) and non-Annex I (developing) countries, or a balance of disciplines?

• Method of review: Would the TAPs use formal scientific peer review, formal review by SBSTA, or no review at all?

• Committee structure: Would the TAPs have a fixed number of members or a fixed steering committee with flexible ability to create subpanels? Or, would the number of TAPs be flexible so as to accommodate new questions that might arise? Further, how many members would the TAPs have?

• Duration: Would the TAPs be permanent, have a fixed duration, or be of contingent duration depending on periodic SBSTA review?

• Line of authority: Would the TAPs report to SBSTA through the IPCC or would they report directly to SBSTA?

• Terms of reference: Would SBSTA establish fixed terms of reference or would the TAPs determine their own terms of reference?

These issues encompassed a wide array of divergent expectations about what makes for credible knowledge and what makes for legitimate policy. Not all of these issues carried equal weight. Some were merely raised as organizational possibilities during SBSTA deliberations. Others achieved the status of formal proposals. Over time, a few alternatives coalesced into a small number of competing proposals. Among contested issues, the most prominent—that of membership—separated many Western countries, who favored membership based on demonstrated disciplinary achievement, from many developing countries, who favored establishing a fixed membership of government-nominated experts with explicit geographic representation and explicit balance between developed and developing country representatives. Although both sides offered strong rationales for their positions, and a number of compromise proposals were put forward, no resolution of this division was ultimately achieved. Negotiators eventually shelved discussion of TAPs for future consideration and turned to an alternative approach to the production of expert advice.

The failure to constitute two new TAPs left participants in SBSTA in a quandary. They were not, in general, willing to abandon the possibility of deriving expert input for the decisions of the climate regime. Nor did they prove willing to revert back to the IPCC as the sole source of scientific and technological advice. Instead, negotiators' deliberations about the organization of expert advice turned in new directions. First, at SBSTA's third meeting, in conjunction with the second meeting of the Conference of the Parties to the Framework Convention in July 1996, participants agreed to establish a "Roster of Experts." The purpose of this roster was to establish a pool of experts on which SBSTA could draw to answer particular questions, should it so choose. The roster works as follows. Governments nominate experts to the roster. No limit has been placed on either the number of experts a government may nominate nor the areas of expertise a government's nominations may cover. Once SBSTA participants have identified an issue of interest, the Framework Convention Secretariat then selects "appropriate" experts to constitute a panel and prepare a response to SBSTA's questions. To date, SBSTA has received reports from five such panels: three on issues of technology transfer, two on issues related to methods for accounting for national emissions of greenhouse gases.

The second response to the demise of the TAPs proposal was to establish two "informal working groups" on methodologies and technology transfer. These working groups hold no formal authority but bring together government representatives interested in the particular issues at hand for informal (i.e., off-the-record) discussions. These groups meet frequently during regular SBSTA meetings (which occur two to three times per year) but are typically attended by only a small fraction of the governments attending SBSTA. Under the auspices of these informal working groups, SBSTA participants have continued to pursue ongoing deliberations and have succeeded in moving forward on several proposals to constitute new scientific advisory arrangements in three important areas of expert involvement in international environmental policymaking: risk assessment, standard methodologies, and technology transfer.

Risk Assessment

Since their failure to establish a common approach to TAPs, SBSTA participants have pursued separate agendas for risk assessment, standard methodologies, and technology transfer. With regard to risk assessment, negotiators have primarily sought to work out an appropriate relationship between SBSTA and the IPCC. In 1991, as described earlier, after the creation of the Intergovernmental Negotiating Committee as the institutional home of the climate negotiations, the IPCC reorganized its activities around the provision of periodic climate risk assessments. However, the creation of SBSTA by the Framework Convention raised a barrage of questions about the ongoing status of the IPCC within the climate regime. Would SBSTA organize its own risk assessments with an eye to competing with the IPCC? If not, what would relations between the two organizations look like? Would the IPCC remain independent of SBSTA or would it become subsidiary to it? Would IPCC reports retain any formal authority within the climate regime? If so, would they be subject to review by SBSTA or not?

These questions led to extensive debate within SBSTA about the process for producing and validating risk assessments within the climate regime. Many governments, largely from the North, viewed the IPCC as the most authoritative, international expert body on climate change and insisted that it continue to produce risk assessments for the climate regime unhindered by interference from SBSTA. However, other governments, predominantly of developing countries, continued to view the IPCC as inattentive to their concerns and as overly dominated by Northern experts and their regionally biased interpretations of climate change. They wanted answers to questions such as the regional distribution of climatic changes about which the IPCC seemed unwilling to make clear statements. Still other governments, mostly from oil-producing states, questioned the validity of the IPCC's conclusions regarding the existence of climate change at all.

The ensuing compromise with regard to the IPCC contained several parts. The IPCC would continue to provide risk assessments of the climate issue every five years. Technical advice in other areas, however, particularly as regards technology transfer (which has always remained a prominent agenda item of developing countries), would be dealt with through alternative, SBSTA-based processes. SBSTA would act as the interface between the IPCC and the climate regime, with the IPCC submitting its reports to SBSTA. SBSTA participants would then decide whether to recommend the reports to other international institutions, in what form to pass them on, and whether or not they wished to supplement the IPCC reports with their own interpretations and conclusions or with reports from other bodies. Finally, SBSTA and the IPCC would establish a joint liaison group, composed of representatives from the Secretariat of the IPCC, the Bureau of the IPCC, the Secretariat of the Framework Convention, and SBSTA. This group would be responsible for establishing working arrangements between the two organizations.

Standard Methodologies
The UN Framework Convention on Climate Change mandates that all signatories compile "national inventories of anthropogenic emissions by sources and removals by sinks of all greenhouse gases not controlled by the Montreal Protocol, using comparable methodologies to be agreed upon by the Conference of Parties" (Mintzer and Leonard 1994, 341). In addition, in pursuing their objectives under the climate regime,

countries have identified a wide array of other areas in which cooperation would be facilitated by common standards or approaches to technical analysis. These include methods for assessing climate impacts and
vulnerability, assessing climate-friendly technologies, assessing the effectiveness of national policies to reduce greenhouse gas emissions, compiling national communications to the Framework Convention process,
assessing activities implemented jointly under emissions trading projects,
and several others.

Debates within SBSTA over the development of international standards have encompassed a variety of different issues. Participants have
disagreed over the content of specific standards, the means by which
standards would be created, the institutions that would be delegated
the task of standardization, the prioritization of standards development,
and the degree to which governments would be legally bound to specific
standards—that is, would standards be for informational purposes
only, would they constitute defaults to be used in the absence of alternative choices by individual governments, or would they be binding on
parties?

One of the most advanced, and also contentious, areas of standards
development has been how to assign responsibility for greenhouse gas
emissions. The negotiation of the Framework Convention itself settled
numerous issues for these standards. As the passage quoted at the beginning of this subsection suggests, the Framework Convention establishes
several principles to guide methodology development: (1) it assigns
responsibility to nations, as opposed to individuals or firms; (2) it establishes that governments will assess their own national emissions, as
opposed to an international body assessing each nation's emissions; (3)
it assesses responsibility on the basis of sources and sinks of greenhouse
gases, not just sources; (4) it establishes that only anthropogenic sources
and sinks, not those seen as natural, will count toward national responsibility; (5) it requires assessing emissions of all greenhouse gases, not
just carbon dioxide, except for chemicals already covered by the Montreal Protocol; (6) it requires that countries use comparable (and, hence,
not necessarily identical) methodologies; (7) and it requires that standard
methods be established by the Conference of the Parties (and not some
other international body).

However, many debates remained to be sorted out by SBSTA. Participants have agreed that the IPCC will be responsible for constructing (and continually updating) a default set of methods which must be ratified by the Conference of Parties. These methods need not be used if a government determines that it has a more accurate method and can specify adequately how its method differs from the IPCC's. However, no criteria has been established for judging either the relative accuracy of methods or the adequacy of documentation provided by a government. In practice, these judgments are left up to the governments themselves. Although SBSTA participants have decided that national inventories will be subject to international review, that review is facilitative and not binding on governments. Several questions directly related to the content of the emissions inventory standards, such as how to deal with bunker fuels from transnational air and sea transport and harvested wood products, have also been included on SBSTA's agenda. The former raises questions about who will be assigned responsibility for emissions that do not occur inside the territory of any particular country. The latter raises questions about who will be held accountable for emissions that result from deforestation in cases in which the actual emission of carbon dioxide into the atmosphere takes place in a different country than the original deforestation. In these cases, negotiators have agreed that important value choices are at stake and that SBSTA, and not the IPCC "technical" groups working out the details of the standards, is the proper forum in which to address them. However, no criteria for determining whether future value choices are sufficiently important to be shifted from the IPCC to SBSTA have been established, leaving resolution of this issue to an ad hoc, case-by-case basis.

Technology Transfer

Finally, debates within SBSTA have addressed a number of issues around technical assistance and technology transfer, although technology transfer has dominated most of the discussions. These issues constitute one of the most difficult topics under deliberation by SBSTA. Technology is considered essential to solving the public policy challenges created by climate change. Establishing effective means of promoting the creation and adoption of new technologies is difficult, however, and particularly so when what is entailed is the transfer of technologies from one country to another. How can technology transfer be carried out effectively? Whose norms will govern technology transfer? Will such transfers take place in the public or the private sector (and whose definitions of public and private will govern policy development)? How will the effectiveness of transfer be assessed, and at what point in the transfer will this be done? How will technology transfer be embedded in wider questions about distributing the costs and benefits of climate change?

Debates over technology transfer have addressed a number of these important underlying issues within the context of more specific proposals—for example, providing information to developing countries about the availability of greenhouse-friendly technologies, assessing the effectiveness and appropriateness of technologies for specific localities, and determining how much credit countries will receive under the treaty for technologies transferred through joint implementation and emissions trading projects.[3] Debates have also taken place over questions such as how countries' needs for technology transfer would be determined.

Over time, SBSTA participants have reached agreement on a number of fronts. A Dutch NGO was contracted to conduct a survey asking developing countries to identify what they believed were their technology needs. This survey has been completed, and SBSTA participants are now working on what to make of the data compiled. An Internet database of industry contact points is in the process of being set up to make information available to developing countries about greenhouse-friendly technologies. SBSTA participants were also able to reach agreement on the desirability of a report detailing current trends in both government-to-government foreign aid related to technology transfer and foreign direct investment. This report was produced by six experts from developing countries selected by the Framework Convention Secretariat from the "Roster of Experts" described earlier. Finally, SBSTA participants have recently agreed to recommend to the Conference of Parties the creation of a new technology information center for climate-friendly technologies.

Coproducing Science and Politics

Debates within SBSTA over risk assessment, international standards, and technology transfer exemplify three challenges that are inherent in any effort to mobilize science to support public policymaking, which can be loosely categorize as those of *contingency*, *trust*, and *moral order*. This categorization stems from recent work in the field of science studies. The *contingency* of science advice has been most fully explored in adversarial political and legal systems such as those of the United States. Studies of expert knowledge in U.S. policymaking have frequently observed the deconstruction of scientific evidence in the course of contests over planned government action. In such contexts, which share many similarities with the contentious negotiations that have taken place within SBSTA, efforts to highlight discrepancies and inconsistencies, emphasize uncertainties, and challenge the adequacy of experimental techniques or the motives of expert advisers are commonly used to discredit scientific testimony and to point out the indeterminacy of scientific findings (Nelkin 1992; Jasanoff 1990).

Although, in principle, science depends on data and models to definitively establish the truth of particular scientific knowledge claims, detailed, empirical studies of the conduct of scientific research have found that, in practice, scientific claims in the making are inevitably subject to varying interpretation. While much science is never subjected to the rigorous public questioning common in organizations like SBSTA, these studies suggest that grounds for skepticism about particular knowledge claims can almost always be found. For example, participants in scientific and public controversies often criticize evidence from experiments and models on grounds of either the validity of the assumptions that go into their interpretation or the skill of the practitioner in carrying them out (Collins 1985). More recently, Miller et al. (1997) have argued that, in addition to scientific data and models, the organization of science and its place among broader institutions in public life are equally subject to the problem of contingency. Policy-analytic tools and methods, information on public values and perceptions, rules of participant selection among experts, policy elites, and the public, and boundaries between disciplines, between science and politics, and between expert and lay domains of authority are all open to multiple interpretations and conflict (see also Gieryn 1999). Certainly many of these subjects have proven contingent within SBSTA's debates.

Likewise, science studies research has shed light on the varying means necessary to secure *trust and credibility* among expert communities, between experts and policy communities, and between elite institutions and lay publics. Scientific knowledge is commonly assumed to derive its credibility from its objective or universal validity. Empirical studies of scientific and political controversies have demonstrated, however, that the credibility of scientific claims often differs across audiences, is interactionally constituted in particular contexts, and frequently depends on reference to deeply embedded cultural norms and practices for securing trust and warranting truth (Shapin 1996, 1994; Jasanoff 1986). Credibility, then, is something that can be achieved only in relevance to particular circumstances and particular expectations regarding the trustworthiness of expert knowledge.

Finally, the organization of scientific advisory processes also raises important issues of *moral or political order*. Science is value-laden in the sense in that it privileges certain voices, certain ways of knowing, and certain interpretations of nature over others. The very act of using science to inform public choices confers power on some actors in the policy process while removing it from others. This empowering and disempowering function of science raises questions of equity, legitimacy, and authority as science becomes more central to shaping public policy and the organization of politics in democratic societies. Science studies research has revealed that questions often assumed to lie within the domain of inquiry into nature—for example, questions regarding the kinds of evidence used to determine the validity of scientific knowledge claims or the expertise appropriate to the resolution of certain problems—often have significant normative dimensions (Jasanoff 1996c, 1990; Wynne 1995). Consequently, normative issues such as discretion, representation, participation, and transparency have emerged as central to the organization of scientific advisory bodies and the admission of scientific testimony in legal and administrative proceedings (Jasanoff 1996c, 1996b, 1990; Porter 1995; Ezrahi 1990).

Although problems of contingency, trust, and moral order are, in principle, intrinsic to all science-based policy enterprises, they have ceased to pose fundamental threats to political legitimacy in industrial countries. Government officials, scientists, and citizens in most Western nations—

even when they disagree on the details of scientific interpretation—are able to draw on culturally specific systems of rhetoric and practice for warranting scientific knowledge in policy contexts, for securing the trustworthiness and credibility of institutions that use science, and for rendering the uses of political power consistent with norms of legitimate governance, such as transparency, openness, and public participation (Brickman, Jasanoff, and Ilgen 1985; Jasanoff 1986; Shapin 1994; Wynne 1982). Meanwhile, public institutions have become active in setting criteria for legitimating scientific evidence, selecting experts, organizing review procedures, demarcating the mandate and authority of scientific and political institutions, and numerous other choices in organizing science advice (Jasanoff 1996c, 1990; Gieryn 1996, 1999). Ideas about the proper organization and evaluation of scientific advisory processes, in other words, have become through continuous use deeply embedded in many national political cultures.

What SBSTA provides is a forum in which participants in the climate regime have been able to overcome these challenges, however minimally. Over time, negotiators have reached a number of settlements that have established practical ground rules for instituting expert advisory processes in international relations. SBSTA's experiences thus illustrate one route by which the coproduction of science and politics occurs in contemporary international relations. The question remains, however, whether or not SBSTA's experiences provide a workable model for other international institutions searching for ways of incorporating science into their policymaking processes.

A Model for Future Global Governing Arrangements?

SBSTA's emergence as a novel kind of institutional space in international relations—one in which governments and NGOs deliberate about the proper organization of expert advisory arrangements in international institutions—raises important questions about how we should evaluate its design and performance. Does SBSTA provide a model for future global governing arrangements? Are policymakers evolving new, globally shared norms and ideas about international scientific advisory processes? Or is the effort to forge such processes simply reinscribing

Western notions of science advice in international contexts, exacerbating tensions in international politics? Do some approaches to designing scientific advisory processes appear to generate more stable or effective institutions? Do others appear to lead more frequently to protracted controversies or poor policy outcomes?

If the global community is going to successfully delegate its problem-solving needs to a cadre of environmental experts, these delegates will have to walk a fine line between two countervailing tendencies: (1) promoting adherence to a broad normative commitment that science, in general, provides an important resource for pursuing the objectives of international regimes; while simultaneously (2) making it possible for participants from very different origins to critically examine specific arrangements for producing, validating, and using scientific knowledge against their own criteria of normative acceptability and instrumental rationality. To accomplish this balancing act, participants in international institutions—scientists, government officials, and citizens alike—need to be aware that expert knowledge and advisory arrangements embed tacit, value-laden assumptions about both nature and society, that the provision of science advice is contingent and often subject to multiple interpretations, and that trust and credibility are interactionally constituted, often according to different criteria among far flung public audiences. As Jasanoff and Wynne (1998, 77) have put it, the successful mobilization of science in the climate regime, in the face of the challenges posed by contingency, trust, and moral order, relies for its authority on "the patient construction of communities of belief that provide legitimacy through inclusion rather than exclusion, through participation rather than mystification, and through transparency rather than black-boxing." Without these reflexive insights, integrating global scientific advisory processes into the norms and practices of policymaking will be made considerably more difficult. Does SBSTA offer opportunities for achieving this goal?

Examined individually, the policy resolutions achieved by SBSTA appear as no more than partial, often temporary, negotiated settlements, seemingly leaving little room for evaluation along these more general lines. Divergent national expectations and interests, perceived high stakes, weak global institutions, and rapidly changing global norms and practices all help render problematic the possibility of universally credible and authoritative approaches to science advice. In such contexts, contingent compromises, reflecting no more than a sorting out of interests around a specific issue at a specific time, seem to be all that SBSTA has to offer. Examined collectively, however, SBSTA's deliberations illustrate several patterns of interaction that offer a more generalizable model for future institutional design.

Deconstruction

Science advice, as we well know, cannot be wholly purified of embedded values. Normative concerns are always at stake in the mobilization of scientific knowledge in public policy contexts (Jasanoff 1996b, 1990). One criteria for evaluating SBSTA is the degree to which its institutional design allows participants to negotiate their tacit, often deeply embedded commitments to alternative models of the proper linkages between science and policy and competing interpretations of human-nature interaction.[4] Does SBSTA help reveal underlying value choices? To whom? For what purposes? Does it help reduce controversy over particular proposals for advisory arrangements in ways that buttress the credibility of the resulting organizations?

SBSTA operates by consensus rules of procedure, effectively providing each participating government with veto power over any proposed arrangement for science advice. Tacit social conventions among diplomats discourage representatives of individual countries from openly opposing the will of the rest (unless, of course, one represents "the world's only superpower"). Yet, this veto power has substantially strengthened the ability of individual governments to make their preferences known, during SBSTA debates and behind the scenes, regarding the proper organization of science advice. During SBSTA's fifth meeting in early 1997, I witnessed the ability of representatives from several small developing countries and from the so-called small island states to oppose strategies for securing expert advice that did not meet their perceived priorities or normative perspectives. Numerous developing countries, for example, opposed creating technical advisory panels organized around well-defined areas of disciplinary expertise for fear that these bodies would come to be dominated by acknowledged Northern experts.

Similarly, when the United States and the European Union wanted to move forward on developing a methodology for assessing joint implementation of projects under an emissions trading system and for verifying treaty compliance, the representative from the Marshall Islands objected on the ground that methods for assessing the risks of sea-level rise were far more important for his country and that they must also receive a high priority.

The ability of individual governments to block the adoption of particular proposals in SBSTA proceedings facilitates their ability to open up discussions of specific expert advisory processes—for example, to raise competing perspectives on how to such advisory processes should be structured to provide credible or authoritative knowledge. SBSTA's institutional design thus opens the possibility of exposing underlying value choices that might otherwise remain hidden. In short, SBSTA provides a forum in which countries that might not otherwise be involved in organizing scientific advisory processes can articulate publicly (backed up with the force of their frequently nonexercised veto) just what kind of expertise would facilitate their own efforts to respond effectively to climate change. Ignoring their voices would weaken the legitimacy of the climate regime, even if the matter never came to a formal vote. At the same time, by helping to reveal important normative disagreements— over such issues as participation, transparency, and priority—earlier rather than later in the design of scientific advisory arrangements, SBSTA may help avoid subsequent losses of public trust and legitimacy.

On the flip side, observers and participants have criticized SBSTA's consensus rules of procedure. SBSTA's "one vote–one veto" rule is largely responsible for its slow progress in establishing working scientific advisory arrangements. Compared to its sister organization under the Convention on Biological Diversity, for example, which has an almost identical treaty mandate, SBSTA has moved at a glacial pace. An important upshot of participants' difficulties in constituting advisory arrangements under SBSTA has been that the progress of negotiations in some areas of the regime has slowed when countries have insisted on waiting for expert advice that SBSTA could not yet provide. However, in an issue area such as climate change in which the norms and practices of global environmental science and management are jointly contested, uncertain, and in rapid flux, forcing governments to address both the normative and the practical dimensions of science advice may have added to the overall credibility of the climate regime. Over the long haul, SBSTA can be accounted successful if its scientific advisory arrangements, although taking longer to get started, produce knowledge that is more widely credible among diverse populations around the globe.

Additionally, in some cases, the failure to submit questions about scientific advisory arrangements within the climate regime to SBSTA for resolution has carried important consequences. For example, on occasion, scientific and technical advisory groups have chosen not to submit difficult and contested issues to SBSTA, for fear of the stalemate that might result, only to have their decisions later come back to haunt them. One area where this has occurred is in the construction of methodologies for counting national emissions of greenhouse gases. Choices among competing methods typically have clear implications for the allocation of national responsibility within the climate regime. Consequently, they are understandably contentious. Participants in some of the IPCC working groups responsible for developing these methodologies have decided on several occasions not to submit questions about alternative methods to SBSTA out of a perception that such submission would entail onerous delays. In at least one case, however, this overly unreflective attitude toward SBSTA backfired when unresolved issues erupted into public controversy and damaged the IPCC's credibility in ways that might have been avoided by early negotiations within SBSTA.

Observers in other contexts have suggested that expert consensus in international negotiations is more easily obtained if expert agreement can be obtained prior to the politicization that occurs in high stakes policy decisions (Thacher 1976; Haas 1990a). This view privileges expert consensus too highly, however, if that consensus is achieved by preventing scientific claims and arrangements from being subjected to sufficient critical scrutiny. Since scientific knowledge inevitably embeds tacit values and assumptions, too little politicization early on in the policy process will often mean that important potential fault lines among participants do not get identified and resolved. Occasionally, these fault lines may later emerge in contested form when political tensions do rise, potentially contributing to losses of public trust and credibility among certain audiences. Too much early politicization, however, can prevent the constitution of advisory arrangements and also potentially detract from the overall legitimacy of the climate regime. From a reflexive perspective which recognizes that expert knowledge and advisory arrangements are implicitly value-laden, SBSTA seems to offer a reasonable compromise between too little and too much politicization. This compromise allows many sources of disagreement to be identified and resolved through widespread and active government involvement in the process of organizing science advice but has not yet hamstrung the regime's activities.

If we accept that one of SBSTA's most important contributions is to help enable states to negotiate among tacit, deeply held commitments to alternative strategies for constituting scientific advisory arrangements, then we can propose further additional features of institutional design that might enhance this function. One suggestion would be to raise the number of developing country participants actively participating in all scientific and technical advisory panels constituted within the climate regime. In coming years, the climate regime seems likely to increasingly shift its focus toward socially and environmentally sustainable strategies for reducing the emission of greenhouse gases and adapting to the consequences of climatic variability. As it does so, local knowledge about heterogeneities in human values, social practices, and natural conditions from place to place will rise in importance relative to knowledge about global systems—for reasons of both effectiveness and legitimacy. Since SBSTA itself will never succeed in fully revealing even a small fraction of the tacit commitments embedded in climate science advice, it is also important that individual advisory arrangements adopt similar critical approaches that empower voices to speak to local heterogeneity in nature and society.

A second suggestion would involve improving the communication channels between SBSTA and the scientific advisory processes it constitutes and interacts with. When value conflicts arise in the context of scientific advisory choices, greater communication would allow SBSTA's participants to provide advice to experts about their needs and views, and even to make explicit political choices should those become necessary. A final suggestion is to continue to enhance the ability of developing countries to participate effectively in SBSTA. Important normative and political issues are at stake in SBSTA's activities, yet many developing countries continue to find participation difficult due to a variety of factors—for example, the small number of representatives (often one) comprising their delegations to SBSTA; limited domestic expertise in important areas of scientific and technological research; and poorly developed channels of communication at home between citizens, delegates, and national expert and policy communities around issues relevant to SBSTA deliberations (see VanDeveer 1998 on East European participation in the scientific advisory arrangements for the acid-rain regime in Europe).

Learning and Confidence Building

The deconstruction of scientific advisory processes, if carried too far, can provide a basis for viewing science as inherently political and therefore render it untrustworthy in the eyes of some participants. To date, however, SBSTA appears to have managed to avoid this problem. Claims, for example, that Northern science represents a novel form of environmental colonialism—such as those made in 1991 by the Indian NGO, Centre for Science and the Environment, and picked up, in turn, in many developing countries during the negotiation of the Framework Convention—have been largely absent from deliberations within SBSTA. Although substantial debates have occurred over how to organize science advice, participants in SBSTA continue to express their support for the relevance of "neutral expertise" to the policy choices of the climate regime, and, as detailed above, they have achieved agreement on several arrangements for providing that expertise.

SBSTA's principal counter to its critical tendencies has been to take recourse in scientific advisory arrangements over which SBSTA participants retain considerable authority. For example, the shift from TAPs to the "Roster of Experts" and "informal working groups" discussed earlier illustrates how SBSTA participants have opted to retain the bulk of deliberations about methodologies and technology transfer within SBSTA's institutional procedures, rather than to delegate such deliberations to permanent advisory bodies which might be, to some degree, autonomous. Likewise, in using the "Roster of Experts," SBSTA participants have decided to consult with outside experts only for specific, collectively agreed-on questions, narrowly delimiting the mandate of requests for information in scope and time. SBSTA participants have also agreed to thoroughly review the "Roster of Experts," the activities of other expert working groups to whom SBSTA has delegated particular tasks (such as methodology construction or the conduct of surveys of technological needs in developing countries), as well as SBSTA's own decisions, on a fairly frequent basis.

This contingent, incremental approach has allowed participants in SBSTA to move forward in providing expert advice in some areas relevant to the climate regime without being held hostage by disagreement in other areas or over the development of more permanent institutions. As a result, participants from various countries have been able to test proposed strategies for acquiring science advice on limited scales, learning which ones satisfy their own criteria of legitimacy and "neutrality" and building confidence in SBSTA and its various advisory arrangements. Through the Roster of Experts, SBSTA has been able to secure expert input on several important issues related to methodologies and technology transfer that participants have generally viewed as credible and reliable. Similarly, SBSTA has been able to work out several arrangements with other international institutions to construct particular international standards relevant to the climate regime, including methods for counting greenhouse gas emissions, assessing climate impacts, and evaluating atmosphere-friendly technologies. Meanwhile, SBSTA's informal working groups have enabled government representatives to continue to deliberate about the need for more permanent arrangements, to present alternative proposals for such arrangements, and to negotiate settlements of contested issues.

The creation of permanent, independent expert advisory institutions might, if achievable, carry certain advantages in terms of long-term consistency and capacity building. The provision of strong, independent sources of action and authority within the climate regime might also diffuse the power of governments. This could, if carried out in appropriate ways, strengthen the voices of a variety of nongovernmental actors. The danger of such institutions, however, particularly in clearly demarcated scientific domains, is that they will unreflectively reiterate narrow Northern perspectives due to the inhomogeneous distribution of scientific and technological capacity around the world. In an arena with widespread global economic and political implications, the world may thus be better served by a slower, incremental approach that ensures that developing country voices (and NGO voices, to some degree) are heard, however weakly. Over time, SBSTA has demonstrated that its relatively unwieldy approach can help governments to work out their differences. Permanent, semi-independent institutions seem likely to be much more difficult to change in response to evolving experience with their use or to changing circumstances than the kinds of arrangements SBSTA has generated so far. Given the rapid changes occurring in global governance, this flexibility may help SBSTA remain effective over a much longer time period and much broader range of responsibilities.

Warranting Credibility
SBSTA's third strength is its ability to make use of a wide variety of systems of rhetoric and practice for warranting knowledge claims. As noted earlier, culturally specific norms and practices for securing trust and warranting truth often vary considerably from country to country. If the provision of science advice in international institutions can draw on these national systems of warrant in appropriate contexts, and can find ways to accommodate or integrate them in international policy-making discourses, it may significantly enhance the credibility of their knowledge claims among diverse publics around the world. In general, Western analysts and policymakers have tended to assume that systems of warrant for scientific knowledge are (and should be) entirely internal to science. Research in science studies has revealed, however, that political norms and institutions may also play an important role in enhancing the credibility of policy-relevant expert knowledge. For example, the rise of social regulation in the United States in the 1970s demanded not only that policymakers make more intensive use of scientific knowledge in public decisions but also that public institutions become involved in setting standards for science in policy contexts. In this way, political institutions became responsible for deciding what would constitute legitimate claims to public knowledge (Jasanoff 1990).

Participants in SBSTA have structured its activities to draw on at least three repertoires of rhetoric and practice for warranting knowledge.

First, SBSTA's institutional design mobilizes the political norms of democratic participation and consensus rule-making to strengthen the credibility of scientific advisory arrangements about which participants are able to secure collective agreement. SBSTA is open to the participation of any government that has signed the Framework Convention. It is also open to the participation of representatives of any NGO that has registered with the Framework Convention Secretariat. This openness, combined with the requirement that decisions in SBSTA be achieved by consensus among governments before any action is taken, helps secure the organization's credibility, at least among governments. Proposed scientific advisory arrangements that are ultimately agreed on thus enjoy strong political backing at the outset, although there is no guarantee that this backing will continue as their work proceeds.

Perhaps the most important example of SBSTA's ability to mobilize political norms to enhance the credibility of scientific advisory arrangements relates to the IPCC. Although it had brought together many leading climate experts to participate in its activities, the IPCC was, by 1993–94, on the verge of becoming irrelevant within the climate regime. Developing countries, distrustful of the panel's processes for including their participation, had led the move to create SBSTA and to oppose any future role for the IPCC in climate policymaking. By acting as a buffer organization between the IPCC and the Conference of the Parties, SBSTA has been able to temper criticism of the IPCC in two ways: (1) by enabling developing countries to pose questions to the IPCC through SBSTA, thus providing a space for developing countries to voice alternative views about the IPCC's choices about what areas of expertise to prioritize; and (2) by requiring that IPCC reports receive SBSTA approval before being brought to the attention of the Conference of the Parties. In this way, the IPCC's tendency to reinscribe Western ideas of nature and governance is diluted, while its authority to make claims about the need for international cooperation in working out new global governing arrangements is strengthened (see also Miller forthcoming).

Second, SBSTA has helped governments mobilize culturally specific systems for warranting knowledge in several areas by devolving responsibility for knowledge production to individual governments. In adopting this strategy, SBSTA has faced an explicit trade-off between tight, uniform international standards, on the one hand, and divergent national expectations about the production of public knowledge on the other. The development of international standards has been credited with a variety of positive outcomes in international cooperation (Chayes and Chayes 1995). Standards can help harmonize state practices, potentially rendering state behavior more transparent and permitting easier and more efficient coordination in global policymaking. Standards can also help states build capacity to implement global agreements. At the same time, however, the use of standards can also backfire. If standards come to be seen as resting on inappropriate assumptions or as reflecting inappropriate power relations, their use can damage the credibility of the regime. And the misapplication of standards to domains where they do not apply can lead to policies that blatantly disregard aspects of perceived reality (see, e.g., Zehr 1994, describing how early satellite data illustrating the ozone hole was regularly ignored as spurious). Finding universally acceptable standards can also prove difficult, as SBSTA's experience has demonstrated, as deeply embedded values and interests shape countries' perspectives on ideas of great importance to standards development, such as causality, agency, and responsibility.

For these reasons, SBSTA has pursued the development of international standards but has opted in nearly all circumstances to avoid making those standards binding on national governments. In developing standards for measuring national emissions of greenhouse gases, for example, SBSTA, as pointed out earlier, adopted international standards developed by the IPCC as default guidelines only. In addition, national governments, and not the IPCC or any other international body, are responsible for compiling inventories of national emissions. Combined, these two provisions have allowed national governments to deviate dramatically from the IPCC standards, and many have subsequently done so where they believe they have a method that is more credible for their own national context. Similarly, although national inventories are subject to international review once submitted to the Conference of the Parties, participants in SBSTA have decided that the outcome of that review will be facilitative only and not legally binding. Consequently, national governments retain almost complete autonomy in the production of national inventories of greenhouse gas emissions. This could be read merely as an exercise of political control over potentially sensitive numbers (designed, even, to allow countries to fudge their data). However, the radically divergent approaches that states have adopted in preparing their inventories suggests that giving countries autonomy also allows them to compile their data in ways that are consistent with nationally specific expectations for the production of public knowledge. Time, and concerted policy actions to hold countries uniformly accountable for their emissions, may enable the nation-specific approach to achieve an eventual and more trustworthy convergence. Or they may reveal deep fault lines between national approaches that reflect important value commitments in need of negotiation.

Finally, SBSTA participants have also, in some areas, succeeded in mobilizing other systems of warrant by retaining a degree of flexibility in SBSTA's approaches to configuring scientific advisory arrangements. For example, as described earlier, the initial proposal to create two new technical advisory panels (TAPs) alongside the IPCC generated strong disagreement over the criteria to be used to determine their membership. Industrialized countries generally favored disciplinary balance. Developing countries generally favored regional balance. This disagreement ultimately proved to be fatal to the proposal. Participants responded by creating a "Roster of Experts" from which the Framework Convention Secretariat selects an "appropriate" panel of experts to address particular issues questions agreed on in SBSTA deliberations. This has created sufficient flexibility in the process to allow several different criteria (geography, area of expertise, developed/developing, and so on) to enter into the selection of experts depending on the circumstances surrounding each specific task. For instance, for its report on current trends in technology transfer, six experts from developing countries were chosen. This flexibility helps to ensure an appropriate procedural response to contingency: committee membership can fluctuate from task to task and credibility for report conclusions can be flexibly promoted through the mobilization of political considerations such as involving representatives from specific countries or regions on the panel.

Conclusion

Conflict and controversy over science are often viewed as manifestations of the introduction of political bias into the interpretation of scientific data and theories. SBSTA itself is often criticized along these lines by participants in its activities. The difficulty participants have had in reaching agreement on arrangements for producing, validating, and using scientific knowledge within the climate regime is frequently taken as evidence of the body's politicization and consequent lack of scientific authority.

To evaluate SBSTA solely in terms of its ability to avoid conflict and to produce uncontested advisory arrangements, however, is to misunderstand both the nature of science advice and SBSTA's significance within the climate regime. The application of science to global affairs cannot be understood as a mere exercise in "speaking truth to power." Rather, challenges of contingency, trust, and moral order pervade efforts to use science to generate shared understandings of global environmental risks and to underpin planetwide arrangements for environmental management. The creation of SBSTA has led to an important innovation for dealing with these challenges within the climate regime by opening up a space in which issues related to uncertainty, credibility, and the politics of science can be debated and negotiated within the context of broader regime activities. Few if any participants in SBSTA's activities currently conceptualize SBSTA with this degree of reflexivity. Nevertheless, SBSTA has become a forum for negotiating temporary settlements of what will count as science, in practical terms, within the climate regime, and how science will relate to the formulation and implementation of global policy.

The detailed account of SBSTA's activities presented in this chapter thus illustrates the more general proposition with which I began the chapter: the principles and practices that define the meaning of concepts like "science" and "governance" are not shared widely around the world. Rather, to the extent that they come to be shared at all, it is as the upshot of ongoing negotiations over specific technical questions in particular institutional settings. Cognitive models of science and politics are thus *coproduced* alongside the specific institutional arrangements that link them to people's activities around the world. What environmental science comes to mean in global society, and how it comes to relate to global governance, will be determined largely through the incremental progress of organizations such as SBSTA.

SBSTA's experiences demonstrate, more than anything else, that policymakers cannot afford to take questions of institutional design for granted when organizing global science advice. SBSTA has made strides toward the establishment of globally shared knowledge about the environment and about policy responses to it by incrementally and contingently working out ways for governments to sort out their often deep-seated political differences. This has been particularly important across North-South divides within the climate regime. SBSTA's (to date limited) successes in this regard stem from its development of institutional practices that facilitate the questioning of particular scientific advisory arrangements; that are incremental and responsive to a broad range of participants' views; and that allow science-based policy legitimation to take advantage of a wide array of alternative rhetorics and practices for warranting knowledge claims in public discourse. These approaches have all helped SBSTA achieve agreement on approaches for acquiring expert advice in the climate regime that are broadly credible among governments from around the world. In this regard, SBSTA is particularly notable for its success in avoiding an unreflective, unintentional reinscription of Western ideas of both climate change and the right relations between science and governance in international institutions. These approaches might fruitfully be adapted in other, nonenvironmental regimes as well, such as the International Monetary Fund, whose dissemination of Western market-based models of financial exchange has achieved less than stellar success in integrating developing countries into the global economy.

SBSTA's institutional design is by no means perfect. However, as many of the examples described here indicate, choosing among scientific advisory arrangements always involves important trade-offs. Consequently, the kind of tentative, flexible, responsive, incremental approaches offered by SBSTA may very well be the best places to begin. This is particularly true in areas like climate change, for which the norms and practices of global governance are themselves contingent, uncertain, and in rapid flux and for which potential solutions will ultimately require commitments on the part of a significant majority of the world's inhabitants and a restructuring of some of the core functions of modern industrial economies. Western scientists and policymakers are familiar with the negotiated, partial, confidence-building approaches that have been necessary in addressing complex political problems like U.S.-Soviet arms control or Middle East peace. SBSTA's experiences suggest that similar approaches are equally, if not more, necessary for organizing science to address environmental policy problems on global scales.

SBSTA's activities can be improved, to be sure, through experience. For example, one clear assumption made in almost all of SBSTA's activities is that all national governments are as responsive to their citizenry as we expect governments in the West to be. In many countries, however, including many Western countries, communication channels and lines of authority and accountability between government officials and various groups within the state vary dramatically in their ability to foster responsive governance. Whether it is through SBSTA or some other institutional process, the success of efforts to promote sustainable development and environmental protection worldwide will ultimately depend on being able to subject these relationships to the same kind of scrutiny that a body like SBSTA has enabled for science advice on climate change.

Appendix: Documentary Sources

UN Documents
Documents of the Subsidiary Body for Scientific and Technological Advice, UN Framework Convention on Climate Change
FCCC/SBSTA/1995/3 Report of the Subsidiary Body for Scientific and Technological Advice on the work of its first session held at Geneva from 28 August to 1 September 1995.

FCCC/SBSTA/1996/2 Establishment of intergovernmental technical advisory panel(s).

FCCC/SBSTA/1996/3 National communications from Parties included in Annex I to the Convention: Report on the guidelines for the preparation of first communications by Annex I Parties.

FCCC/SBSTA/1996/4 Initial report on an inventory and assessment of technologies to mitigate and adapt to climate change.

FCCC/SBSTA/1996/5 Activities implemented jointly under the pilot phase: Options for reporting guidelines.

FCCC/SBSTA/1996/6 Scientific assessments: Cooperation with the Intergovernmental Panel on Climate Change.

FCCC/SBSTA/1996/7 Scientific assessments: Consideration of the second assessment report of the Intergovernmental Panel on Climate Change.

FCCC/SBSTA/1996/8 Report of the Subsidiary Body for Scientific and Technological Advice on its second session.

FCCC/SBSTA/1996/9 Communications from Parties included in Annex I to the Convention: Guidelines, schedule and process for consideration.

FCCC/SBSTA/1996/9/Add.1 Communications from Parties included in Annex I to the Convention: Guidelines, schedule and process for consideration, Addendum.

FCCC/SBSTA/1996/10 Progress report on issues in the programme of work of the Subsidiary Body for Scientific and Technological Advice.

FCCC/SBSTA/1996/10/Add.1 Progress report on issues in the programme of work of the Subsidiary Body for Scientific and Technological Advice, Addendum.

FCCC/SBSTA/1996/13 Report of the Subsidiary Body for Scientific and Technological Advice on the work of its third session Geneva, 9–16 July 1996.

FCCC/SBSTA/1996/15 Activities implemented jointly under the pilot phase. Uniform reporting format. Note by the secretariat.

FCCC/SBSTA/1996/16 Methodological issues: longer-term programme of work.

FCCC/SBSTA/1996/16/Add.1 Methodological issues: longer-term programme of work, Addendum.

FCCC/SBSTA/1996/17 Activities implemented jointly under the pilot phase. Update. Note by the secretariat.

FCCC/SBSTA/1996/18 Cooperation with the Intergovernmental Panel on Climate Change. Progress report.

FCCC/SBSTA/1996/19 Activities implemented jointly under the pilot phase. List of methodological issues. Note by the secretariat.

FCCC/SBSTA/1996/20 Report of the Subsidiary Body for Scientific and Technological Advice on the work of its fourth session.

FCCC/SBSTA/1996/Misc.1 Activities implemented jointly under the pilot phase. Views from Parties on a framework for reporting. Note by the secretariat.

FCCC/SBSTA/1996/Misc.3 Establishment of intergovernmental technical advisory panel(s). Comments from Parties. Positions of the Group of 77 and China, and of the United States of America. Note by the secretariat.

FCCC/SBSTA/1996/Misc.4 Scientific assessments. (a) Consideration of the Second Assessment Report of the Intergovernmental Panel on Climate Change. (b) Research and Observation issues. National Communications. Establishment of a roster of experts. Development and transfer of technologies. Comments from Parties. Note by the secretariat.

FCCC/SBSTA/1996/Misc.5 Methodological Issues. Comments from Parties and an international organization. Note by the secretariat.

FCCC/SBSTA/1996/Misc.5/Add.1 Methodological Issues. Comments from Parties and an international organization. Note by the secretariat.

FCCC/SBSTA/1997/2 Cooperation with Relevant International Organizations. Progress report on research and systematic observation. Note by the secretariat.

FCCC/SBSTA/1997/3 Activities implemented jointly under the pilot phase. Uniform reporting format. Note by the secretariat.

FCCC/SBSTA/1997/4 Report of the Subsidiary Body for Scientific and Technological Advice on the work of its fifth session, Bonn 25–28 February 1997.

FCCC/SBSTA/1997/6 Report of the Subsidiary Body for Scientific and Technological Advice on the work of its sixth session, Bonn 28 July–5 August 1997.

FCCC/SBSTA/1997/8 Cooperation with Relevant International Organizations. Monitoring of greenhouse gases in the atmosphere. Note by the secretariat.

FCCC/SBSTA/1997/9 Methodological issues. Progress report.

FCCC/SBSTA/1997/10 Development and transfer of technologies. Progress report.

FCCC/SBSTA/1997/11 Roster of experts. Experience of the secretariat in its use.

FCCC/SBSTA/1997/12 Activities implemented jointly under the Pilot Phase. Synthesis report on activities implemented jointly. Note by the secretariat.

FCCC/SBSTA/1997/14 Report of the Subsidiary Body for Scientific and Technological Advice on the work of its seventh session, Bonn 20–28 October 1997.

FCCC/SBSTA/1997/Misc.1 Technology and technology information needs. Comments from a Party. Note by the secretariat.

FCCC/SBSTA/1997/Misc.2 Cooperation with the Intergovernmental Panel on Climate Change. Long-term emissions profiles. Comments from Parties. Note by the secretariat.

FCCC/SBSTA/1997/Misc.3 Activities implemented jointly under the pilot phase. Uniform reporting format. Methodological issues. Comments from Parties.

FCCC/SBSTA/1997/Misc.4 Cooperation with the Intergovernmental Panel on Climate Change. Structure and content of the Third Assessment Report by the IPCC. Note by the secretariat.

FCCC/SBSTA/1997/Misc.5 Activities implemented jointly under the pilot phase. Submission by the Group of 77 and China. Note by the secretariat.

FCCC/SBSTA/1997/INF.5 Development and Transfer of Technologies. Proposal from a Party. Draft decision regarding the transfer of technology. Note by the secretariat.

FCCC/SBSTA/1997/INF.6 Roster of experts: Nominations to the roster.

Technical Papers

FCCC/TP/1997/1 Trends of Financial Flows and Terms and Conditions Employed by Multilateral Lending Institutions. First Technical Paper on Terms of transfer of technology and know-how.

FCCC/TP/1997/2 Methodological issues. Temperature adjustments. Technical Paper.

FCCC/TP/1997/3 Technological issues. Adaptation Technologies. Technical Paper.

FCCC/TP/1997/5 Methodological issues. Synthesis of information from National Communications of annex I Parties on sources and sinks in the land-use change and forestry sector. Technical Paper.

FCCC/TP/1998/1 Technical paper on terms of transfer of technology and know-how. Barriers and opportunities related to the transfer of technology.

Earth Negotiation Bulletin

The *Earth Negotiation Bulletin* provides summaries of ongoing international negotiations related to environmental issues. Copies of the *Bulletin* can be obtained at the website of the International Institute for Sustainable Development (http://www.iisd.ca/linkages). Individual authors and editors are listed in an appendix to each issue of the *Bulletin*. Negotiations related to the UN Framework Convention on Climate Change are summarized in Volume 12. Specific issues relevant to the activities of SBSTA include:

Volume 12. Number 23. 1st Session SBSTA & SBI. August 28–September 01, 1995. Geneva, Switzerland.

Volume 12. Number 26. 2nd Session SBSTA & SBI. February 27–March 04, 1996. Geneva, Switzerland.

Volume 12. Number 39. AGBM 5, SBI 4, SBSTA 4, AG13 3. December 09–18, 1996. Geneva, Switzerland.

Volume 12. Number 40. 5th Session SBSTA & SBI, AG13 4. February 25–28, 1997. Bonn, Germany.

Volume 12. Number 46. 6th Session SBSTA & SBI, AG13 5. July 28, 1997. Bonn, Germany.

Volume 12. Number 47. 6th Session SBSTA & SBI, AG13 5. July 29, 1997. Bonn, Germany.

Volume 12. Number 48. 6th Session SBSTA & SBI, AG13. 5 July 30, 1997. Bonn, Germany.

Volume 12. Number 49. 6th Session SBSTA & SBI, AG13. 5 July 31, 1997. Bonn, Germany.

Volume 12. Number 56. 7th Session SBSTA & SBI. October 20, 1997. Bonn, Germany.

Volume 12. Number 57. 7th Session SBSTA & SBI. October 21, 1997. Bonn, Germany.

Volume 12. Number 58. 7th Session SBSTA & SBI. October 22, 1997. Bonn, Germany.

Volume 12. Number 60. 8th Session of the AGBM, SBSTA 7. October 23, 1997. Bonn, Germany.

Volume 12. Number 61. 8th Session of the AGBM, SBSTA 7. October 20, 1997. Bonn, Germany.

Volume 12. Number 62. 8th Session of the AGBM, SBSTA 7. October 20, 1997. Bonn, Germany.

Volume 12. Number 63. 8th Session of the AGBM, SBSTA 7. October 20, 1997. Bonn, Germany.

Volume 12. Number 64. 8th Session of the AGBM, SBSTA 7. October 20, 1997. Bonn, Germany.

Volume 12. Number 66. 8th Session of the AGBM, SBSTA 7, SBI 7. October 20–31, 1997. Bonn, Germany.

Notes

1. The word *coproduction* typically refers to the mutual construction and reinforcement of nature and culture or society. In this chapter, I use *coproduction* to refer to the mutual construction and reinforcement of ideas about the world in which people live (whether they choose to view that world in social, natural, or some other terms) and the organization and practices of institutions that enable people to act in that world.

2. I use the term *climate regime* to refer to the institutions authorized by the UN Framework Convention on Climate Change and created to implement it: the Conference of Parties, Subsidiary Body for Implementation, Subsidiary Body for Scientific and Technological Advice, and the Advisory Group on the Berlin Mandate.

3. *Joint implementation* refers to projects carried out jointly by an Annex I (industrialized) country and a non–Annex I (developing) country for the purposes of reducing emissions in the developing country to offset obligations under the Framework Convention by the industrialized country.

4. The term *deconstruction*, as noted previously, is frequently used to mean efforts to highlight discrepancies and inconsistencies, emphasize uncertainties, and challenge the adequacy of experimental techniques or the motives of expert advisers. These tactics are commonly used to discredit scientific testimony and to point out the indeterminacy of scientific findings. Here, I expand this use of the term to include efforts to challenge the adequacy or appropriateness of expert advisory arrangements.

9

Climate Change and Global Environmental Justice

Dale Jamieson

The centerpiece of the 1992 Rio Earth Summit was the signing of the United Nations Framework Convention on Climate Change (FCCC). The signatories to the Convention, numbering more than 160 countries, committed themselves to the goal of achieving "stabilization of greenhouse gas concentrations in the atmosphere at a level that would prevent dangerous anthropogenic interference with the climate system." To begin to reach this objective, the Annex I countries agreed to voluntarily stabilize greenhouse gas emissions (GHGs) at 1990 levels by the year 2000.[1] It soon became clear that while the European Union was likely to keep its commitment, the United States, Australia, New Zealand, Japan, Canada, and Norway would not. In 1995 the parties to the FCCC adopted the Berlin Mandate: they pledged that by the end of 1997 they would reach an agreement establishing binding "quantified, limitation, and reduction objectives" for Annex I countries, and that no new obligations would be imposed on other countries during the compliance period. From December 1 to 10, 1997, the parties met in Kyoto, Japan, to try to negotiate the agreement.

In the run-up to Kyoto there was serious conflict between the developed and developing countries, as well as within both groups. The United States wanted an agreement that required stabilization at 1990 levels between 2008 and 2012. The European Union pressed for 15 percent reductions by 2010. On July 7, 1997, Foreign Minister Alexander Downer stated that "the only target that Australia could agree to at Kyoto would be one that allowed reasonable growth in our greenhouse emissions" (Australian Conservation Foundation 1998). The Alliance of Small Island States (AOSIS), whose very existence is threatened by sea-level rise, proposed stronger measures than anyone would accept. India and China were mainly concerned to avoid undertaking new commitments but some Latin American countries, such as Chile and Argentina, signaled a willingness to do more.

Less than five months before the Kyoto meeting, the U.S. Senate unanimously passed the "Byrd Resolution," directing the president not to sign any agreement that requires the United States to limit or reduce GHGs, unless the same agreement also "mandates new specific scheduled commitments to limit or reduce greenhouse gas emissions for Developing Country Parties within the same compliance period." The Clinton Administration, which had agreed to the Berlin Mandate, also supported the Byrd Resolution, declaring that it would strengthen the American hand at Kyoto. When President Clinton announced the American negotiating position on October 22, 1997, he stated that "the United States will not assume binding obligations unless key developing nations meaningfully participate in this effort" (Clinton 1997).

The Kyoto meeting was a "make-or-break" moment in the development of the international climate regime. Had the parties to the convention not been able to reach an agreement in Kyoto it is likely that the global effort to limit GHG emissions would have fallen apart. But after pulling some "all-nighters," the delegates to the conference managed to hammer out an agreement.

Four provisions are central to the Kyoto Protocol. First, the Annex I countries agreed to differentiated, binding targets that would reduce GHG emissions to about 5 percent below 1990 levels sometime between 2008 and 2012. Second, performance in meeting these targets will be assessed on the basis of "sinks" (e.g., tree planting) as well as "sources," and virtually all GHGs, not only carbon dioxide, will be taken into account. Third, emissions trading among Annex I countries and between Annex I countries and developing countries will be permitted. Finally, the Protocol reaffirmed that developing countries will not be subject to binding emissions limitations during the compliance period of this Protocol.

Hovering in the background are enormous uncertainties about how emissions can be monitored and how changes in sinks can be measured. But many important issues were left unresolved even in principle by the last-minute compromise reached in Kyoto. For example, the Protocol does not specify sanctions for nations that do not keep their commitments. Nor does it address the extent of emissions trading and exactly how it will be implemented. Since it is now clear that the United States as well as some other Annex 1 countries plan to reach their targets primarily by purchasing credits rather than by reducing emissions, this issue is rapidly moving to the center of the debate. While some progress has been made on these issues in subsequent meetings in Buenos Aires in November 1998, and Bonn in November 1999, most major decisions have been deferred.

Debates about climate change are as much about the distribution of wealth, power, and authority as they are about whether or not scientists have accurately depicted the natural and human systems that contribute to climate change. How we as individuals should act in the face of the rapid anthropogenic environmental changes that are now sweeping the globe with disastrous consequences for many of our contemporaries, future generations, and nonhuman nature is one of the most interesting and important ethical issues that climate change confronts us with. But just as important are the ethical questions that underlie our collective responses to climate change. For what is at issue in these debates are the moral principles that will govern future global environmental governance. As Miller and Edwards point out in chapter 1, rising atmospheric concentrations of carbon dioxide and other greenhouse gases raise important questions not only about the nature and extent of anthropogenic degradation of the global environment but also about "how people of vastly unequal technological capacity and means are going to live together on the planet."

In this chapter I will address some competing conceptions of global environmental justice that lie at the heart of the North-South debate about climate change. I argue that the post-Kyoto process must find ways of addressing contentious normative issues, including those bound up with scientific representations of nature, if we are going to be able to mobilize support among diverse and far-flung publics for the kinds of foundational social and economic changes that will be needed to seriously address climate change. As Sheila Jasanoff (1997a, 242) has observed, scientific knowledge can only achieve public credibility and "political authority when the boundaries of the relevant moral-political space [are] redrawn so as to accommodate the interests of all parties."

I begin with a brief discussion of the relations between scientific knowledge and conceptions of justice. I then characterize two competing conceptions of global environmental justice and go on to suggest a proposal of my own that incorporates various elements of them. Finally I speculate on how the climate change issue is likely to develop and draw some conclusions.

Scientific Knowledge and Conceptions of Justice

The question of relationships between scientific knowledge and conceptions of justice is a large one. The simplest thing to say (echoing Hume) is that an "ought" cannot be derived from an "is," and so insofar as science is a purely descriptive enterprise it can never tell us how the world ought to be. While this is true, it does not take us as far as it may seem. Logical derivation is one relation that may hold between beliefs, but it is not the only one. Moreover, it is hard to defend the idea that science is a purely descriptive enterprise. Scientific knowledge informs normative commitments, undermines them, and shapes them in all sorts of other ways both gross and subtle.[2] While I cannot provide a full account of the relations between science and conceptions of justice here, I will discuss some examples of these relations in the context of climate change.

One principle at work is this: In order for the language of justice to come into play, there must at least be the serious possibility of people significantly affecting each other's interests. What this means in the context of climate change is that if human activity does not pose a threat to climate stability, then questions of global environmental justice do not arise with respect to GHG emissions. This principle explains in part why the discussion of climate change has focused so centrally on science.[3] A strong case for skepticism about the reality of climate change undercuts the debate about what are morally acceptable (or required) policy responses.

A second general principle at work is that "*ought* implies *can*." People or nations cannot be morally obliged to do what they cannot do. Thus there can be no moral injunction that requires humans to fly or cats to be vegetarian. Similarly, some argue that the United States cannot reasonably be required to conform to the Kyoto reductions since it would not be possible for the United States to do so. This argument, of course, trades on treating a lack of political will as if it were equivalent to some strong form of impossibility such as physical impossibility.

Science enters the debate about justice in various other, more subtle, ways as well. Scientific debates about "the facts" are often proxies for debates about moral responsibility. For example, a recent study suggests that North America is such a large carbon sink that the United States may be neutral with respect to its contributions to concentrations of atmospheric carbon dioxide (Fan et al. 1998). Many scientists are disposed not to believe these results. Some worry that "groups opposed to the Kyoto treaty will seize on the estimate to argue that the United States doesn't need to reduce its emissions to comply with the accord" (Kaiser 1998, 386). Sure enough, Steven Crookshank of the American Petroleum Institute says that this study "calls into question the scientific basis on which we're making these decisions, when we still don't know if the United States is even emitting any carbon in the net" (Kaiser 1998, 387).

Scientific and moral authority are sometimes explicitly linked. The science most influential in shaping the discussion relating to the FCCC has been that collected and organized by the Intergovernmental Panel on Climate Change (IPCC). In 1988 the IPCC was convened by the World Meteorological Organization (WMO) and the United Nations Environmental Program (UNEP). The IPCC published major reports in 1990 and 1996, and has also produced a stream of occasional studies, assessments, and discussion papers. Hundreds of scientists from all over the world have been involved with the IPCC in various ways. The IPCC process is one of the most ambitious attempts ever mounted to mobilize science for the purposes of making international law and policy.

Despite its scientific credentials, different constituencies see the IPCC in different ways. Since most of the scientific expertise relating to climate change is in the industrialized world, and indeed in only a few highly specialized research centers (Hadley in the United Kingdom; CCCMA in Canada; Max Planck Institute in Germany; GISS, GFDL, Livermore, Lamont-Doherty, and NCAR in the United States; Macquarie University in Australia, among others), the IPCC's attempts at inclusiveness are seen by some as a noble dream and by others as a cruel deception. Academics who write in the tradition of science studies have seen the IPCC as a paradigmatic institution for creating apparently natural facts through the artful use of social processes (Jasanoff and Wynne 1998). Environmentalists typically view the IPCC as the voice of reason and dispassionate objectivity in the cacophony of greed and self-interest characteristic of environmental politics. But leftists who oppose "globalism" and rightists who are against "internationalism" have seen the IPCC as a malevolent conspiracy. Even some who champion the IPCC see it as a conspiracy, but one that is benign (Lahsen 1999). The credibility and authority of the IPCC rests, in part, on perceptions of its moral and political authority.

These observations may help to explain cross-national differences in the acceptance of scientific claims about climate change, especially those emanating from the IPCC. From the beginning the IPCC's conclusions were more warmly embraced in Europe than in the United States. This is part of the reason why during the 1990s most European countries were reducing GHGs while the United States was still increasing them.[4] By the early 1990s the European debate had already moved away from the science of climate change to discussion of policy responses.

Developing-country views of climate science illustrate how the credibility of science is conditioned by the moral and political structures of global society. One of the peculiarities of the climate change debate in the United States is that much of the science that supports the case for climate change has been produced in that country, yet skepticism about climate change is more influential in the United States than in any other industrial country. This has certainly damaged the credibility of the case for climate change in the developing world. Americans who have promoted the idea of climate change have not been able to convince many of their fellow citizens (e.g., the Republican Congress) of its reality, much less have they succeeded in motivating them to take significant actions in response. When this apparently self-refuting nature of American climate-change science is combined with the standing skepticism that many Third World people have about knowledge produced in the developed world, it is not surprising that many developing countries have refused to take the climate change issue seriously even though they may have the most to lose if climate change is in fact underway.

A final example of how perceptions of moral authority can condition particular scientific claims emerges from recent activities in the United States that have fragmented the unity of industrial opposition and helped move the American debate over climate change away from the question of whether climate is changing toward the question of what we should try to do about it. In a speech at Stanford University in 1997, Sir John Browne, CEO of British Petroleum, acknowledged the threat of climate change and went on to sketch some steps to address it. This was the first major crack in the united front that major corporations had erected to defend their interests against possible climate stabilization policies. In his October 22 speech later that year, President Clinton began to reframe the domestic political discussion of climate change when he declared that "the problem is real" (Clinton 1997). And although most of the American business community opposed the Kyoto Agreement, their coalition continued to fragment throughout 1998, the warmest year since at least 1856. In May 1998, the Pew Center on Global Climate Change was launched with the support of such corporations as American Electric Power, BP America, and Weyerhauser. The Pew Center explicitly accepts "the views of most scientists" and views Kyoto as the first step in addressing climate change. Throughout the unusually warm summer of 1998 the President and Vice President repeatedly linked various weather anomalies to climate change. On October 20, 1998, executives from General Motors, Monsanto, and British Petroleum held a joint press conference with the World Resources Institute in which they declared that "climate change is a cause for concern, and precautionary action is justified now" (*Greenwire* 1998). In 1999 such companies as Ford Motor Company followed Shell Oil, United Technologies, and BP Amoco in withdrawing from the Global Climate Coalition, the industry lobbying group that campaigns against climate change measures. Shell, United Technologies, and BP Amoco went on to join the Pew Center's "Business Environmental Leadership Council." Although a stubborn band of contrarians continues to have the ear of the Republican Congress, warmer temperatures and the breakup of the industrial coalition have fractured the moral authority behind critiques of climate science, and the discussion has dramatically shifted away from climate models toward economic models.

Global Environmental Justice: The View from Above

Differences between the United States and developing countries are grounded in competing views of global environmental justice. In this section I concentrate primarily on the American position on climate change.

In the United States the debate about reducing GHG emissions centers on self-interest and national interest rather than on appeals to morality. Studies have been thrown around about how compliance with the Kyoto Protocol would devastate the economy or put Americans at a competitive disadvantage. The debate has been dominated by hypersensitivity to the domestic politics of which sectors would win and which would lose as a result of controlling GHGs. There is also a backlash, at least in Congress, against government regulation, and the Kyoto Protocol is seen as underwriting government intrusion in the economy. In addition to these factors, the United States is in a period of confusion and ambivalence about its role in the world. This is well symbolized by a series of votes in April 1999, in which the U.S. House of Representatives both failed to support the NATO air campaign against Serbia, and voted twice as much money for carrying out the campaign as the president had requested.

In his October 22, 1997, speech President Clinton did make a moral appeal when he stated that "if there are dislocations caused by the changing patterns of energy use in America, we have a moral obligation to respond to those to help the workers and the enterprises affected." And although the president noted that "the United States has less than 5 percent of the world's people, enjoys 22 percent of the world's wealth, but emits more than 25 percent of the world's greenhouse gases," he did not cast this observation in the language of morality. Instead he went on to say that "we must begin now to take out our insurance policy on the future" (Clinton 1997). While American opposition to Kyoto might therefore be viewed as resting on national interests, these conceptions of interests are constructed against a backdrop of moral claims about such issues as efficiency, feasibility, and fairness.

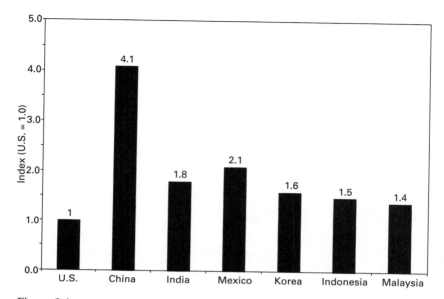

Figure 9.1
1995 Energy/GDP ratios for the United States and several developing countries.
Source: President's Council of Economic Advisers, 1998, 29.

The appeal to efficiency rests on the fact that developed countries generally produce more GDP per unit of energy than developing countries,[5] as can be seen from figure 9.1. Any policy that restricts energy use on the part of the developed countries will only lead to a more inefficient global economy, which will be bad not only for people in the industrial countries but for the world as a whole. The CEO of Exxon went so far as to visit China in the run-up to Kyoto, where he argued that restrictions on GHG emissions for developed countries was bad for China, since anything that is bad for the economies of the developed world will ultimately hurt China's export industries. He was not found to be terribly convincing.

A second argument appeals to feasibility. Just as the United States could not remain half slave and half free, so restricting the emissions of only a small number of countries and leaving the emissions of most countries uncontrolled cannot possibly be an effective permanent solution. Two arguments are given for why this is so. First, in the post-Kyoto world, energy-intensive industries in the developed countries will simply

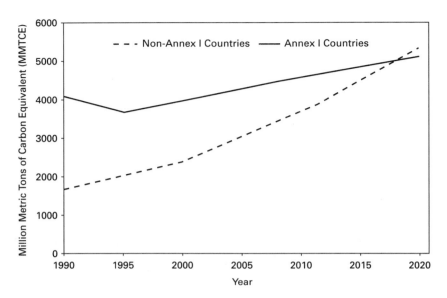

Figure 9.2
Projected emissions of Annex I and Non-Annex I countries without new abatement policies. Data represent million metric tons of carbon equivalent (MMTCE) from fossil fuel combustion. *Source*: President's Council of Economic Advisers, 1998, 11.

move offshore to escape emissions limitations. The same GHG emissions will occur, but the developed countries will no longer benefit from them. The second argument is that by 2020 the developing countries will be the largest emitters of GHGs, and over the next century their emissions will swamp those of the Annex I countries, as can be seen from figure 9.2, which graphs GHGs through time in million metric tons carbon equivalent (MMTCE).

The third argument against the Kyoto Protocol that is commonly given has a more distinctively moral tone: "It just ain't fair to single out a handful of countries for emissions control when the rest of the world goes unregulated." If climate change really is a global problem, then everyone in the world should do their part in solving it. The Kyoto Protocol violates a principle of equality that requires all countries to be treated the same, or (perhaps more subtly) a principle regarding the fair sharing of burdens.

Outside the United States these arguments are not seen as particularly persuasive. Probably most Europeans see the developed world as getting off easy in the Kyoto Protocol. Even countries like Australia, Norway, and New Zealand, which argued for increases in their emissions, did not for the most part argue on the basis of general principles like efficiency or equality. Instead they claimed that they were special cases and should be exempted from requirements imposed on larger, richer industrial countries with more diversified economies.

In my opinion, the most serious of these arguments given for the view from above is the second one. Anyone who appreciates the seriousness of climate change should recognize the importance of eventually bringing at least some of the developing countries into an emissions control regime. The question is how to do this in a way that is both fair and effective. Indeed, proposals not seen as fair by developing nations are not likely to be effective in bringing them into an emissions control regime (for a contrary view, see Victor 1999).

Global Environmental Justice: The View from Below

Like the Annex I countries, the developing countries are a large and diverse group of nations. I will focus mainly on the picture of global environmental justice that is commonly expressed in the climate negotiations by countries such as India and China. Roughly, their view is that if there is a problem about climate change, it has been caused by the developed countries; therefore, they are the ones who should be addressing it. This is really a version of the "polluter pays" principle. In this case the developed countries are the polluters, so they should pay to mitigate the damages.

This position can be bolstered by appeal to various principles of equality. One version of such a principle would ask which countries can afford to address GHG emissions without damaging the serious interests of their citizens. The answer would be the same one that the delegates to Kyoto arrived at: the developed countries. Another version of the appeal to equality would assert that reducing inequality between nations is in itself a good thing. So the provision of any global public good that requires developed but not developing countries to pay is a good thing.

The developing countries point out that throughout the nineteenth and twentieth centuries the nations of Europe and North America became rich by developing powerful industrial economies driven by fossil fuels. Four out of five pounds of carbon dioxide currently in the atmosphere were emitted by these countries. (Carbon dioxide has a residency time of 120 years in the atmosphere.) Although in recent years emissions have been growing in the less developed countries, the United States is still by far the world's largest emitter, with annual emissions one-third greater than China's and two-thirds greater than India's. Each year since 1990 the actual increase in American carbon emissions has been greater than that of any country except China and greater than the total carbon emissions of Brazil. On a per capita basis, Americans are responsible for more than seven times the emissions of the Chinese and eighteen times the emissions of the Indians. Moreover, it is unfair to equate the "necessary" emissions of the developing world, in their quest for minimally decent lives for their citizens, with the "luxury" emissions of the industrialized world produced by gas-guzzling second cars and overheated and overcooled homes and offices.[6]

It is easy to see why most developing countries are unwilling even to think about controlling their emissions until they see some real progress on the part of the developed countries in reducing theirs. The rich countries are the ones that have caused the problem, continue to emit more than their fair share per capita, and have the resources to develop and adopt alternative technologies. Moreover, they promised to stabilize their emissions at Rio in 1992 but failed to keep their promise. Even worse, it now appears that at least some Annex I countries are planning to fulfill their Kyoto agreements through clever bookkeeping rather than through emissions reductions. The Kyoto loopholes that make this possible involve counting sinks as well as sources in assessing performance, emissions trading, and reductions keyed to 1990 emissions levels.

The Australian government plans to take advantage of the first loophole. By planting trees (many of them plantation pines that will be harvested for timber) in areas in which native forest was cleared before 1990, Australia may earn enough credits for increasing its sinks to significantly increase its emissions. The United States plans to achieve 75 to 85 percent of its emissions reductions by purchasing credits from other countries, many of them from the former communist countries (President's Council of Economic Advisers 1998). The latter's GHG emissions have declined precipitously since 1990 as a result of their collapsing economies, leaving them with many salable permits. However, these permits really reflect what is sometimes called "hot air" (rather than emissions that would actually occur if they were not sold). Thus emissions trading not only allows some countries to avoid reducing their own emissions at home, but also results in GHG emissions that would otherwise not occur. Paradoxically the Kyoto Protocol has the potential for increasing GHG emissions rather than reducing them. This enrages environmentalists and many in the European Union. It also does little to convince the developing world that the rich countries are serious about climate change.

A Modest (?) Proposal

Progress within the climate regime seems likely to continue to elude negotiators so long as North and South continue to remain deadlocked over such basic moral issues. Is it possible to envision a way forward? In this section, I offer a modest proposal. The proposal is modest in that it conjoins two ideas that are very much alive in the policy world, each of which has influential supporters. However, the conjunction that constitutes this proposal has not been forcefully advocated because those who support one conjunct typically oppose the other. The proposal has been discussed by a number of authors, however, in varying degrees of detail (see Rayner et al. 1999 and the literature cited therein). This modest proposal is important both for its intrinsic fairness, and because it provides an illustration of how an explicit focus on normative principles might help shift grounds of debate, thus opening new opportunities for settlements that can bridge rather than exacerbate existing differences.

In my view both developed and developing countries have a point. The emphasis on efficiency being promoted by the United States is potentially good for the world as a whole. But the emphasis on equality on the part of the developing countries seems to me to be morally unassailable. The main problem with emissions trading as it is developing is that no serious thought is being given to what might be called the end game and start game: the total global emissions that we should permit and how permissions to emit should be allocated. I propose that we give the Americans what they want—an unrestricted market in permits to emit GHGs—but that we distribute these permits according to some plausible principle of justice.

What would be such a principle? I can think of the following general possibilities:

1. Distribute permissions on a per capita basis

2. Distribute permissions on the basis of productivity

3. Distribute permissions on the basis of existing emissions

4. Distribute permissions on the basis of some other principle

5. Distribute permissions on the basis of some combination of these principles

Principles 4 and 5 are principles of last resort, and principle 3 is implausible.[7] The existing pattern of emissions primarily reflects temporal priority in the development process, rather than any moral entitlement. In general, it is hard to see why temporal priority in exploiting a commons should generate any presumptive claim to continue the exploitation. Suppose that I started grazing a large herd of cows on some land that we own together before you were able to afford any cows of your own. Now that you have a few cows you want to graze them on our land. But if you do, some of my cows will have to be taken off the land and as a result I will be slightly less rich. Therefore, I demand compensation. Surely you would be right in saying that since we own the land in common you have a right to your fair share. The fact that you have not been able to exercise that right does not mean that you forfeited it.

Principle 2 has a point. Surely we would not want to allocate emissions permissions toward unproductive uses. If the world can only stand so many GHG emissions, we have an interest in seeing that they are allocated toward efficient uses. But what this point bears on is how emissions should be allocated, not on how they should initially be distributed. Markets will allocate permissions toward beneficial uses. But it is hard to see why those who are in a position to make the most productive use of GHGs should therefore have the right to emit them for free. This is certainly not a principle that we would accept in any domestic economy. Perhaps if you owned my land, you would use it more productively than I do. For this reason you have an incentive to buy my land, but this does not warrant your getting it for free.

In my opinion the most plausible distributive principle is one that simply asserts that every person has a right to the same level of GHG emissions as every other person. It is hard to see why being American or Australian would thereby give someone a right to more emissions, or why being Brazilian or Chinese should give someone less of a right. The problem with this proposal is that it provides an incentive for pronatalist policies. A nation can generate more permissions to emit simply by generating more people. But this problem is easily addressed. For other purposes the FCCC has seen the importance of establishing baseline years. There is no magic in 1990 as the reference year for emission stabilization. But if 1990 is a good year for that purpose, let us just say that every nation should be granted equal per capita emissions permissions, indexed to its 1990 population. If you do not like 1990, then index to another year. It is important to my proposal that per capita emissions be indexed to some year, but exactly which year is open to negotiation.

Four problems (at least) remain. First, in indexing emissions to 1990 populations I am in effect giving the developed countries their historical emissions for free. But don't the same considerations that suggest that everyone who was alive in 1990 should have equal permissions apply to everyone who has ever lived? There is some force to this objection. But knowledge of the consequences of GHG emissions does seem to some extent to be morally relevant. Suppose that when my mother grazed her cows on our common property, the world was very different. Neither of us thought of what we were doing as eroding common property. Indeed, neither of us thought of the area on which the cows were grazing as property at all. I benefited from the activities of my mother, but neither your mother nor mine were aware of any harms being produced. If my mother had been cleverer perhaps she would have asked your mother for the exclusive right to graze cows on this piece of land. Perhaps your mother would have granted it because she had no cows and did not think of land—much less this land—as property, much less her property. Suppose that I say that since we now have different understandings, I am going to set matters right, and that henceforth you have an equal right to graze cows on our land. I acknowledge that if I am to graze more cows than you I will have to buy the right.

I think many people would say that I have done enough by changing my behavior in the light of present knowledge. Perhaps others would say that there is still some sort of unacknowledged debt that I owe you because of the benefits I reaped from my mother's behavior. But what I think is not plausible to say is that what my mother did in her ignorance is morally equivalent to my denying your right to use our land to the same extent that I do. For this reason I do not think that historical emissions should be treated in the same way as present and future emissions. The results of historical emissions are also so much a part of the fabric of the world that we now presuppose that it is difficult to turn the clock back. At a practical level both Australia and the United States have had a difficult time in determining what compensation they owe their indigenous peoples. Determining the effects of unequal appropriation of the atmosphere would be even more difficult.

The second problem is that some would insist that it matters where GHG emissions occur, not because of their impact on climate, but because of their effects on quality of life. High quality of life, it is argued, is associated with high levels of GHG emissions. But what this objection brings out is that a bad market in emissions permissions would be worse than no market at all. In a properly functioning market, nations would only sell their emissions permissions if the value of the offer was worth more to them than the permission to emit. But while no international market in emissions permissions could be expected to run perfectly, there is no reason to think that they cannot run well enough to improve the welfare of both buyers and sellers.

A third objection would come from some developing countries. My proposal brings them into the regime before developed countries have taken the first steps to reduce emissions. True enough, but the developing countries have the most to lose from climate change. If the regime is to be effective they will have to enter at some point anyway, and the terms that I have proposed are the most favorable ones they will get by a long shot.

This leads to the problems of monitoring, enforcement, and compliance. These are difficult problems for any climate regime. Perhaps they are more difficult for the regime that I suggest than for others, but I think that it is clear that any meaningful emissions control will require vast improvement in these areas.

The scheme that I suggest has many advantages. It would stabilize emissions in a way that would be both efficient and fair. It would also entail a net transfer of resources from the developed to the developing countries, thus reducing global inequality. Too bad that it does not have much chance of being adopted.

Back Down to Earth: Prevention vs. Adaptation

By the end of September, 2000, the Kyoto Protocol had been signed by eighty-four countries and ratified by twenty-nine. What happens in the United States is especially important since, in order to take effect, the Protocol must be ratified by fifty-five countries that together emitted 55 percent of GHGs in 1990. Since the United States was responsible for about 38 percent of the world's emissions in 1990, the Protocol is unlikely to become law without American ratification (Bolin 1998).

The United States signed the Protocol in November 1998, but the Clinton administration has stated that it will not submit the Protocol to the Senate for ratification until it gets "key developing countries to mean-ingfully participate." At the Fourth Conference of the Parties in Buenos Aires, Nauru, Niue, Argentina, and Kazakhstan volunteered to reduce the rate of their increases in GHGs on the same timetable as the Annex I countries. A good case can also be made for supposing that many developing countries, including China, are taking steps to reduce the rate of their increases of GHG emissions (Knight 1998). But clearly the American government is unimpressed. The administration's standard of "meaningful participation" is weaker than the Senate's requirement that there be an agreement than mandates specific limitations or reductions for developing countries, but the administration wants more from the developing countries than it got in Buenos Aires before it is willing to argue the case for ratification. Senator Hagel, the leading Republican sponsor of the Byrd resolution, was openly dismissive of Argentina's commitment to reduce its rate of increase in GHG emissions. Ultimately, however, what is important about the Byrd Resolution is not its language but rather its unanimous passage, which indicated just how much opposition there was to the Kyoto Protocol in the U.S. Senate. The 1998–2000 Republican Senate was unlikely to ratify the Protocol under any circumstances, and even if the Democrats had taken control of the Senate in the elections of 2000 it would still have faced serious opposition. However, even if the United States succeeds in extracting more concessions from developing countries and the Kyoto Protocol is submitted and ratified in the future, it is far from clear that the United States would be able to keep its commitments. The heavy reliance on emissions trading in the administration's plan is in part a response to the enormous political, economic, and social obstacles to trying to get Americans to change their energy-profligate ways. The *New York Times* has reported that "since the early 1970s, as the average household has shrunk by a sixth, the average new home has grown by a third" (October 22, 1998, A-1). And because American GHG emissions have been on an upward trajectory since the country's 1992 promise to reduce emissions, meeting the Kyoto target will require a reduction of more than 30 percent from "business-as-usual" scenarios.

Even if every country kept its Kyoto commitments, the effect on the atmosphere would be relatively slight. In 2010 the difference between a world of perfect conformity to the Kyoto Protocol and a world with no such agreement at all is 1.5 parts per million (ppm) of carbon dioxide, the difference between an atmospheric concentration of 382 ppm and 383.5 ppm (Bolin 1998). To put the point in perspective, in the 1990s we have had a 0.25°C warming from the 1961–1990 baseline; perfect conformity to the Kyoto agreement would reduce the expected warming in 2050 from 1.4°C to 1.395°C (Parry 1998).

But much more may be at stake than 1.5 ppm of carbon dioxide or .005°C of warming. In the positive scenario the FCCC process is like the ozone regime. The initial highly publicized agreements at Vienna and Montreal were not sufficient for repairing the ozone layer, or even to prevent further depletion. The really strong agreements to phase out CFCs happened later, in London and Copenhagen, with much less publicity. From this perspective what is important is to bring the countries of the world into an emissions-control regime by getting them to agree to take the first steps, and then to revisit this agreement as the science develops and the consequences of a warming are felt. Moreover, mandatory limits, even if they are relatively weak at present, send signals to the markets that the era of fossil fuels is over. New investment will move away from fossil fuels toward renewables, bringing about technical innovation and lower prices.

However, the more likely outcome (in my opinion) is that the attempt to control GHG emissions will fall apart. Diplomats may continue to fly around the world having meetings and coining acronyms, but none of this will matter much on the ground. Or perhaps in a fit of international honesty we may just give it all up. Either way, the world will turn toward adaptation. Indeed, influential voices in the research community are calling for just this shift in focus (Rayner and Malone 1997; Pielke 1998; Parry 1998).

Traditionally it is said that there are three options in responding to climate change: prevention, mitigation, and adaptation. But if the science is at all credible, then for some time prevention has not been an option. The debate is over mitigation. Will the world succeed in significantly mitigating climate change, or will we have a global policy of adaptation?

There is a lot to be said for adaptation. Most countries are not currently well adapted to the variability that is part of a stable climate regime.[8] But a policy of adaptation without mitigation runs serious practical and moral risks. The practical risk is that a GHG forcing may drive the climate system into some unanticipated, radically different state, to which it is difficult to adapt.[9] The moral risk is that a policy of adaptation will be one that hits the developing countries hardest. For a global policy of adaptation is an expression of the "the polluted pay" principle rather than the "polluter pays" principle. This is because adaptation policies are typically national or subnational and require resources and knowledge. Since the developed countries have resources and knowledge, they will succeed in adapting to climate change. Since the developing countries do not have (the right sort of) resources and knowledge, they will suffer the worst effects of climate change.

We could try to internationalize adaptation by creating a global fund that countries contribute to on the basis of their GHG emissions and make withdrawals from on the basis of the climate change impacts that they suffer. Indeed, something like this is supported by those in the research community who champion adaptation. Even more grandly we could envision, like Al Gore in chapter 15 of his book *Earth in the Balance*, a "Global Marshall Plan" aimed at "heal[ing] the global environment" (Gore 1992). If such a plan were focused on reducing the vulnerability of susceptible people to climate-related extremes, there is even some reason to believe that it would have some resonance with Western publics. The vocabularies of "at risk populations" and "humanitarian assistance" are already at play in the West, and the United States, for example, funds the International Research Institute for Climate Prediction in an effort to help developing countries respond to El Niño–like events (Glantz and Jamieson, forthcoming). Still, the rich countries, especially the United States, have the political equivalent of attention deficit disorder. Grand promises are made and big money is promised when a hurricane devastates Honduras (for example), but all this is forgotten when the next humanitarian crisis erupts. A "Global Marshall Plan" would require a level of non-crisis-sustained commitment that most Western societies seem incapable of maintaining. My gloomy conclusion is that if we had the moral and political resources to internationalize adaptation, the Kyoto Protocol would succeed and we could effectively mitigate the effects of climate change. The positive scenario envisioned by the optimists would prevail, and we would not need to focus our attention solely on adaptation.

Conclusion

In this chapter I have examined one of the ethical problems posed by global climate change. I have argued for one approach to mitigating climate change that would be both just and efficient. Unfortunately I am not sanguine about its prospects. A more likely option is an approach to mitigation that is not very responsive to concerns about justice, or what is even more likely, a slide toward adaptation, an outcome that is likely to be the most unjust of all.[10]

Notes

1. The Annex I countries are the members of the European Union (EU), the other members of the Organization for Economic Cooperation and Development

(OECD), and the "countries in transition," a euphemism for the European countries that were formerly communist. The Annex I countries are also known as "the rich countries."

2. For an excellent discussion of how scientific knowledge can undermine normative beliefs, see Rachels 1990.

3. But it is not the whole of the explanation. Americans have a deep cultural tendency to displace their moral and political differences onto scientific discourse. I have discussed this with respect to climate change in Jamieson 1990, 1991, 1992.

4. The usual political reasons were in play as well, such as Thatcher's desire to break the power of the coal miners' union.

5. All figures in this chapter are from the President's Council of Economic Advisers 1998.

6. For the claims and arguments reported in this paragraph see the work of the Centre for Science and Environment, especially their report *Politics in the Post-Kyoto World*, available at http://www.oneworld.org/cse/html/cmp/cmp334.htm.

7. Henry Shue (1995) has proposed a version of 4, and Eileen Clausen and Lisa McNeilly (1998) have proposed a version of 5. These are serious proposals that would have to be considered in a fuller treatment of this subject. For further discussion of these and other proposals see the papers collected in Toth 1999, and various papers by Michael Grubb, notably Grubb 1995.

8. The climate impacts community has argued this position for many years. For a sample of this literature, see the work of Michael Glantz and his colleagues in the Environmental and Societal Impacts Group at the National Center for Atmospheric Research (http://www.esig.ucar.edu).

9. The idea that a GHG warming may trigger a "climate surprise" is associated with the work of Wallace S. Broecker and his students and colleagues.

10. Earlier versions of this material were presented in lectures at the Society for the Social Studies of Science, the University of California at Davis, the Fondazione Eni Enrico Mattei in Venice, Italy, Politea in Milan, Italy, the Queensland University of Technology in Brisbane, Australia, Webster University in St. Louis, Princeton University, and Amherst College. In addition to those who took part in these discussions, I thank Roger Pielke Jr. (National Center for Atmospheric Research), James White (University of Colorado), various anonymous referees, and especially the editors of this volume, Clark A. Miller and Paul N. Edwards, for their hard work on various versions of the manuscript.

10

Image and Imagination: The Formation of Global Environmental Consciousness

Sheila Jasanoff

As the mood of the West turned retrospective and millennial in the final years of the twentieth century, it became clear that the images by which Western societies were defining the meaning of this stretch of history had shifted their form and emphasis—from pictures of division and conflict for the first three-quarters of the century to those of interconnectedness at its end. War and destruction dominate the frames through which we look at most of the past hundred years: the disjointed march of troops from nowhere to nowhere on the battlefields of the Great War, the emaciated bodies and charred cities of the second and wider World War, the eruptions of American firepower in the fields and villages of Vietnam, and the mass evacuations that presaged the killing fields of Cambodia. These images have not faded from our collective consciousness. Rather, they have gained secondary and tertiary currency through the commemorative efforts of contemporary historians, novelists, filmmakers, and museologists, all intent, it seems, on finding at this emotionally charged calendrical moment the appropriate visual languages to memorialize the century's vast conflicts.[1] One need think only of the controversies surrounding the U.S. Vietnam memorial and their resolution through Maya Lin's inspired and reflective roll call of names, the attack against perceived revisionism in the *Enola Gay* exhibit at the U.S. National Air and Space Museum (Harwit 1996), Stephen Spielberg's embrace of black-and-white cinematography in his 1993 opus *Schindler's List*, and the lengthy, emotional debates about how best to commemorate the Holocaust in Berlin, the once and future capital of reunified Germany.

Sometime during the last three decades of the century, however, images of connection, of dissolving boundaries, began to supplement, and at times crowd out, division in our visual and imaginative space. President Richard Nixon's controversial visit to China in 1972 was perhaps a starting point, providing compelling television footage of one of the world's most committed cold warriors visiting the shrines and monuments of the very nation he had fought so doggedly to isolate from communion with the West. The watershed year of 1989 brought additional stirring images, with the fall of the Berlin wall on November 9 signaling the official end of bipolar tensions and, to some, even "the end of history" (Fukuyama 1992). And on January 1, 1999, months before war-torn Kosovo gripped the television screens, the pictures of a new common currency, the Euro—as yet available only in virtual form—made concrete the extraordinary, voluntary ceding of sovereignty through which eleven European nations sought to erase the scars and trenches of the two world wars that had split their continent.[2]

One image perhaps more than any other has come to symbolize the Western world's heightened perceptions of connectedness at the end of the millennium: that of the earth suspended in a void, captured by the cameras of the U.S. space program beginning with the Lunar Orbiter in 1966 and culminating with *Apollo 17*, the last mission to land men on the moon.[3] The image confronts Americans today at every turn, from the revolving globe used as a background for so many televised, and now networked, news programs to the logo that wordlessly asserts the global reach of credit cards, airlines, automobile manufacturers, telephone companies, bookstores, academic programs, and virtually every other product or service that travels. It is also an image that catches the spirit of contemporary environmentalism, one of late modernity's signature social movements. The picture of the earth hanging in space not only renders visible and immediate the object of environmentalists' concern, but it resonates with the themes of finiteness and fragility, and of human dependence on the biosphere, that have provided growing impetus for environmental mobilization since the 1960s. It is as well a deeply political image, subordinating as it does the notional boundaries of sovereign power in favor of swirling clouds that do not respect the lines configured by human conquest or legislation. It is in this respect a fitting emblem of Western environmentalism's transnational ambitions.

While many have pointed to the image of Earth from space as an artifact that fundamentally altered human consciousness, there have been few systematic attempts to explore how, when, or to what extent such a transformation occurred, let alone how this potent visual resource interacted with other, more commonly recognized political forces (for example, scientific knowledge, economic interests, or hegemonic power) in the formation of shared environmental awareness (for an exception, see Sachs 1994). There are several reasons why it is important to fill this gap. To begin with, the televised distribution of standardized visual symbols, and visual language more generally, is creating a global communicative resource whose political implications demand closer exploration. Images may transcend cultural lines in ways that words cannot, thereby helping to create communities of meaning and shared responses or demands that cut across ordinary linguistic and governmental divides. More generally, there is growing interest in the social sciences in the power of visual representation to sway both belief and action (Scott 1998). Sight moreover, like any sense, is now seen as something that has to be manipulated and disciplined in order for people in the aggregate to see things in the same ways.[4] The politics involved in constructing common vision has accordingly begun to draw attention.

This chapter, then, is a study of the reception of the image of planet Earth in American and, through U.S. mediation, international environmental politics. At a theoretical level, this project can be seen as contributing to the interpretive turn in international relations theory by attempting to understand better the role of ideas in promoting transnational cooperation and conflict (Keohane 1988; Haggard and Simmons 1987; Haas 1990b; Litfin 1994). More specifically, it extends earlier work on epistemic communities by probing, within visual culture, one possible source of shared beliefs about the environment.[5] At the same time, the project also fits comfortably within the core research agenda of science and technology studies: it explores the creation of new knowledge about nature and its diffusion to varied audiences through technologically mediated visual representations.[6] The chapter's organization reflects these paired objectives. I begin by reviewing major strands in international relations theory and science and technology studies that deal with image making and its power to foster shared social and political awareness. I then successively discuss the emergence of Earth consciousness in postwar politics, the early history of responses to the image of planet Earth, and its later thematization and uptake in the discourses of risk, politics, economics, and ethics. The chapter ends with reflections on the merits of the planetary image as a resource for global action to protect the environment.

Common Vision, Concerted Action

What makes people from different societies believe that they should act to further common goals, even if these goals require them to sacrifice or postpone perceived economic and social interests? In a world in which political will has classically been exercised through national institutions, how can we account for the rise of transnational coalitions, such as the contemporary environmental movement, that seem to articulate their objectives in defiance of the positions of nation states? How, more generally, do people form commitments to collective action on a global scale, and from where do they derive notions of an international common good that are strong enough to override the intense but parochial pull of national self-interest?

A promising place from which to begin addressing these questions is Benedict Anderson's influential work *Imagined Communities* (1991), which sought to explain how nationality became modernity's most compelling social identifier. Why, Anderson asked, has nationality proved to be such a peculiarly robust form of ideology, resisting for instance Marxism's brave attempts to reclassify world politics in terms of shared class allegiances? Why are people willing to go to war, courting death in defense of nationhood, and why do they agree to do this even when, as in the case of Indonesia, the entity that commands their loyalty is a loosely connected string of islands with no plausible claims to linguistic or cultural unity? Rejecting geographic determinism as inadequate, Anderson defined the nation as "an imagined political community—and imagined as both inherently limited and sovereign" (Anderson 1991, 6). The move from physical to imagined demarcations proved intensely liberating to theoreticians of the state. Anderson and his many followers were able to probe the mechanisms by which people come to think that they belong to something so invisibly put together as a nation, and which, in short, endow the concept of nationhood with meaning. The turn to "imagined communities" made it possible to encompass within a single theoretical frame such disparate manifestations of nationhood as Austria-Hungary, Indonesia, and the ultimately failed construct of postpartition Pakistan, its brackets not firmly enough welded through a shared Islamic faith to withstand the divisive force of intervening Hindu India.

Print capitalism plays a central role in Anderson's story of nationalism. Newspapers, he argued, exerted a profound pull on social imagination, making it possible for people in far-flung places to read about and react to the same events at the same time. The printed page became the instrument through which people who previously had no connection with each other could now *imagine* that they were part of a single community, experiencing and participating in a single national drama. Readers were bound together by the newspaper's inbuilt clock, which inexorably marked off the days (and, through morning and evening papers, even times of days), juxtaposed happenings from around the world in a collage of adventitiously related events, and rendered them obsolete the very next day with another collection of stories, equally random though united by the same seemingly inevitable logic.

Anderson originally ascribed to the controllers of the printed page an almost unlimited capacity to mobilize nationalism, but in the book's second edition he added a chapter, more Foucauldian in inspiration, on three other institutions of power—the census, the map, and the museum—through which modern states have tried to discipline their citizens' nationalistic imaginations. Through these institutions, enterprising states could manufacture or erase boundaries and histories, connections and divisions. A telling exercise in image making occurred at the fifteenth anniversary celebration of Cambodia's independence in November 1968, in honor of which

Norodom Sihanouk had a large wood and papier-mâché replica of the great Bayon temple of Angkor displayed in the national sports stadium in Phnom Penh. The replica was exceptionally coarse and crude, but it served its purpose—instant recognizability via a history of colonial-era logoization. "Ah, our Bayon"— but with the memory of French colonial restorers wholly banished. French-reconstructed Angkor Wat, again in "jigsaw" form, became . . . the central symbol of the successive flags of Sihanouk's royalist, Lon Nol's militarist, and Pol Pot's Jacobin regimes. (Anderson 1991, 183)

Nationalism as "logoization," imagination overwritten by image making—all possible, so Anderson argued, through a means of production that permitted images to be removed from context, made infinitely reproducible, and so implanted in people's minds as seedlings of national fellow-feeling.

This account of political community building strikingly resonates with work in science and technology studies that attempts to explain how scientific representations of the natural world acquire a hold on people's beliefs. No one perhaps has done more to illuminate this process than Bruno Latour, the French ethnographer and philosopher of science, for whom the investigation of scientific knowledge in the making has long been coextensive with trying to understand what gives scientific images and inscriptions their special persuasive power (Latour and Woolgar 1979). In one of his best-known expositions of the subject, Latour argues that the difference between "savage," or prescientific, and "civilized," or scientific, knowledge lies not so much in how people perceive reality but in the ability of modern science to *circulate* its perceptions by rendering them "mobile, flat, reproducible, still, and of varying scales" (Latour 1990, 45). Latour called the resulting inscriptions "immutable mobiles" because—unlike the *objects* that science observes (countries, planets, microbes)—*representations* of them (maps, photographic plates, Petri dishes) can move around in fixed forms created by the exertions of scientists.

To this point, there is a startling family resemblance between Anderson's logoized nations and Latour's immobilized inscriptions: both move, both can be mobilized, both are torn away from the specific circumstances of their production, gaining greater power through this erasure of history and context. Yet the two writers are profoundly at odds in their understanding of the forces that create mobility. For Anderson, capital is the prime mover. Without its support, states could not control the printing presses that produce the maps and images that impress themselves, in turn, on the awaiting minds of protonationalist citizens. Latour turns this argument on its head, insisting that it is the mundane craftsmanship of visualization that moves things and people, and so gives rise to power. In passages that read almost as if they were written to counter Anderson, or equally Foucault, Latour says that we continually misunderstand the relationship between science and power

because we take for granted that there exist, somewhere in society, macroactors that naturally dominate the scene: Corporation, State, Productive Forces, Cultures, Imperialism, "Mentalités," etc. . . . Far from being the key to the understanding of science and technology, these entities are the very things a new understanding of science and technology should explain. The large-scale actors to which sociologists of science are keen to attach "interests" are immaterial in practice so long as precise mechanisms to explain their origin or extraction and their changes of scale have not been proposed. (Latour 1990, 56–57)

"Capitalism," for Latour then becomes a special case of accumulation—that of money:

Thus capitalism is not to be used to explain the evolution of science and technology. It seems to me that it should be quite the contrary. Once science and technology are rephrased in terms of immutable mobiles it might be possible to explain economic capitalism as another process of mobilization and conscription. (Latour 1990, 59)

Latour's commitment therefore is to elaborating the details of scientific practice (not, characteristically, of science funding), the workaday routines of sampling and observation, recording and classification through which an entire Amazonian forest, for example, can be transformed into a tractable, movable catalog of soil types and plant varieties, and ultimately even a theory of causation to explain whether the forest is advancing or retreating (Latour 1995).

Does the aggregated power of money control the technologies of representation, which then function as levers of ideology? Or does mastery of the craftwork of representation cumulatively give rise to power, lodged, as Latour (1990, 59) memorably puts it, in "centers of calculation" that in effect *make* the world by circulating particular interpretations of it? Or is the relationship between image and imagination altogether more complex, requiring visual stimuli to resonate with cultures of interpretation that help define their ultimate meanings in time and place? Let us turn to an actual case. In this chapter we will investigate more exactly how a single image, that of planet Earth, became an inhabitor of Western consciousness and an icon, more particularly, of U.S. environmentalism. In doing so, we will not (following a typically Latourian program) seek chiefly to retrace the networks of rocket and satellite production, nor the labyrinths of the military-industrial establishment, that importantly enabled the making of the original image. Rather, we will focus on a less easily encapsulated dimension of the story, that of the image's transmission, uptake, and interpretation within disparate communities of discourse and action.[7] Unlike Anderson and Latour, both of whom link their studies of power primarily to the technologies of image making and circulation, I aim to look more closely at the imagination of the viewers—and the self-conscious consumers—of this potent symbol of human and natural interconnectedness. Without the participation of these ordinarily unsung actors, images would not be invested with the meanings that motivate political action.

Viewing Planet Earth

American commentators have frequently written of the transforming impact of the picture of Earth suspended in space, as captured on film by successive *Apollo* mission astronauts. The late astronomer Carl Sagan, whose televised program *Cosmos* won him something akin to cult status in the 1980s, was one of those who helped to popularize this theme:

While almost everyone is taught that the Earth is a sphere with all of us somehow glued to it by gravity, the reality of our circumstance did not really begin to sink in until the famous frame-filling *Apollo* photograph of the whole Earth—the one taken by the *Apollo 17* astronauts on the last journey of humans to the Moon. (Sagan 1994, 5)

Sagan selected a particular image as his icon, the one that shows Earth in the round, with no clouds concealing (to a culture familiar with conventions of mapping) the readily imaginable outlines of states in Arabia and the horn of Africa.[8] Environmentalists as a rule are inclined to agree with Sagan's judgment about the importance of that image. Writing in 1990, the ecologist Daniel Botkin said:

It is more than 20 years since the phrase "spaceship Earth" was coined and made popular and 20 years since the Apollo astronauts took this famous photograph of the Earth from space—a blue globe, enveloped by swirling white clouds, against a black background—creating an image of a small island of life floating in an ocean of empty space. (Botkin 1990, 5)

A remarkably similar point was made some years earlier by the World Commission on Environment and Development (WCED) in its influential report, *Our Common Future*:

In the middle of the 20th century, we saw our planet from space for the first time. Historians may eventually find that this vision had a greater impact on thought than did the Copernican revolution of the 16th century, which upset humans' self-image by revealing that the Earth is not the centre of the universe. From space, we see a small and fragile ball dominated not by human activity and edifice but by a pattern of clouds, oceans, greenery, and soils. Humanity's inability to fit its activities into that pattern is changing planetary systems fundamentally. (World Commission on Environment and Development 1987, 308)

The idea of a "scientific revolution" held particular appeal for others who commented on the *Apollo* picture. Laurence Tribe, at one time a critic of technology's instrumental rationality and later a constitutional scholar at Harvard Law School, remarked that this image—"the earth as a dramatically finite and surprisingly delicate blue-green globe" (Tribe 1973, 620)—had ushered us toward "the fourth discontinuity." This was a moment that displaced the human ego by making it conscious of the physical limitations of the place it inhabits. This decentering effect, Tribe and others have said, was on a par with three great intellectual discontinuities of the past: the Copernican revolution, which displaced the earth from the center of the universe; the Darwinian revolution, which displaced human beings from the pinnacle of the tree of creation; and the Freudian revolution, which exposed the workings of the unconscious mind and made humankind aware that we are not, after all, masters in our own house.

Continuing the theme of scientific revolutions, some suggested that the picture of our lonely planet brought about nothing less than a paradigm shift in ways of thinking about how the world works. Lynton Caldwell, a leading figure in the new environmentalism of the 1970s, explicitly took this position:

The change from the belief that the sun, moon, and stars revolved around the earth to the Copernican view of the earth's place in the solar system was a paradigm shift. The change marked by [the aftermath of *Apollo*] is from the view of an earth unlimited in abundance and created for man's exclusive use to a concept of the earth as a domain of life or biosphere for which mankind is a temporary resident custodian. . . . The newer view sees it as an ultimately unified system . . . that may supply man's needs as long as he observes the system's rules. (Caldwell 1990, 21)

Elsewhere, Caldwell linked the image to the internationalization of environmental policy:

The first landing on the moon on 20 July 1969 and pictures of the Earth from outer space brought to many people a realization that their environment had many of the characteristics of a closed system. "Spaceship Earth" became a metaphor, and "Only One Earth" was the motto of the 1972 United Nations Conference on the Human Environment. (Caldwell 1992, 67)

Another author, Joseph Campbell, suggested that the making and broadcasting of the *Apollo 17* trip had "transformed, deepened, and extended human consciousness to a degree and in a manner that amount to the opening of a new spiritual era" (Campbell 1972, 239).

All these observations credit the planetary image with inducing a sudden, radical, and far-ranging shift in political consciousness, as human beings redefined their understanding of what it means to live together on the earth. But the widespread acceptance of this reading by environmentalists runs counter to much of what we know about cultural responses to imagery. Whether in the history of science or in the history of art, it seems that images become persuasive only when ways of looking at them have been carefully prepared in advance, through the creation of a stylized visual idiom or an interpretive tradition that knows how to respond to particular types of images (see, for example, Jones and Galison 1998; Alpers 1983; also Sachs 1994). The meaning of pictures is inseparable from the context that supplies the idioms of interpretation. What, then, were the historical, political, and cultural circumstances in which the vision of Planet Earth acquired its now-canonical readings? Did the image give rise to demonstrably new forms of understanding about the environment and associated concepts of governance, or did it simply reinforce older habits and patterns of political association?

Narrative Traditions

Many of the themes invoked in connection with the *Apollo* image predated the photographs that gave them unforgettable embodiment. In particular, reflections on the earth's finiteness, its fragility, its limited resources, the interconnectedness of its physical and biological systems, and the flimsiness of its geopolitical boundaries were all current in Western thought and writing well before the astronauts of *Apollo 17* brought back the most famous icon of the floating planet.[9] Thus, the maverick American engineer and inventor R. Buckminster Fuller (1969), who prided himself on having viewed the earth imaginatively before there were astronauts, coined the term *Spaceship Earth* to describe humanity's flight on a superbly designed, self-contained vehicle lacking only an intelligible operating manual.[10] Fuller's metaphor quickly became popular in enlightened political circles. The influential British economist and environmentalist Barbara Ward (1966; see also Ward and Dubos 1972) borrowed the term, finding it a congenial hook on which to hang her own ideas about transnational harmony, sustainable development, and the need for global redistribution of wealth. Ward's friend and philosophical ally, the noted liberal Democrat Adlai Stevenson, then U.S. Ambassador to the United Nations, observed in a speech before the UN Economic and Social Council in July 1965, "We travel together, passengers on a little spaceship, dependent on its vulnerable resources of air and soil; all committed for our safety to its security and peace; preserved from annihilation only by the care, the work and, I will say, the love we give our fragile craft" (Stevenson 1979, 821).

Although Fuller and others spoke of the earth as a spaceship, it was the University of Michigan economist Kenneth Boulding who explicitly connected the planet's roundness with a global imagination of the environmental predicament (Boulding 1966, 3–14). Air travel, Boulding observed, had begun to accustom people since World War II with "the notion of the spherical earth and a closed sphere of human activity" (Boulding 1966, 3).[11] The sense of the planet as an enclosed system had gradually replaced the image of the frontier, with its wide open spaces and promise of endless resources. The new era, Boulding argued, would require a new kind of economic discipline: a "spaceman economy," in which "man must find his place in a cyclical ecological system," replacing the earlier "cowboy economy," which countenanced "reckless, exploitative, romantic, and violent behavior," especially with respect to resource consumption (Boulding 1966, 9).

Others, too, had begun to note that the triad of population, consumption, and pollution might place irreversible stresses on the planet's health. Among the earliest and most influential voices in the United States was that of Rachel Carson, whose 1962 book *Silent Spring* offered a part-scientific, part-elegiac exposition of how the indiscriminate use of chemical pesticides was silencing bird populations throughout North America and gravely threatening all earthly life (Carson 1962). The Club of Rome, a prestigious association of scientists and intellectuals, went further. Using newly developed techniques of computer modeling, its controversial report, *The Limits to Growth*, predicted a sudden and drastic collapse of the Earth's economic, social, and environmental systems (Meadows et al. 1972). Despite many methodological criticisms, the report's catastrophist vision persisted as one of the enduring themes of modern environmentalism (Cotgrove 1982; Ashley 1983; Bloomfield 1986).

Perceptions of the earth as a closed system inspired less calamitous scientific stories as well. Most widely discussed perhaps is the so-called Gaia hypothesis, originated in the early 1970s by James Lovelock, a physicist working for the National Aeronautics and Space Administration (NASA), and further developed with Lynn Margulis, a microbiologist known for her theory of the origins of eukaryotic cells. Versions of the hypothesis range from a weak (and uncontroversial) form that merely posits complex linkages between biological and nonbiological activity at the earth's surface to stronger claims that the earth's atmosphere maintains a steady state for the express purpose of sustaining life, and that biological organisms actively manipulate their environment to this end (Lovelock 1979; Margulis and Lovelock 1976; Schneider 1991). It is not the validity or theoretical coherence of Lovelock's and Margulis's scientific ideas that is significant for our purposes, but rather their integrative, planetary vision. An outgrowth of NASA's interest in the possible existence of life on Mars, the Gaia hypothesis illustrates how the technology of space exploration fostered a global science of biogeochemical interactions. Even before the space age allowed humanity actually to look the earth in the face, scientific imaginations were constructing narratives on a global scale about what was happening at the earth-atmosphere interface.

If the conquest of flight ushered in an age of environmental claustrophobia, it also made geopolitical divisions seem more vulnerable. By the late 1940s, the world's major powers were already arrayed into the sharp dualities of the Cold War. The earth as globe dominated the visual renditions of this new political dispensation. The standard view adopted by the superpowers looked down on the world from the North Pole. From this standpoint, image makers were free to decide only how much of the Southern Hemisphere they would include in their field of vision. A 1947 report sponsored by the Council on Foreign Relations, for example, displayed the polar perspective as adapted from an official chart used by the U.S. Air Force. The map was cut off at the 30th parallel, showing the top of north Africa and fringes of Iran and India, but nothing at all of Latin America. "Strategists," the report observed, "term the area between the 30th and 65th parallels the key zone since all modern wars have started there" (Baldwin 1947). The same projection appears as a logo to this day on publications of Harvard University's Belfer Center for Science and International Affairs, an institution whose identity was molded during the Cold War. Image and imagination, still powerfully fused, deny the emergence of a more complex, less bipolar political order.

Yet the implications of the global perspective have always been thoroughly ambiguous. The view from the pole could be read, on one hand, as an invitation to strengthen state sovereignty. A 1948 report by the President's Air Policy Commission took just this tack, using the polar projection to underscore the threat of aerial war. Surveying the world from a position near Point Barrow, Alaska, and looking 7,000 miles along the earth's surface in all directions, the Commission graphically illustrated the emergence of "a new element through which this country may be attacked—the air." The report called for a stronger air force as "the best conceivable defense" against air attack (President's Air Policy Commission 1948, 11). But the polar gaze equally supported messages of peaceful coexistence in the postwar world, as evidenced by the United Nations logo adopted in the same period. Although the North Pole again occupied the image's center, no nation, however far from the central viewing point, was excluded from the UN's encompassing vision. Even Australia appears outlined in full at the outer margins of the UN world.

Finally, no account of the interpretive conventions that have grown up around the earth image would be complete without the voices of the astronauts who saw as eyewitnesses what the rest of humanity experienced only through pictures and television. Unwitting seers, confronting a vision for which little in life had prepared them, these observers tell what it was like to see the earth before anyone else had appropriated the images. Some were stirred to uncharacteristic eloquence. William Anders, a member of the December 1968 *Apollo 8* mission that for the first time saw the earth whole, set the tone for many later interpreters. Imagining the earth as a "little Christmas-tree ornament against an infinite black backdrop of space," Anders commented on its "fragility and finiteness":

I find it somewhat ironic that we went up there for the moon, but probably it was the Earth and the perspective of it that most impressed hard-bitten test pilots like us—and I guess the rest of the world—the most. Because the pictures of the first Earthrise and the first full Earth floating in space, I think, have been a major contribution in helping people get a better feeling for the Earth's place in our lives and in the universe. You realize that Earth is about as physically significant as one grain of sand on a beach. But it's our only home. (In Folger, Richardson, and Zimmer 1994, 38)

One recognizes in Anders's groping phrases some familiar strains of contemporary environmentalist discourse: fragility, finiteness, insignificance, the unavoidable dependence of human life on this planet ("our only home"). Yet barely three and a half years later, when NASA was winding up the first phase of lunar exploration, the "hard-bitten" edge was back, and some members of the *Apollo 17* crew seemed able to take the spectacular earthscape for granted. Eugene Cernan records the following conversation with his fellow-traveler Harrison "Jack" Schmitt:

C: "Oh, man—Hey, Jack, just stop. You owe yourself 30 seconds to look up over the South Massif and look at the Earth."
S: "What? The Earth?!"
C: "Just look up there."
S: "Aaah! You've seen one Earth, you've seen them all." (Chaikin 1994)

If even astronauts on the moon could so quickly accustom themselves to one of the twentieth century's grandest displays, then it hardly seems probable that the rest of humanity, preoccupied with innumerable local cares and conflicts, was drawn by the image to an all-new, enduring global ecoconsciousness. Did its mere dissemination—as capitalist logo or scientific "immutable mobile"—compel people to reimagine their political affiliations on a worldwide scale? Not so. As we will see, connections between the global image and an imagined global community evolved along more intricate pathways, as human actors and institutions strove in disparate ways to assimilate, or exploit, the photographic evidence of their common destiny. Associated shifts from local or national to global environmental thinking—although they *can* be documented— have been neither seamless nor smooth, but rather subtle, sporadic, partial, and unevenly distributed among the world's political communities.

Varieties of Global Experience

Global environmental consciousness, I am suggesting, did not coalesce all at once in response to a striking visual stimulus, but took shape gradually in diverse domains of social and political practice during the final decades of the twentieth century. Strands of increasing global awareness can be traced in the discourses of risk, politics, commerce, and ethics. In each context, we observe a selective uptake of themes prevalent in older narratives of earthwatching, but reinforced and given new persuasive power through association with the *Apollo* photographs.

Framing Risks Globally

Most observers of American environmental politics agree that something happened to alter its character in the decade roughly marked by the publication of *Silent Spring* in 1962 and the celebration of the first Earth Day, a nationwide event involving some 300,000 citizens, on April 22, 1970. Often termed the *new environmentalism*, the movement born in this period of ferment diverged from earlier forms of environmental activism in its focus on people and their habitats rather than on the preservation of nature for its own sake. It was founded, according to one analyst, on "a broader conception of the place of man in the biosphere, a more sophisticated understanding of that relationship, and a note of crisis that was greater and broader than it had been in the earlier conservation movement" (McCormick 1989, 48).

Its targets, however, were initially local. They concerned first and foremost the effects of pollution on common people's homes and lives. Rachel Carson's ground-breaking vision, as noted earlier, took as its central theme the harms caused by pervasive use of chemical pesticides. Although she imagined a whole world deprived of birdsong (her book began with a fable of poisoned landscapes and dying animals and vegetation), some of her most telling vignettes involved ordinary citizens reporting on changes in their immediate surroundings: the disappearance of robins in one town, the decimation of swallows in another. These local insults added up to a problem of concededly national proportions. In 1970, President Nixon ordered the creation of the Environmental Protection Agency largely on the ground that a new organization was needed to coordinate a nationwide fight against pollution. But local issues continued to predominate in environmental politics. In 1978, for example, toxic chemicals found in the basements of homes in the Love Canal area of Niagara Falls, New York, precipitated an intense flurry of pollution-centered legislative and regulatory activity (see for example Levine 1982). This and similar episodes fueled the era's most powerful form of environmental mobilization, the NIMBY, a social movement whose tightly bounded, *local* imagination was synonymous with the slogan "not in my backyard." Chemical pollution of communities remained a guiding theme of U.S. environmentalism well into the 1980s. ·

It was not until the later 1980s that a global conception of environmental protection rooted itself in Western consciousness. Almost imperceptibly, the causes and extent of environmental degradation (for example, deforestation, desertification, ozone depletion) began to be defined across political domains far larger than individual communities and even more encompassing than nation states. Attention began to shift from end-of-pipe controls on specific polluting facilities to questions of prevention and lifestyle change. A new term—*sustainability*—came into common use in policy discourse, bridging what had previously seemed an irreconcilable contradiction between environmental protection and human development. Appropriately enough, the book that heralded this new era was not the work of a single author with a distinctively personal vision, but of an international committee of experts, the World Commission on Environment and Development, chaired by Norway's prime minister Gro Harlem Brundtland. Both in its title and by explicit reference (as quoted above), *Our Common Future* conveyed, and helped crystallize, a sense of the whole/human condition, framed by the planetary image and global in its prescriptive scope. No longer would it be sufficient for environmental activism to concentrate its energies primarily on the invasion of individual backyards by chemical pollution.

Global Environmental Politics

From NIMBY to a global politics of the environment, however, was no easy step. The transition arguably began with the age of space exploration, but its progress is not yet complete. That the earth image impels many observers to "think globally" has been apparent ever since the early satellite launches. Barbara Ward, for example, imagined the solidarity that U.S. astronauts must feel with their Soviet counterparts and speculated that their feat would erase the cold war's central political conflict:

When the astronauts spin through more than a dozen sunrises and sunsets in a single day and night; when the whole globe lies below them with California one minute and Japan the next; when, as they return from space, they feel spontaneoulsy, with the first Soviet spaceman: "How beautiful it is, *our* earth"; it is inconceivable that no modification of consciousness or imagination occurs, no sense that quarrels are meaningless before the majestic yet vulnerable reality of a single planet carrying a single human species through infinite space. (Ward 1966, 146)

Her comments interestingly foreshadowed Anders's testimony that the earth is "as physically significant as one grain of sand" and yet "it's our only home."

Perhaps self-consciously echoing Anders, Ward and her distinguished environmentalist colleague Rene Dubos published a book in 1972 with the title *Only One Earth*. This became the official theme of that year's Stockholm conference on the environment. Carl Sagan in turn gave the image an explicitly political spin, while reiterating the theme of humanity's insignificance:

We are too small and *our statecraft too feeble* to be seen by a spacecraft between the Earth and the Moon. From this vantage point, our obsession with nationalism is nowhere in evidence. The Apollo pictures of the whole Earth conveyed to

multitudes something well known to astronomers: On the scale of worlds—
to say nothing of stars or galaxies—humans are inconsequential, a thin film of
life on an obscure and solitary lump of rock and metal. (Sagan 1994, 5–6;
emphasis added)

The World Commission on Environment and Development similarly
juxtaposed the transitory geopolitical constructions of the globe against
an enduring ecological view: "From space, we see a small and fragile ball
dominated not by human activity and edifice but by a pattern of clouds,
oceans, greenery, and soils" (World Commission on Environment and
Development 1987, 308).

While astronauts, astronomers, and international experts identified
the earth image with coexistence and political interdependence, the use
and enjoyment of environmental resources remained for many other
actors tightly bound to national interests. A case in point was President
Ronald Reagan's astonishing and emphatic rejection of the draft Law of
the Sea (LOS) convention following his election in 1980. Here was an
accord governing the oceans that had enjoyed bipartisan support in
the Nixon, Ford, and Carter administrations and had seemed virtually
ready for adoption by the late 1970s. It presented U.S. negotiators with
an apparently uncomplicated trade-off: increased navigational freedom,
simplifying and counteracting a patchwork of jurisdictional claims by
coastal states, in return for a decrease in the right to mine seabed
resources, including manganese nodules, which developing countries
saw as the common heritage of mankind (Sebenius 1984). For those
committed to the negotiation, the two issues were inextricably linked.
Concession on seabed mining was the necessary price for avoiding the
expanding territorial ambitions of coastal states. Yet as the negotiations
went on for decades, fissures appeared in the U.S. position corresponding to changing perceptions of costs and benefits among some of the
parties to the negotiation. American mining interests, in particular,
came to believe that the draft treaty was penalizing them more than
was warranted by corresponding gains on the side of navigation. As
support crumbled for a comprehensive solution, combining navigation and mining, a new Republican administration began to think the
unthinkable and pulled the United States out of the almost-completed
negotiations.

The fate of the LOS conference in the 1980s can be seen in retrospect as a triumph of persistent nationalist claims to global environmental resources over a nascent internationalist worldview. This point was well understood by Richard Darman, vice chair of the U.S. delegation to the 1977 session of the third LOS conference and a perceptive participant-observer of the treaty process. In an article in *Foreign Affairs*, Darman argued that the conference was being driven, to the detriment of U.S. interests, both pragmatic and ideological, by a community of

internationalist lawyer-codifiers. The internationalists' tendency to favor collective over individual action is combined with the codifiers' tendency to wish to see the world in neat static terms. Above and beyond practical considerations, there is an aesthetic antipathy toward the "disaster" of nonuniformity, and a general distrust of the possible benignness of self-regulating, dynamic processes. (Darman 1978, 381)

Darman conceded that foiling the internationalist vision of a single regime for navigational and seabed governance carried a risk. The oceans occupying two-thirds of the earth's surface might be "carved up" along geopolitical lines in ways that would benefit developed countries over developing ones. Political boundaries would in effect be drawn where none had previously existed. This, in turn, would create a problem of equity that might affect U.S. strategic interests. But this issue, Darman urged, should be addressed on its own terms through mechanisms such as loan guarantees and technology transfer. International equity did not require the United States to recognize the oceans as the common property of humankind or to countenance the development of new international institutions whose mandates would inevitably erode national sovereignty. Equity problems, in other words, could be redressed without having to accept the case for global ownership or control of the oceans.

When the Clinton administration signed a revised LOS convention in July 1994, the official explanation declared a victory for free-market principles over objectionable centralized planning and for sovereignty over loss of national control. A government fact sheet on LOS posted on the World Wide Web notes that the United States has won a guaranteed seat on key committees, increased power to block adverse decisions, and credit for exploration already undertaken by American companies (U.S. Department of State 1996). Clearly, framing the seas as an economic resource ran counter to the planetary imagination. A view founded on the necessity—even the rightness—of competition among nations blocked the emergence of transnational management institutions and a genuinely global politics of resource allocation.

A similar resurgence of national sovereignty can be observed under the Convention on Biological Diversity (CBD), even though this treaty, too, was initially conceived as an instrument for effectuating allegedly global environmental interests. By the late 1980s, alarm about rapid, world-wide extinctions of species and associated activism by leading biologists had created a demand for international action to protect the earth's scarce biological resources (Takacs 1996). In response, the United Nations Environment Program (UNEP) initiated in 1988 a series of expert and intergovernmental deliberations with the aim of preparing a legal instrument for the conservation and sustainable use of biodiversity. Consistent with these goals, the convention sought to address economic and social concerns along with scientific ones. The experts convened by UNEP were asked to take into account "the need to share costs and benefits between developed and developing countries" as well as "ways and means to support innovation by local people."

The text of the CBD was adopted in Nairobi in May 1992 and opened for signature in June of the same year at the United Nations Conference on Environment and Development (the Rio "Earth Summit"). It entered into force in December 1993, ninety days after the thirtieth ratification. From the beginning, however, international efforts to balance conservation against development, equity against economics, and global management against national sovereignty proved to be highly contentious. A test of the convention's attempted global framing of biodiversity arose in 1999 at a meeting in Cartagena, Colombia, to approve an international biosafety protocol governing genetically modified organisms. Acrimony between developing countries and major grain exporters caused the meeting to break down without any agreement being reached. A new round of negotiations in Montreal in early 2000 proved more productive, although the agreement reached there represented more a working conpromise among contrary interests than a global accord on basic presumptions. As in the case of LOS, environmentalism's global ambitions bowed to pressure from economic interests defined at the national level.

One of the few environmental regimes in which the political ideal of "One Earth" arguably has come closer to fruition is that governing climate change (popularly better known as "global warming"). Here, there has been a convergence between the *scientific* construction of a problem that transcends national boundaries (see Miller, chapter 8, this volume) and a *normative* construction that recognizes the rights of developing as well as developed countries to be protected against the worst consequences of greenhouse-gas accumulation in the atmosphere. Even on the scientific side, considerable work was needed to define climate change as something other than the sum of local weather patterns—in other words, as a problem of "whole-earth" dimensions. Prerequisites included the formation of the Intergovernmental Panel on Climate Change and its hard-fought acquisition of credibility as a body capable of representing the best scientific judgment with respect to climate change. This evolutionary story stands markedly at odds with the conventional account of the earth image as herald and harbinger of a sudden paradigm shift in environmental consciousness. The emergence of climate change as a global phenomenon, moreover, coincided, as we will see, with the appearance of an ethical discourse that had no precursor in the politics of either LOS or biodiversity.

Commerce's Global Ambition

U.S. environmentalists were not alone in sizing up the symbolic potential of the earth image. The picture of the planet became almost an instant classic in the visual repertoire of advertising, at first retaining its connections to themes of environmental stewardship, but gradually shedding these in favor of something more like "universalism" or simply "global reach."[12] Commerce, environment, and the earth in space were first linked together in 1968 in the *Whole Earth Catalog*, an entrepreneurial venture that both articulated and capitalized on the values of the new environmental movement. The cover picture showed the North American land mass almost entirely obstructed by a large white cloud; below appeared the caption "THE UNIVERSE: from planet Earth on a sunny day." Inside, the *Catalog*'s offerings emphasized environmental restoration, community building, simplicity, authenticity, and medical self-help. Supplements published over the next five years all carried the same cover picture accompanied by the same message. A new version planned for the millennium continued several themes of the late 1960s. Consumers were still offered "tools for producing knowledge, reporting and broadcasting the news as you see it, and creating communities according to your own values and ideals" (Rheingold 1994).

Contemporary advertisers of services and products no longer link globality so explicitly to environment-friendly lifestyles. Instead, the planetary symbol tends to stress the deployer's capacity to move people and products (and, in the case of television, images) effortlessly around the globe. Not surprisingly, the picture has become an important property for CNN, the cable news service that built its empire by bringing viewers face to face with events from the furthest reaches of the earth—with an immediacy prized equally in the White House and in the headquarters of some of the United States's most intransigent enemies. Not only for CNN, but for many other advertisers, it is the imagined shrinking of time and distance between the consumer and the object of consumption that has become the image's most alluring message.

Advertisements also illustrate the infinite interpretive flexibility of an image that has achieved iconic status throughout Western culture. Just as the Mona Lisa, the world's most famous painting, has been adapted, interpreted, and sometimes subverted, to suit every taste,[13] so too has the planet's portrait been manipulated in varied ways to create a universally accessible visual counterpoint to messages of persuasion and seduction. One commonplace strategy is to focus on the part of the globe on which the advertiser's commercial activities are specifically targeted. Another is to superimpose the image on something else—for example, a burning, spherical candle or a pair of clasped hands—thereby hybridizing the instantly comprehensible sign of global interconnection with other, less normalized messages (energy crisis, regional business partnerships, company logos, and the like), which then are empowered to travel, as it were, on the shoulders of the earth.

These advertising pictures do not claim the power of direct representation, the seemingly literal transcription of a new reality celebrated by environmental writers and scientists. Consider, once again, Sagan's evocative reading of the *Apollo 17* image, in which he first zoomed in on its dense, geopolitical meanings before pulling back (as noted above) to a more abstracted, apolitical, indeed dehumanized gaze:

There's Antarctica at what Americans and Europeans so readily regard as the bottom, and then all of Africa stretching up above it: You can see Ethiopia, Tanzania, and Kenya, where the earliest humans lived. At top right are Saudi Arabia and what Europeans call the Near East. Just barely peeking out at the top is the Mediterranean Sea, around which so much of our global civilization emerged. You can make out the blue of the ocean, the yellow-red of the Sahara and the Arabian desert, the brown-green of forest and grassland. (Sagan 1994, 5)

In commercial discourse, by contrast, Earth has no fixed, human-made coordinates. It is, as often as not, a dream image, as in an advertisement for Thai Airways, which shows an incredibly remote planet held at the eye of a huge silvery needle against a black velvet sky and a ribbon of deep purple, the advertiser's signature color. Through their very ubiquity, however, these modified pictures reinforce the status of the underlying "original" image as a common cultural resource; they appear and disappear as figments of our common imagination, even as they cater to our culturally calibrated desires.

A Planetary Ethics

Modern environmentalism includes at its core a widely acknowledged, if only imperfectly realized, ethical imperative to renegotiate human beings' relationship with nature in the light of new scientific understandings. More than a generation ago, Boulding observed in his article on Spaceship Earth that we were as yet "very far from having made the moral, political, and psychological adjustments which are implied in [the] transition from the illimitable plane to the closed sphere" (Boulding 1966, 4). Now, more than two decades after the first landing on the moon and the first photographic portrayals of the earth from space, it is possible to identify at least three strains of ethical discourse that appear specifically to derive their force from a global, as opposed to a national or local, framing of humanity's environmental predicament.

The first is a gradual extension of the precautionary principle into transnational environmental policy. This legal precept originated in German law as one of five fundamental principles governing environmental decisions. Briefly stated, the precautionary principle asks for restraint on human activities that could harm the environment when there is not enough evidence to determine for sure whether such harm will occur. American environmental law has opted on the whole for a seemingly more pragmatic, risk-based approach that allows development to proceed when the benefits are calculated to exceed the probable harm.[14]

With respect to environmental hazards of global scope, however, the utilitarian calculus of risks and benefits is harder to sustain than in the context of localized pollution problems from a waste dump or chemical factory. Uncertainties loom larger, and, within the contested frameworks of global politics, practices of analysis and reassurance cannot be as readily stabilized through well-worn channels of expert deliberation. The result has been to introduce what some international theorists have termed a *bias shift* away from problem solving toward a set of actions geared more toward prevention (Ruggie 1986). For example, as Karen Litfin has argued, the Montreal accord on the control of ozone-depleting substances would not have adopted nearly so stringent a set of targets had it not been for the discovery of the Antarctic ozone hole, which atmospheric science had not predicted and for which there was no obvious nonanthropogenic explanation (Litfin 1994).

The second strand of an emergent global ethical discourse centers on the idea of sustainability and more specifically on concepts of steward-ship for future generations. The World Commission on Environment and Development built the norm of stewardship into its very definition of sustainability in *Our Common Future*, endorsing only those patterns of development that would leave future generations no worse off than their forebears in the present. Lynton Caldwell, for one, explicitly ties this shift to the *Apollo* image, which in his telling induced a move "from the view of an earth unlimited in abundance and created for man's exclusive use to a concept of the earth as a domain of life or biosphere for which mankind is a temporary resident custodian" (Caldwell 1990, 21). The elaboration of intergenerational ethics as a legal principle likewise rests on a recognition of "the planet" as the appropriate spatial framing for sustainable environmental action. Edith Brown Weiss's important treatise on the legal foundations of intergenerational ethics begins on a note familiar to all earthwatchers: "The human species inhabits a small, relatively new, and so far as we know, unique planet—Earth. It is also a fragile planet" (Weiss 1989, 1).

The third ethical strand has to do with the obligations of the developed North to the developing South in matters of environmental policy. The recognition of such an obligation is not in itself new, as is evident from our earlier discussion of the Law of the Sea negotiations. Thus, even while espousing a position of unilateralism and national self-interest, Darman rejected the prospect of a highly inequitable regime for exploiting global seabed resources. Ethics, however, was embedded within the discourse of rational choice, where it became simply one more item to tote up along with other national interests. Acting ethically was no more important in principle than respecting the needs of the mining industry or of commercial shipping. Efforts to treat equity as a higher-order variable within LOS proved unsuccessful. Indeed, in his analysis of the U.S. withdrawal from LOS, the negotiation theorist James Sebenius has argued that developing nations' attempt to promote a supervening, *transnational* ethical discourse—that of the New International Economic Order (NIEO)—was a prime reason for the formation of a "blocking coalition" and the eventual breakdown of the conference (Sebenius 1991).

Claims about equity have received a more sympathetic hearing under regimes that have (unlike LOS and CBD) successfully constructed environmental problems on a transnational level, most notably ozone and climate change. Thus, an Indian environmental group, the Centre for Science and Environment, successfully deconstructed the tacit normative assumptions incorporated within early efforts to create objective measures of the "global warming potential" of greenhouse gases (Agarwal and Narain 1991). Perhaps more important, the very kinds of equity arguments that were rejected by Northern nations when put forward by the South under the label *NIEO* now seem to carry greater moral as well as political weight. Terms like *vulnerability* and *equity* have entered the language of global environmental accords. Both the ozone and climate change regimes explicitly recognize the special economic and ethical claims of developing nations through legal provisions ensuring delayed implementation, funding, and technology transfer. It is tempting to conclude that framing pieces of the natural world in global terms—such as the *ozone hole* or *climate*—has facilitated an ethical discourse that also operates at the global level, without needing to be subordinated to the older calculus of national interests.

Seeing Things Together

The power of words to compel action has been a subject for philosophical and political analysis from Plato down to modern times. The power of images may be no less profound, especially in this era of mass visual communication, but it has yet to receive the same sustained scrutiny from social theorists. My object in this chapter has been to trace the complex pathways by which one image—that of planet Earth—has come to inhabit our political consciousness as an icon of global environmentalism.

A closer look at the image's reception suggests that its connections with environmental thought and action have been anything but straight-forward. The picture, to begin with, picked up and reinforced themes of the earth's fragility and finiteness that had begun to percolate through policy discourses decades before the space age began. In this way, it may have subtly helped to shift the perception of environmental risk from issues of purely local scope to longer-term concerns for human survival. Yet although it was appropriated early on to support arguments for global environmental governance (witness the "Only One Earth" theme of the 1972 Stockholm conference), such thinking failed to move entrenched national interests in resource management regimes ranging from seabed mining to the protection of biodiversity. The image's wide-spread exploitation by commercial enterprises has underscored its iconic properties but arguably blunted its moral and political connotations. Possibly the most important consequence to flow from the planetary image is the visual anchor it has provided for emerging, globally articu-lated ethical concepts, such as the precautionary principle, sustainabil-ity, and intergenerational equity.

Global stewardship remains nonetheless a deeply contested concept. Battles over the Law of the Sea and biodiversity throw into sharp relief some of the dangers that people around the world—from the South as well as the North—perceive in allowing environmental risks, and their control, to be framed globally. The idea of international governance, especially in matters of natural resource management, carries for many the threatening specters of bureaucratic inflexibility, loss of sovereignty, and even new forms of colonial domination drawing their legitimacy from science. The planetary image, moreover, conveys a serene (some might say contemptuous) disregard for the human condition. Not only does it appear to erase the territorial claims of nation states, but it also renders invisible the day-to-day environmental insults suffered by billions of the world's poorest citizens: dirty air, polluted water, inadequate sanitation, infectious diseases, damaged crops, loss of green spaces, and the decay of built environments. Indeed, people themselves are eliminated from this image of environmentalism, as in some of the darker fantasies of ecofascism. As a dazzling offshoot of the twentieth century's most destructive technological impulses, the photograph that preeminently symbolizes planetary togetherness ironically undermines its own authority in the eyes of skeptics. It promises an imagined community as encompassing as the earth itself, but is this a community in which those without the power to patrol the heavens, to map and perhaps to devastate the earth, can ever meaningfully participate?

I would like nevertheless to end this chapter on a note of mild optimism. Seldom in the course of preparing an academic publication have I encountered so much spontaneous interest among my U.S. conversation partners as in discussing the topic of this piece. Almost everyone I spoke with, it seemed, had his or her own favorite associations with the Earth image; many admitted to possessing a variant of it on some prized but mundane object, such as a T-shirt, a tote bag, or a poster. If general circulation models and integrated assessments belong to the "high" scientific language of global environmentalism, then for most Americans the picture of planet Earth surely belongs to its vernacular. It is not, if it ever was, an arcane "immutable mobile" that simply extends the dominance of instrumentally rational ways of perceiving the environment. Nor is it a decontextualized, impersonal logo through which an unscrupulous, hegemonic power is asserting its reach across the globe. Thoroughly domesticated and sustaining multiple meanings, the image may, after all, rekindle an associationism through which America's too self-centered political culture can embrace in imagination those billions of others who also regard the Earth as their only home.

Notes

Many people helped in the preparation of this chapter by contributing pictures, articles, citations, and fascinating personal experiences with the image of Earth. I have benefited as well as from astute comments on presentations at MIT, UC Berkeley, and Harvard. I owe particular thanks to Xandra Rarden for her superb research, Michael Dennis for insights on the Cold War's polar perspective, Bernward Joerges for a most helpful critical reading, and Clark Miller for his unfailing patience and encouragement.

1. For the appearance of memory as a major theme in historiography, see Nora 1987. On war and memory, see Fussell 1975 and Winter 1995. Representations of the two world wars multiplied in both high and popular culture in the last decades of the century, as exemplified by Michael Ondaatje's *The English Patient* (in both novel and film versions), Pat Barker's *Regeneration* trilogy, and Stephen Spielberg's *Schindler's List* and *Saving Private Ryan*. For a comparative account of attempts to memorialize World War II in Germany and Japan, see Buruma 1995.

2. As of January 1999, only four (Britain, Denmark, Greece, Sweden) of fifteen members of the European Union had not joined the European Monetary Union.

3. There are, in fact, a large number of pictures of the earth from space, as documented and archived by NASA's Johnson Space Program; these may be viewed at NASA's website. The most famous (as discussed below) is the *Apollo 17* picture of the whole planet, showing the horn of Africa and Saudi Arabia. Its popularity can be attributed to several factors, including the size and fullness of the planet, the absence of clouds, and the clarity and color of the visible land masses. Like all of NASA's pictures, this one is not covered by copyright and can be downloaded from the web.

4. For instance, in a courtroom, the jury's ability to see things is framed by the judge and discursively constituted by expert witnesses. See, in this regard, Goodwin 1994 (writing about the videotape in the Rodney King trial) and Jasanoff 1998.

5. See, for example, Haas 1989, 1990. See also my own argument that political analysis needs to take more seriously the ways shared epistemes are created and achieve standing in the political realm, in Jasanoff 1996a, 173–197.

6. Representation has long been a topic of major interest in science and technology studies and there is a vast literature dealing with scientific representations in particular. For an introduction, see Lynch and Woolgar 1990.

7. By focusing on *regimes* of interpretation, I do not mean to suggest that individual perceptions are unimportant. There is clearly interesting research to be done on ways in which people in varying national or social surroundings have

made sense of the image of the earth. This type of ethnographic work, however, lies outside the scope of this chapter.

8. One can only speculate on the reasons for this particular choice. It is, as noted, one of the relatively few earth images that shows the full globe relatively unencumbered by clouds. It therefore conforms well to the ways in which cultures familiar for some five centuries with the artifacts of mapping, both spherical and two-dimensional, *expect* to see the earth.

9. It should be noted that representations of the earth seen as if from a distant vantage point in space were not unknown in the Western mapping tradition. For example, the *Celestial Atlas of Harmony*, a magnificent series of engravings by the seventeenth-century Polish cartographer Andreas Cellarius, displays the earth set amidst the other bodies of the solar system as conceived by the astronomers Ptolemy, Copernicus, and Tycho Brahe. In several of these illustrations, the earth appears as a delicately suspended, beautiful, blue-green globe.

10. Fuller noted that few people actually sense themselves to be in a spaceship because most have seen only small portions of the earth's surface; even veteran pilots, he observed, had not viewed more than about one-hundredth of the earth. In a bow to Fuller, *Spaceship Earth* is the name given to the giant geosphere (a full rather than a half-sphere or geodesic dome) that marks the entrance to the Future World exhibit at Walt Disney World's Epcot Center in Florida.

11. Boulding (1966, 3) specifically contrasted the new visual perception of the earth as a sphere with the earlier imaging of earth as "an illimitable cylinder, essentially a plane wrapped around a globe."

12. This change in meaning may help to account for the image's widespread use as a symbol of the new millennium in the final years of the twentieth century. A detailed exploration of this phenomenon would be rewarding, but it unfortunately cannot be attempted within the scope of this chapter.

13. "She has also been used to advertise cheese, oranges, gramophone needles, cigars and ladies' shoes in Italy, Spain, Holland and England. Her name is an unfailing password everywhere. The German post office has issued the painting on a stamp. At the same time, she has given rise to many iconoclastic manifestations on postcards or cartoons, but they were friendly jokes and could also be taken for tokens of admiration" (Ottino della Chiesa 1985, 105).

14. To be sure, the precautionary principle also requires a balancing of caution against other desired policy objectives. Nonetheless, the principle's very framing places a greater emphasis on prevention than the discourse of risk analysis. For a persuasive critique of risk-based environmental regulation, see Winner 1986, 138–154.

References

Abelson, P. H. (1980). "Scientific Communication." *Science* 209 (4452): 60–62.

Acheson, D. ([1947, May 8], 1969). "The Marshall Plan; Relief and Reconstruction Are Chiefly Matters of American Self-Interest." In W. LaFeber, *America and the Cold War: Twenty Years of Revolution and Response, 1947–1967*. New York: Wiley.

Adler, E. (1992). "The Emergence of Cooperation: National Epistemic Communities and the International Evolution of the Idea of Nuclear Arms Control." *International Organization* 46 (1): 101–146.

Agarwal, A., and S. Narain. (1991). *Global Warming in an Unequal World*. New Delhi, India: Center for Science and the Environment.

Alcamo, J. (1994). *IMAGE 2.0: Integrated Modeling of Global Climate Change*. Boston: Kluwer.

Alpers, S. (1983). *The Art of Describing: Dutch Art in the 17th Century*. Chicago: University of Chicago Press.

Anderson, B. (1991). *Imagined Communities*. Rev. ed. London: Verso.

Arakawa, A. (1966). "Computational Design for Long-Term Numerical Integration of the Equations of Fluid Motion: Two-Dimensional Incompressible Flow. Part 1." *Journal of Computational Physics* 1: 119–143.

Arrhenius, S. (1896). "On the Influence of Carbonic Acid in the Air upon the Temperature of the Ground." *Philosophical Magazine and Journal of Science* 41: 237–276.

Ashford, O. M. (1959). "International Geophysical Year." *WMO Bulletin* 8 (2): 60–61.

Ashley, R. K. (1983). "The Eye of Power: The Politics of World Modeling." *International Organization* 37 (3): 495–535.

Aspray, W. (1990). *John von Neumann and the Origins of Modern Computing*. Cambridge, MA: MIT Press.

Atlas, D. (1975). "President's Page: Selling Atmospheric Science." *Bulletin of the American Meteorological Society* 56: 688–689.

Australian Conservation Foundation. (1998). *Global Warming Information Sheet* 8. Available at: http://www.acfonline.org.au/campaigns/globalwarming/discussion/8facts.htm.

Baldwin H. W. (1947). *The Price of Power.* New York: Harper and Brothers.

Barker, P. (1991). *Regeneration.* New York: Plume.

Barker, P. (1993). *The Eye in the Door.* New York: Viking.

Barker, P. (1995). *The Ghost Road.* New York: Dutton.

Barnett, T. P., and M. E. Schlesinger. (1987). "Detecting Changes in Global Climate Induced by Greenhouse Gases." *Journal of Geophysical Research* 92: 14772–14780.

Baumgartner, T., and A. Midttun, eds. (1987). *The Politics of Energy Forecasting.* Oxford: Oxford University Press.

Beck, U. (1992). *Risk Society: Towards a New Modernity.* Newbury Park, CA: Sage.

Benedick, R. E. (1991). *Ozone Diplomacy: New Directions in Safeguarding the Planet.* Cambridge, MA: Harvard University Press.

Bergthorsson, P., B. R. Döös, S. Frylkond, O. Haog, and R. Lindquist. (1955). "Routine Forecasting with the Barotropic Model." *Tellus* 7 (2): 272–276.

Berkner, L. V. (1950). *Science and Foreign Relations.* General Foreign Policy Series no. 30. Washington, DC: U.S. Department of State.

Berkner, L. V. (1959). *Reminiscences of the International Geophysical Year.* No. 354. New York: Columbia University Oral History Research Office.

Berry, E. X., and R. W. Beadle, eds. (1974). "Project METROMEX: A Review of Results." *Bulletin of the American Meteorological Society* 55, 86–121.

Bimber, B. (1996). *The Politics of Expertise in Congress.* Albany, NY: SUNY Press.

Bloomfield, B. P. (1986). *Modelling the World: The Social Constructions of Systems Analysis.* Oxford: Blackwell.

Bodansky, D. (1993). "The United Nations Framework Convention on Climate Change: A Commentary." *Yale Journal of International Law* 18: 451–558.

Bodansky, D. (1994). "Global Warming: The Role of International Law." In D. D. Caron, F. S. Chapin, J. Donoghue, M. Firestone, J. Harte, L. E. Wells, and R. Stewardson, eds., *Ecological and Social Dimensions of Global Change,* 297–315. Berkeley: Institute of International Studies, University of California, Berkeley.

Boehmer-Christiansen, S. (1994). "Global Climate Protection Policy: The Limits of Scientific Advice." *Global Environmental Change* 4 (2): 140–159 and (3): 185–200.

Boer, G. J. (1992). "Some Results from and an Intercomparison of the Climates Simulated by 14 Atmospheric General Circulation Models." *Journal of Geophysical Research* 97: 12771–12786.

Boffey, P. (1976). "International Biological Program: Was It Worth the Cost and Effort?" *Science* 193: 866–868.

Bolin, B. (1994). "Science and Policy Making." *Ambio* 23 (1): 28.

Bolin, B. (1996). Letter to Ben Santer (provided by Bert Bolin).

Bolin, B. (1998, January 16). "The Kyoto Negotiations on Climate Change: A Science Perspective." *Science* 279: 330–332.

Bolin, B., J. Houghton, and L. G. Meira-Filho. (1996, June 25). "Letter to Ben Santer." *Wall Street Journal*, p. A15.

Bornstein, R. F. (1991). "The Predictive Validity of Peer Review: A Neglected Issue." *Behavioral and Brain Science* 14 (1): 138–139.

Botkin, D. (1990). *Discordant Harmonies: A New Ecology for the Twenty-First Century.* New York: Oxford University Press.

Boulding, K. E. (1966). "The Economics of the Coming Spaceship Earth." In H. Jarrett, ed., *Environmental Quality in a Growing Economy*, 3–14. Baltimore: Johns Hopkins University Press.

Brickman, R., S. Jasanoff, and T. Ilgen. (1985). *Controlling Chemicals: The Politics of Regulation in Europe and the United States.* Ithaca, NY: Cornell University Press.

Broecker, W. C. (1997). "Will Our Ride into the Greenhouse Future be a Smooth One?" *GSA Today* 7 (5): 1–6.

Brown, G. E. (1996, October 23). *Environmental Science Under Siege: Fringe Science and the 104th Congress.* Report, Democratic Caucus of the Committee on Science, U.S. House of Representatives, Washington, DC. Available at http://www.house.gov/science_democrats/archive/envrpt96.htm.

Bruce, J., H. Lee, and E. Haites, eds. (1996). *Climate Change 1995: Economic and Social Dimensions of Climate Change.* Vol. 3 of 3. Cambridge, England: Cambridge University Press.

Bryson, R. (1975). "The Lessons of Climatic History." *Environmental Conservation* 2: 163–170.

Burks, A. W. (1975). "Models of Deterministic Systems." *Mathematical Systems Theory* 8: 295–308.

Burks, A. W., and A. R. Burks. (1981). "The ENIAC: First General Purpose Computer." *Annals of the History of Computing* 3 (4): 310–389.

Burley, A.-M. (1993). "Regulating the World: Multilateralism, International Law, and the Projection of the New Deal Regulatory State." In John Gerard Ruggie, ed., *Multilateralism: The Anatomy of an Institution.* New York: Columbia University Press.

Buruma, I. (1995). *Wages of Guilt.* London: Vintage.

Bush, V. (1945). *Science: The Endless Frontier.* Washington, DC: U.S. Government Printing Office.

Byers, H. B. (1974). "History of Weather Modification." In W. N. Hess, ed., *Weather and Climate Modification*, 3–44. New York: Wiley.

Cacuci, D. G. (1981a). "Sensitivity Theory for Nonlinear Systems: I. Nonlinear Functional Analysis Approach." *Journal of Mathematical Physics* 22: 2784–2802.

Cacuci, D. G. (1981b). "Sensitivity Theory for Nonlinear Systems: II. Extensions to Additional Classes of Responses." *Journal of Mathematical Physics* 22: 2803–2812.

Caldwell, L. K. (1990). *International Environmental Policy: Emergence and Dimensions*. Durham, NC: Duke University Press.

Caldwell, L. K. (1992). "Globalizing Environmentalism: Threshold of a New Phase in International Relations." In R. Dunlap and A. Mertig, eds., *American Environmentalism: The U.S. Environmental Movement, 1970–1990*, 63–76. Philadelphia: Taylor and Francis.

Campbell, J. (1972). *Myths to Live By*. New York: Viking Press.

Cane, M. A. (1997, March). "ENSO and Its Prediction: How Well Can We Forecast It?" *Internet Journal of African Studies* 2. Available at http://www.esig.ucar.edu/ijas/index.html.

Carson, R. (1962). *Silent Spring*. New York: Fawcett Crest.

Carter, L. J. (1966, January 28). "Weather Modification: Panels Want Greater Federal Effort." *Science* 151: 428–431.

Carter, L. J. (1973, June 29). "Weather Modification: Colorado Heeds Voters in Valley Dispute." *Science* 180: 1347–1350.

Cartwright, N. (1983). *How the Laws of Physics Lie*. Oxford: Clarendon Press.

Chaikin, A. (1994). "The Last Men on the Moon." *Popular Science* 245 (3): 70–74, 88.

Changnon, S. A. (1973). "Weather Modification in 1972: Up or Down?" *Bulletin of the American Meteorological Society* 54: 642–646.

Changnon, S. A. (1975). "The Paradox of Planned Weather Modification." *Bulletin of the American Meteorological Society* 56: 27–37.

Changnon, S. A. (1977). "On the Status of Hail Suppression." *Bulletin of the American Meteorological Society* 58: 20–28.

Changnon, S. A. (1986). "Review of the Tenth Conference on Planned and Inadvertent Weather Modification." *Bulletin of the American Meteorological Society* 67: 1501–1506.

Charney, J. G. (1949). "A Physical Basis for Numerical Prediction of Large-Scale Motions in the Atmosphere." *Journal of Meteorology* 6: 371–385.

Charney, J. G. (1955a). "Numerical Tendency Computations from the Barotropic Vorticity Equation." *Tellus* 7: 248–257.

Charney, J. G. (1955b). "The Use of the Numerical Primitive Equations of Motion in Numerical Prediction." *Tellus* 7: 22–26.

Charney, J. G., and A. Eliassen. (1964). "On the Growth of the Hurricane Depression." *Journal of the Atmospheric Sciences* 21: 68–75.

Charney, J. G., R. Fjörtoft, and J. von Neumann. (1950). "Numerical Integration of the Barotropic Vorticity Equation." *Tellus* 2: 237–254.

Chayes, A., and A. H. Chayes. (1995). *The New Sovereignty: Compliance with International Regulatory Agreements.* Cambridge, MA: Harvard University Press.

Chedin, A., and N. A. Scott. (1983). "The Improved Initialization Inversion Procedure '3I.'" In *First International TOVS Study Conference* (COSPAR-IAMAP-WMO LMD): 117.

Chervin, R. M. (1990). "High Performance Computing and the Grand Challenge of Climate Modeling." *Computers in Physics*, 4 (3): 234–239.

Christy, J. R., R. W. Spencer, and R. T. McNider. (1995). "Reducing Noise in the MSU Daily Lower Tropospheric Global Temperature Data Set." *Journal of Climate* 8: 888–896.

Chubin, D. E., and E. J. Hackett. (1990). *Peerless Science: Peer Review and U.S. Science Policy.* Albany: State University of New York Press.

Cichetti, D. V. (1991). "The Reliability of Peer Review for Manuscript and Grant Submissions: A Cross-Disciplinary Investigation." *Behavioral and Brain Sciences* 14: 119–135.

Claussen, E., and L. McNeilly. (1998). *Equity and Global Climate Change: The Complex Elements of Global Fairness.* Arlington, VA: Pew Center on Global Climate Change.

Clinton, W. J. (1997, October 22). "Remarks by the President on Global Climate Change." Washington, DC: National Geographic Society. Available at http://www.pub.whitehouse.gov/retrieve-documents.html.

Cohen, W. I. (1991). *America in the Age of Soviet Power, 1945–1991.* Vol. 4, W. I. Cohen, ed., *The Cambridge History of American Foreign Relations.* Cambridge, England: Cambridge University Press.

Cole, S. (1992). *Making Science: Between Nature and Society.* Cambridge, MA: Harvard University Press.

Collins, H., and T. Pinch. (1993). *The Golem: What Everyone Should Know about Science.* Cambridge, England: Cambridge University Press.

Collins, H. M. (1985). *Changing Order: Replication and Induction in Science.* London: Sage.

Collins, L. C. J. R., Jr. (1969). "Automated Data Processing at the United States Air Force Environmental Technical Applications Center." In World Meteorological Organization, *Data Processing for Climatological Purposes.* Technical note no. 100. Geneva, Switzerland: World Meteorological Organization.

Comité Spécial de l'Année Géophysique Internationale. (1958). "The Fourth Meeting of the CSAGI." In M. Nicolet, ed., *Annals of the International Geophysical Year*, Vol. IIA, 297–395. New York: Pergamon Press.

Committee on Science, Engineering, and Public Policy of the U.S. National Academy of Sciences (COSEPUP). (1992). *Policy Implications of Greenhouse Warming*. Washington, DC: National Academy Press.

Cotgrove, S. (1982). *Catastrophe or Cornucopia: The Environment, Politics and the Future*. Chichester, England: Wiley.

Courant, R., K. O. Friedrichs, and H. Lewy. (1928). "Über die partiellen Differenzengleichungen der mathematischen Physik." *Mathematische Annalen* 100: 32–74.

Curti, M., and K. Birr. (1954). *Prelude to Point Four: American Technical Missions Overseas 1838–1938*. Madison: University of Wisconsin Press.

Cussins, C. M. (1998). " 'Quit Sniveling, Cryo-Baby. We'll Work Out Which One's Your Mama!' " In R. Davis-Floyd and J. Dumit, eds., *Cyborg Babies: From Techno-Sex to Techno-Tots*. New York: Routledge.

Daley, R. (1991). *Atmospheric Data Analysis*. Cambridge, England: Cambridge University Press.

Damon, P. E., and S. M. Kunen. (1976). "Global Cooling?" *Science* 193: 447–453.

Daniel, H. (1973). *One Hundred Years of International Co-operation in Meteorology (1873–1973)*. Report no. 345. Geneva, Switzerland: World Meteorological Organization.

Daniel, H.-D. (1993). *Guardians of Science: Fairness and Reliability of Peer Review*, trans. W. E. Russey. New York: VCH.

Darman, R. G. (1978). "The Law of the Sea: Rethinking U.S. Interests." *Foreign Affairs* 56 (2): 381.

Davies, A., and O. M. Ashford. (1990). *Forty Years of Progress and Achievement: A Historical Review of WMO*. Geneva, Switzerland: World Meteorological Organization.

Dear, P. (1995). *Discipline and Experience: The Mathematical Way in the Scientific Revolution*. Chicago: University of Chicago Press.

Dennis, A. S. (1972). "Conference Summary: Third National Conference on Weather Modification." *Bulletin of the American Meteorological Society* 53: 878–879.

Dennis, A. S. (1980). *Weather Modification by Cloud Seeding*. New York: Academic Press.

Dennis, M. (1994). " 'Our First Line of Defense': Two University Laboratories in the Postwar American State." *Isis* 85 (3): 427–455.

Donoghue, J. (1994). "International Law and Policy-Making about Global Change." In D. D. Caron, F. S. Chapin, J. Donoghue, M. Firestone, J. Harte, L. E. Wells, and R. Stewardson, eds., *Ecological and Social Dimensions of Global Change*. Berkeley: Institute of International Studies, University of California.

Douglas, M. (1986). *How Institutions Think*. Syracuse, NY: Syracuse University Press.

Dowlatabadi, H., and M. G. Morgan. (1993). "A Model Framework for Integrated Studies of the Climate Problem." *Energy Policy* 21 (3): 209–221.

Downey, G. L. (1992). "Agency and Structure in Negotiating Knowledge." In M. Douglas and D. Hull, eds., *How Classification Works*. Edinburgh: Edinburgh University Press.

Droessler, E. (1968). "Conference Summary of the First National Conference on Weather Modification." *Bulletin of the American Meteorological Society* 49: 982–987.

Droessler, E. G. (1972). "Weather Modification: Review and Perspective." *Bulletin of the American Meteorological Society* 53: 345–348.

Droessler, E. G. (1975). "Weather Modification: Some Proposals for Action." *Bulletin of the American Meteorological Society* 56: 676–678.

Droessler, E. G. (1983). "Remarks on Weather Modification." *Bulletin of the American Meteorological Society* 64: 966–967.

ECMWF Re-Analysis Project. (1995). "ERA Information Bulletin." Available at http://nic.fb4.noaa.gov:8000/research/erainfo.txt.

Edwards, P. N. (1996a). *The Closed World: Computers and the Politics of Discourse in Cold War America*. Cambridge, MA: MIT Press.

Edwards, P. N. (1996b). "Global Comprehensive Models in Politics and Policy-making." *Climatic Change* 32: 149–161.

Edwards, P. N. (1999). "Global Climate Science, Uncertainty and Politics: Data-Laden Models, Model-Filtered Data." *Science as Culture* 8 (4): 437–472.

Edwards, P. N. (2000). "A Brief History of Atmospheric General Circulation Modeling." In D. A. Randall, ed., *General Circulation Model Development, Past, Present, and Future*, 67–87. San Diago: Academic Press.

Edwards, P. N., and M. Lahsen. (Forthcoming). "Climate Science and Politics in the United States." In P. N. Edwards and C. Miller, eds., *Planetary Management and National Political Cultures*. Cambridge, MA: MIT Press.

Edwards, P. N., and C. A. Miller, eds. (Forthcoming). *Planetary Management and National Political Cultures*. Cambridge, MA: MIT Press.

Elliott, R. D. (1974). "Experience of the Private Sector." In W. N. Hess, ed., *Weather and Climate Modification*, 45–89. New York: Wiley.

Epstein, R. (1987). *A History of Econometrics*. Amsterdam: North-Holland.

Epstein, S. (1996). *Impure Science: AIDS, Activism, and the Politics of Knowledge*. Berkeley: University of California Press.

Ezrahi, Y. (1990). *The Descent of Icarus: Science and the Transformation of Contemporary Democracy*. Cambridge, MA: Harvard University Press.

Fairhead, J., and M. Leach. (1996). "Rethinking the Forest-Savanna Mosaic: Colonial Science and its Relics in West Africa." In M. Leach and R. Mearns, eds., *The Lie of the Land: Challenging Received Wisdom on the African Environment*. London: International African Institute.

Fan, S., M. Gloor, J. Mahlman, S. Pacala, J. Sarmiento, T. Takahashi, and P. Tans. (1998, October 16). "A Large Terrestrial Carbon Sink in North America Implied by Atmospheric and Oceanic Carbon Dioxide Data and Models." *Science* 282: 442–445.

Farhar, B. C. (1974). "The Impact of the Rapid City Flood on Public Opinion about Weather Modification." *Bulletin of the American Meteorological Society* 55: 759–764.

Ferguson, J. (1990). *The Anti-Politics Machine: "Development," Depoliticization, and Bureaucratic Power in Lesotho.* Cambridge, England: Cambridge University Press.

Filippov, V. V. (1969). "Quality Control Procedures for Meteorological Data." In World Meteorological Organization, *Data Processing for Climatological Purposes.* Technical note no. 100. Geneva, Switzerland: World Meteorological Organization.

"First Decade of Weather Satellites." (1970). *Bulletin of the American Meteorological Society* 51: 376–377.

Fleagle, R., J. A. Crutehfield, R. W. Johnson, and M. F. Abd. (1974). *Weather Modification in the Public Interest.* Seattle: American Meteorological Society and University of Washington Press.

Fleck, L. ([1935] 1979). *Genesis and Development of a Scientific Fact.* Chicago: University of Chicago Press.

Fleming, J. R. (1998). *Historical Perspectives on Climate Change.* New York: Oxford University Press.

Folger, T., S. Richardson, and C. Zimmer. (1994). "Remembering Apollo: Astronauts Recall Their Flights to the Moon." *Discover* 15 (7): 38.

Folland, C. K., and D. E. Parker. (1995). "Correction of Instrumental Biases in Historical Sea Surface Temperature Data." *Quarterly Journal of the Royal Meteorological Society* 121: 319–367.

Foucault, M. (1977). *Discipline and Punish*, trans. A. Sheridan. New York: Vintage Books.

Fukuyama, F. (1992). *The End of History or the Last Man.* New York: Free Press.

Fuller, R. B. (1969). *Operating Manual for Spaceship Earth.* Carbondale: Southern Illinois University Press.

Fussell, P. (1975). *The Great War and Modern Memory.* New York: Oxford University Press.

Galison, P. (1997). *Image and Logic: A Material Culture of Microphysics.* Chicago: University of Chicago Press.

Gentry, R. C. (1974). "Hurricane Modification." In W. N. Hess, ed., *Weather and Climate Modification*, 497–521. New York: Wiley.

Gibson, J. W. (1986). *The Perfect War: The War We Couldn't Lose and How We Did.* New York: Atlantic Monthly Press.

Giddens, A. (1990). *The Consequences of Modernity.* Stanford, CA: Stanford University Press.

Gieryn, T. F. (1996). "Boundaries of Science." In S. Jasanoff, G. E. Markle, J. C. Peterson, and T. Pinch, eds., *The Handbook of Science and Technology Studies*, 393–443. Thousand Oaks, CA: Sage.

Gieryn, T. F. (1999). *Cultural Boundaries of Science: Credibility on the Line.* Chicago: University of Chicago Press.

Gilbert, G. N., and M. Mulkay. (1984). *Opening Pandora's Box: A Sociological Analysis of Scientists' Discourse.* Cambridge, England: Cambridge University Press.

Gilchrist, B., and G. P. Cressman. (1954). "An Experiment in Objective Analysis." *Tellus* 6: 309–318.

Gilead, M. (1954). "The Technical Assistance Program in Meteorology in Israel: A Review and Reappraisal." *WMO Bulletin* 3 (2): 67–69.

Gilman, D. L., J. R. Hibbs, and P. L. Laskin. (1965). "Special Report: Weather and Climate Modification." *Bulletin of the American Meteorological Society* 46: 637–640.

Glantz, M. (1996). *Currents of Change: El Niño's Impact on Climate and Society.* Cambridge, England: Cambridge University Press.

Glantz, M. H., and D. Jamieson. (Forthcoming). "Societal Responses to Hurricane Mitch and Intra- versus Inter-generational Equity Issues: Whose Norms Should Apply?" *Risk Analysis.*

Global Environmental Assessment Team. (1997). *A Critical Evaluation of Global Environmental Assessments.* Calverton, MD: IGES/CARE.

Goldstine, H. (1972). *The Computer from Pascal to von Neumann.* Princeton: Princeton University Press.

Goodwin, C. (1994). "Professional Vision." *American Anthropology* 96: 606–633.

Gore, A. (1992). *Earth in the Balance: Ecology and the Human Spirit.* New York: Houghton Mifflin.

Graham, L. (1997). *The Ghost of the Executed Engineer: Technology and the Fall of the Soviet Union.* Cambridge, MA: Harvard University Press.

Greenaway, F. (1996). *Science International.* Cambridge, England: Cambridge University Press.

Greenwire. (1998, October 27). Electronic daily environmental news journal. Available at: http://nationaljournal.com/greenwire.

Gribbin, J. (1974, November 15). "Weather Warning: You Are Now Experiencing a Climatic Change." *Nature* 252: 182–183.

Gross, P., and N. Leavitt. (1994). *Higher Superstition: The Academic Left and Its Quarrels with Science.* Baltimore, MD: Johns Hopkins University Press.

Grove, R. (1997). *Ecology, Climate and Empire: Colonialism and Global Environmental History, 1400–1940*. Cambridge, England: White Horse Press.

Grubb, M. (1995). "Seeking Fair Weather: Ethics and the International Debate on Climate Change." *International Affairs* 71 (3): 463–496.

Haas, E. B., M. P. Williams, and D. Babai. (1978). *Scientists and World Order: The Uses of Technical Knowledge in International Organizations*. Berkeley: University of California Press.

Haas, P. M. (1989). "Do Regimes Matter? Epistemic Communities and Mediterranean Pollution Control." *International Organization* 43:377–403.

Haas, P. M. (1990a). "Obtaining International Environmental Protection through Epistemic Consensus." *Millennium* 19 (3): 347–364.

Haas, P. M. (1990b). *Saving the Mediterranean: The Politics of International Environmental Cooperation*. New York: Columbia University Press.

Haas, P. M. (1992). "Introduction: Epistemic Communities and International Policy Coordination." *International Organization* 46 (1): 1–36.

Haas, P. M., and E. B. Haas. (1995). "Learning to Learn: Improving International Governance." *Global Governance* 1 (3): 255–284.

Haas, P. M., R. Keohane, and M. Levy, eds. (1993). *Institutions for the Earth: Sources of Effective International Environmental Protection*. Cambridge, MA: MIT Press.

Hack, J. J. (1992). "Climate System Simulation: Basic Numerical and Computational Concepts." In K. E. Trenberth, ed., *Climate System Modeling*. Cambridge, England: Cambridge University Press.

Hacking, I. (1992). "Style for Historians and Philosophers." *Studies in History and Philosophy of Science* 23 (1): 1–20.

Haggard, S., and B. A. Simmons. (1987). "Theories of International Regimes." *International Organization* 41: 491–517.

Hajer, M. A. (1995). *The Politics of Environmental Discourse: Ecological Modernization and the Policy Process*. New York: Oxford University Press.

Hall, M. C. G. (1982). "Sensitivity Analysis of a Radiative-Convective Model by the Adjoint Method." *Journal of the Atmospheric Sciences* 39: 2038–2048.

Hammond, A. (1973, August 17). "Hurricane Prediction and Control: Impact of Large Computers." *Science* 181: 643–645.

Hammond, A. L. (1974). "Modeling the Climate: A New Sense of Urgency." *Science* 185: 1145–1147.

Hansen, J. E., A. Lacis, R. Ruedy, and M. Sato. (1992). "Potential Climatic Impact of Mount Pinatubo Eruption." *Geophysics Research Letters* 19: 215–218.

Hansen, J. E., W. Rossow, and I. Fung, eds. (1993). *Long-Term Monitoring of Global Climate Forcings and Feedbacks*. NASA Conference Publication no. 3234. Greenbelt, MD: Goddard Space Flight Center.

Hansen, J. E., M. Sato, A. Lacis, and R. Ruedy. (1997). "The Missing Climate Forcing." *Philosophical Transactions of the Royal Society of London B* 352: 231–240.

Hanson, N. R. (1958). *Patterns of Discovery*. Cambridge, England: Cambridge University Press.

Hart, D. M., and D. G. Victor. (1993). "Scientific Elites and the Making of U.S. Policy for Climate Change Research, 1957–74." *Social Studies of Science* 23: 643–680.

Harvey, D. (1986). *The Condition of Postmodernity*. Oxford: Blackwell.

Harwit, M. (1996). *An Exhibit Denied: Lobbying the History of Enola Gay*. New York: Copernicus.

Hays, S. P. (1989). "Three Decades of Environmental Politics: The Historical Context." In M. J. Lacey, ed., *Government and Environmental Politics*, 19–79. Washington, DC: Woodrow Wilson Center Press.

Heims, S. J. (1980). *John von Neumann and Norbert Wiener*. Cambridge MA: MIT Press.

Hess, W. N., ed. (1974). *Weather and Climate Modification*. New York: Wiley.

Hide, R. (1953). "Some Experiments on Thermal Convection in a Rotating Liquid." *Quarterly Journal of the Royal Meteorological Society* 79: 161.

Hilgartner, S. (1997). "The 'Sokal Affair' in Context." *Science, Technology & Human Values* 22: 506–522.

Hope, C. (1993). "Policy Analysis of the Greenhouse Effect: An Application of the PAGE Model." *Energy Policy* 21 (3): 327–338.

Houghton, J. T., L. G. Meira-Filho, J. Bruce, H. Lee, B. A. Callander, E. Haites, N. Harris, and K. Maskell, eds. (1995). *Climate Change 1994: Radiative Forcing of Climate Change and an Evaluation of the IPCC IS92 Emission Scenarios*. Cambridge, England: Cambridge University Press.

Houghton, J. T., L. G. Meira-Filho, B. A. Callander, N. Hams, A. Kattenberg, and K. Maskell, eds. (1996). *Climate Change 1995: The Science of Climate Change*. Vol. 1 of 3. Cambridge, England: Cambridge University Press.

Houghton, J. T., F. W. Taylor, and C. D. Rodgers. (1984). *Remote Sounding of Atmospheres*. Cambridge, England: Cambridge University Press.

Howell, W. E. (1977). "Environmental Impacts of Precipitation Management: Results and Inferences from Project Skywater." *Bulletin of the American Meteorological Society* 58 (1977): 488–501.

Hurrell, J. W., and K. E. Trenberth. (1997). "Spurious Trends in Satellite MSU Temperatures from Merging Different Satellite Records." *Nature* 386 (6621): 164–166.

Huschke, R. E. (1963). "A Brief History of Weather Modification Since 1946." *Bulletin of the American Meteorological Society* 44: 425–429.

Ikenberry, G. J. (1992). "A World Economy Restored: Expert Consensus and the Anglo-American Postwar Settlement." *International Organization* 46 (1): 289–322.

Intergovernmental Panel on Climate Change. (1993). "IPCC Procedures for Preparation, Review, Acceptance, Approval, and Publication of its Reports." Produced by IPCC Secretariat. Available at http://www.usgcrp.gov/ipcc/html/rulespro.html.

International Meteorological Organization. (1947). *Conference of Directors, Washington, September 22–October 11, 1947*. Geneva, Switzerland: World Meteorological Organization.

Iriye, A. (1997). *Cultural Internationalism and World Order*. Baltimore: Johns Hopkins University Press.

Jamieson, D. (1990). "Managing the Future: Public Policy, Scientific Uncertainty, and Global Warming." In D. Scherer, ed., *Upstream/Downstream: Essays in Environmental Ethics*, 67–89. Philadelphia: Temple University Press.

Jamieson, D. (1991). "The Epistemology of Climate Change: Some Morals for Managers." *Society and Natural Resources* 4: 319–329.

Jamieson, D. (1992). "Ethics, Public Policy, and Global Warming." *Science, Technology, and Human Values* 17 (2): 139–153.

Jasanoff, S. (1986). *Risk Management and Political Culture*. New York: Russell Sage Foundation.

Jasanoff, S. (1990). *The Fifth Branch: Science Advisors as Policymakers*. Cambridge, MA: Harvard University Press.

Jasanoff, S. (1991). "Acceptable Evidence in a Pluralistic Society." In D. G. Mayo and R. D. Hollander, eds., *Acceptable Evidence: Science and Values in Risk Management*. New York: Oxford University Press.

Jasanoff, S. (1993, March). "India at the Crossroads of Global Environmental Change." *Global Environmental Change*, 3: 32–52.

Jasanoff, S. (1995). "Product, Process, or Programme: Three Cultures and the Regulation of Biotechnology." In M. Bauer, ed., *Resistance to New Technology*. Cambridge, England: Cambridge University Press.

Jasanoff, S. (1996a). "Is Science Socially Constructed—And Can it Still Inform Public Policy?" *Science and Engineering Ethics* 2 (3): 263–276.

Jasanoff, S. (1996b). *Science at the Bar: Law, Science, and Technology in America*. Cambridge, MA: Harvard University Press.

Jasanoff, S. (1996c). "Science and Norms in International Environmental Regimes." In F. O. Hampson and J. Reppy, eds., *Earthly Goods: Environmental Change and Social Justice*, 173–197. Ithaca, NY: Cornell University Press.

Jasanoff, S., ed. (1997a). *Comparative Science and Technology Policy*. Cheltenham, England: Elgar.

Jasanoff, S. (1997b). "Compelling Knowledge in Public Decisions." In L. A. Brooks and S. VanDeveer, eds., *Saving the Seas: Values, Scientists, and International Governance*, 229–252. College Park: University of Maryland Press.

Jasanoff, S. (1998). "The Eye of Everyman: Witnessing DNA in the Simpson Trial." *Social Studies of Science* 28: 713–740.

Jasanoff, S. (1999). "The Songlines of Risk." *Environmental Values* 8 (2): 135–152.

Jasanoff, S. (forthcoming). "Introduction: Empowering States." In S. Jasanoff, ed., *States of Knowledge: Science, Power, and Political Culture*.

Jasanoff, S., and B. Wynne. (1998). "Science and Decisionmaking." In S. Rayner and E. Malone, eds., *Human Choice and Climate Change: Vol. 1. The Societal Framework*, 1–87. Columbus, OH: Battelle Press.

Jasanoff, S., G. E. Markle, J. C. Peterson, and T. Pinch, eds. (1996). *Handbook of Science and Technology Studies*. Thousand Oaks, CA: Sage.

Jencks, C. (1984). *The Language of Post-Modern Architecture*. New York: Rizzoli.

Jenne, R. (1998, November 1). Interviewed by P. N. Edwards at National Center for Atmospheric Research. Oral history interview.

Johnson, G. L., J. M. Davis, T. R. Karl, A. L. MeNab, K. P. Gallo, J. D. Tarpley, and P. Bloomfield. (1994). "Estimating Urban Temperature Bias Using Polar-Orbiting Satellite Data." *Journal of Applied Meteorology* 33: 358–369.

Jones, C. A., and P. Galison. (1998). *Picturing Science, Producing Art*. New York: Routledge.

Jones, S. H. S. (1959). "The Inception and Development of the International Geophysical Year." In S. Chapman, ed., *Annals of the International Geophysical Year*, 383–414. New York: Pergamon Press.

Judson, H. F. (1994). "Structural Transformation of the Sciences and the End of Peer Review." *Journal of the American Medical Association* 272: 92–94.

Kalser, J. (1998, October 16). "Possibly Vast Greenhouse Gas Sponge Ignites Controversy." *Science* 282: 386–387.

Karl, T. R., R. W. Knight, and J. R. Christy. (1994). "Global and Hemispheric Temperature Trends: Uncertainties Related to Inadequate Spatial Sampling." *Journal of Climate* 7 (7): 1144–1163.

Kassirer, J. P., and E. W. Campion. (1994). "Peer Review: Crude and Understudied, But Indispensable." *Journal of the American Medical Association* 272: 96–97.

Katz, M. B. (1978). *Questions of Uniqueness and Resolution in Reconstruction from Projections*. Lecture Notes in Biomathematics, Vol. 26. Berlin: Springer-Verlag.

Kaufmann, W. J. III, and L. L. Smarr. (1993). *Supercomputing and the Transformation of Science*. New York: Scientific American Library.

Kay, L. (1993). *The Molecular Vision of Life: Caltech, the Rockefeller Foundation, and the Rise of the New Biology.* Oxford: Oxford University Press.

Kellogg, W. W., and S. H. Schneider. (1974, December 27). "Climate Stabilization: For Better or for Worse?" *Science* 186: 1163–1172.

Kennan, G. F. (1951). *American Diplomacy 1900–1950.* Chicago: University of Chicago Press.

Kennan, G. F. (1997). "Diplomacy without Diplomats?" *Foreign Affairs* 76 (5): 198–213.

Keohane, R. O. (1988). "International Institutions: Two Approaches." *International Studies Quarterly* 32: 379–397.

Keohane, R. O. (1996). "Analyzing the Effectiveness of International Environmental Institutions." In R. Keohane and M. Levy, eds., *Institutions for Environmental Aid: Pitfalls and Promise.* Cambridge, MA: MIT Press.

Keohane, R. O., P. M. Haas, and M. Levy. (1993). "The Effectiveness of International Institutions." In P. Haas, R. O. Keohane, and M. Levy, eds., *Institutions for the Earth: The Sources of Effective International Environmental Protection,* 3–26. Cambridge, MA: MIT Press.

Keohane, R. O., and M. Levy, eds. (1996). *Institutions for Environmental Aid: Pitfalls and Promise.* Cambridge, MA: MIT Press.

Kerr, R. A. (1982, July 16). "Test Fails to Confirm Cloud Seeding Effect." *Science* 217: 234–236.

Kerr, R. A. (1997, May 9). "Climate Change: Model Gets It Right—Without Fudge Factors." *Science* 276: 1041.

Kiehl, J. T. (1992). "Atmospheric General Circulation Modeling." In K. E. Trenberth, ed., *Climate System Modeling.* Cambridge, England: Cambridge University Press.

Knight, D. (1998, November 10). "Developing Countries Cutting Greenhouse Emissions." *ECONET Highlights.* Available at http://www.igc.org/igc/en/hg/LDC-greenhouse.html.

Knorr-Cetina, K. (1991). "Epistemic Cultures: Forms of Reason in Science." *History of Political Economy* 23: 105–122.

Koertge, N. (1998). *A House Built on Sand: Exposing Postmodernist Myths about Science.* Oxford: Oxford University Press.

Kohler, R. (1992). *Lords of the Fly.* Chicago: University of Chicago Press.

Krasner, S., ed. (1982). *International Regimes.* Ithaca, NY: Cornell University Press.

Kuhn, T. S. (1962). *The Structure of Scientific Revolutions.* Chicago: University of Chicago Press.

Kwa, C. (1987). "Representations of Nature Mediating Between Ecology and Science Policy: The Case of the International Biological Program." *Social Studies of Science* 17: 413–442.

Kwa, C. (1993). "Modeling the Grasslands." *Historical Studies in the Physical and Biological Sciences* 24: 125–155; addendum in *HSPBS* 25 (1994): 184–186.

Kwa, C. (1994). "Modelling Technologies of Control." *Science as Culture* 4 (20): 363–391.

Lahsen, M. (1999). "The Detection and Attribution of Conspiracies: The Controversy Over Chapter 8." In G. E. Marcus, ed., *Paranoia within Reason: a Casebook on Conspiracy as Explanation.* Chicago: University of Chicago Press.

Landsberg, H. (1970). "Man-Made Climatic Changes." *Science* 170: 1265–1274.

Landsea, C. W. (1996, rev. 2000). "Hurricanes, Typhoons, and Tropical Cyclones: Part C. Tropical Cyclone Myths" [FAQ]. Atlantic Oceanographic and Meteorological Laboratory, National Oceanic and Atmospheric Administration. Available at http://www.aoml.noaa.gov/hrd/tcfaq/tcfaqC.html.

Langer, E. (1963, August 9). "Weather Bureau." *Science* 141: 508–509.

Lansford, H. (1973). "Weather Modification: The Public Will Decide." *Bulletin of the American Meteorological Society* 54: 658–660.

Lansing, J. S. (1991). *Priests and Programmers: Technologies of Power in the Engineered Landscape of Bali.* Princeton: Princeton University Press.

Latour, B. (1987). *Science in Action.* Cambridge MA: Harvard University Press.

Latour, B. (1988). *The Pasteurization of France.* Cambridge, MA: Harvard University Press.

Latour, B. (1990). "Drawing Things Together." In M. Lynch and S. Woolgar, eds., *Representation in Scientific Practice.* Cambridge, MA: MIT Press.

Latour, B. (1993). *We Have Never Been Modern.* New York: Harvester Wheatsheaf.

Latour, B. (1995). "The Pédofil of Boa Vista," trans. B. Simon and K. Verreson. *Common Knowledge* 4 (1): 144–187.

Latour, B., and S. Woolgar. (1979). *Laboratory Life: The Social Construction of Scientific Facts.* London: Sage.

Laudan, L. (1990). "Demystifying Underdetermination." In C. W. Savage, ed., *Scientific Theories,* 267–297. Minneapolis: University of Minnesota Press.

Law, J., ed. (1986). *Power, Action and Belief: A New Sociology of Knowledge?* London: Routledge and Kegan Paul.

Leary, N. (1999, July 19). "Non–Peer Reviewed Sources." E-mail to Intergovernmental Panel on Climate Change (IPCC) authors. Technical Support Unit, IPCC Working Group II.

Lee, D. (1973). "Requiem for Large-Scale Models." *Journal of the American Institute of Planners* 39, 117–142.

Levine, A. (1982). *Love Canal: Science, Politics, and People.* Lexington, MA: Lexington Books.

Lipschutz, R. D., and K. Conca, eds. (1993). *The State and Social Power in Global Environmental Politics.* New York: Columbia University Press.

Lipschutz, R. D., and J. Mayer. (1996). *Global Civil Society and Global Environmental Governance.* Albany, NY: SUNY Press.

Litfin, K. (1994). *Ozone Discourses: Science and Politics in Global Environmental Cooperation.* New York: Columbia University Press.

Litfin, K., ed. (1998). *The Greening of Sovereignty in World Politics.* Cambridge, MA: MIT Press.

Logan, J. D. (1987). *Applied Mathematics: A Contemporary Approach.* New York: Wiley.

"Long-Range Weather Forecasts by Computer." (1954, June 12). *Science News Letter,* p. 376.

Lovelock, J. E. (1979). *Gaia: A New Look at Life on Earth.* Oxford: Oxford University Press.

Lowi, T. (1969). *The End of Liberalism: The Second Republic of the United States.* New York: Norton.

Lynch, M. (1990). "The Externalized Retina: Selection and Mathematization in the Visual Documentation of Objects in the Life Sciences." In M. Lynch and S. Woolgar, eds., *Representation in Scientific Practice,* 153–186. Cambridge, MA: MIT Press.

Lynch, M., and S. Woolgar. (1990). *Representation in Scientific Practice.* Cambridge, MA: MIT Press.

MacKenzie, D. (1990). *Inventing Accuracy.* Cambridge, MA: MIT Press.

MacKenzie, D. (1999). "The Science Wars and the Past's Quiet Voices." *Social Studies of Science* 29: 199–214.

Mahoney, M. J. (1977). "Publication Prejudices: An Experimental Study of Confirmatory Bias in the Peer Review System." *Cognitive Therapy and Research* 1: 161–175.

Maienschein, J. (1991). "Epistemic Styles in German and American Embryology." *Science in Context* 4 (2): 407–427.

Mallows, C. L., and J. W. Tukey. (1982). "An Overview of Techniques of Data Analysis, Emphasizing its Exploratory Aspects." In J. T. de Oliveira and B. Epstein, eds., *Some Recent Advances in Statistics,* 111–172. New York: Academic Press.

Malone, T. F. (1967, May 19). "Weather Modification: Implications of the New Horizons in Research." *Science* 156: 897–901.

Manabe, S. (1967). "General Circulation of the Atmosphere." *Transactions of the American Geophysical Union* 48 (2): 427–431.

Manabe, S. (1997). "Early Development in the Study of Greenhouse Warming: The Emergence of Climate Models." *Ambio* 26 (1): 47–51.

Manabe, S., J. Smagorinsky, and R. F. Strickler. (1965, December). "Simulated Climatology of General Circulation with a Hydrologic Cycle." *Monthly Weather Review* 93: 769–798.

Manabe, S., and R. J. Stouffer. (1994). "Multiple-Century Response of a Coupled Ocean-Atmosphere Model to an Increase of Atmospheric Carbon Dioxide." *Journal of Climate* 7: 5–23.

Manabe, S., R. Stouffer, and M. Spelman. (1994). "Response of a Coupled Ocean-Atmosphere Model to Increasing Atmospheric Carbon Dioxide." *Ambio* 23 (1): 44–49.

Mann, M. E., R. S. Bradley, and M. K. Hughes. (1999). "Northern Hemisphere Temperatures during the Past Millennium: Inferences, Uncertainties, and Limitations." *Geophysical Research Letters* 29 (6): 759.

Margulis, L., and J. Lovelock. (1976, June/July). "Is Mars a Spaceship, Too?" *Natural History*: 86–90.

Marsh, H. W., and S. Ball. (1989). "The Peer Review Process Used to Evaluate Manuscripts Submitted to Academic Journals: Interjudgmental Reliability." *Journal of Experimental Education* 57 (2): 151–169.

Masood, E. (1996). "Climate Report 'Subject to Scientific Cleansing.'" *Nature* 381 (6583): 546.

McCormick, J. (1989). *Reclaiming Paradise: The Global Environmental Movement*. Bloomington: Indiana University Press.

McDonald, J. E. (1956). "Conference on the Scientific Basis of Weather Modification Studies, April 10–12, Institute of Atmospheric Physics, University of Arizona." *Bulletin of the American Meteorological Society* 37: 318.

McDougall, W. A. (1997). *Promised Land, Crusader State: The American Encounter with the World since 1776*. Boston: Houghton Mifflin.

McGuffie, K., and A. Henderson-Sellers. (1997). *A Climate Modelling Primer*. 2nd ed. Chichester, England: Wiley.

Meadows, D. H., D. L. Meadows, S. Rauderg, and W. W. Behrens. (1972). *The Limits to Growth*. New York: Universe Books.

Meehl, G. A. (1990). "Seasonal Cycle Forcing of El Niño–Southern Oscillation in a Global Coupled Ocean-Atmosphere GCM." *Journal of Climate* 3: 72–98.

Meehl, G. A. (1992). "Global Coupled Models: Atmosphere, Ocean, Sea Ice." In K. E. Trenberth, ed., *Climate System Modeling*, 555–581. Cambridge, England: Cambridge University Press.

Merton, R. K. (1973). *The Sociology of Science: Theoretical and Empirical Investigations*. Chicago: University of Chicago Press.

Michelson, A. A. (1881). "The Relative Motion of the Earth and the Luminiferous Ether." *American Journal of Science* 22: 120–129.

Michelson, A. A. (1882). "Sur le mouvement relatif de la terre et de l'éther." *Comptes Rendus* 94: 520.

Michelson, A. A., and E. W. Morley. (1887). "On the Relative Motion of the Earth and the Luminiferous Ether." *American Journal of Science* 34: 333–345.

Miller, C. (Forthcoming). "Undermining the Postwar Settlement: The Reconstruction of Climate Science and Global Order." In S. Jasanoff, ed., *States of Knowledge: Science, Power, and Political Culture.*

Miller, C., S. Jasanoff, M. Long, W. Clark, N. Dickson, A. Iles, and T. Parris. (1997). "Shaping Knowledge, Defining Uncertainty: The Dynamic Role of Assessments." In Global Environmental Assessment Project, eds., *A Critical Evaluation of Global Environmental Assessments: The Climate Experience.* Calverton, MD: CARE.

Miller, D. C. (1922). "Ether-Drift Experiments at Mount Wilson Solar Observatory." *Physical Review* 19: 407–408.

Miller, D. C. (1925a). "Ether-Drift Experiments at Mount Wilson." *Proceedings of the National Academy of Sciences of the United States of America* 11: 306–314.

Miller, D. C. (1925b). "Ether-Drift Experiments at Mount Wilson." *Science* 61: 617–621.

Miller, D. C. (1928). "Conference on the Michelson-Morley Experiment: Held at the Mount Wilson Observatory, Pasadena, California, February 4 and 5, 1927." *Astrophysical Journal* 68: 352–267.

Miller, D. C. (1933). "The Ether-Drift Experiment and the Determination of the Absolute Motion of the Earth." *Reviews of Modern Physics* 5: 204–242.

Mintzer, I., and J. A. Leonard, eds. (1994). *Negotiating Climate Change: The Inside Story of the Framework Convention.* Cambridge, England: Cambridge University Press.

Mitchell, J. F. B., C. A. Senior, and W. J. Ingram. (1989, September 14). "CO_2 and Climate: A Missing Feedback?" *Nature* 341: 132–134.

Moran, G. (1998). *Silencing Scientists and Scholars in Other Fields: Power, Paradigm Controls, Peer Review, and Scholarly Communication.* Greenwich, CT.: Ablex.

Morgan, M. G., and D. W. Keith. (1995). "Subjective Judgments by Climate Experts." *Environmental Science and Technology* 29: 468A–476A.

Morrisette, P. (1989). "The Evolution of Policy Responses to Stratospheric Ozone Depletion." *Natural Resources Journal* 29 (3): 793–820.

Moss, R., and S. H. Schneider. (1997). "Session Synthesis Essay: Characterizing and Communicating Scientific Uncertainty: Building on the IPCC Second Assessment." In S. J. Hassol and J. Katzenberger, eds., *Elements of Change 1996.* Aspen, CO: Aspen Global Change Institute. Available at http://www.gcrio.org/ASPEN/science/eoc96/AGCIEOC96SSSII/AGCIEOC96SynthesisSSSII.html.

Namias, J. (1970). "Climatic Anomaly over the United States during the 1960s." *Science* 170: 741–743.

Namias, J. (1986). "Autobiography." In J. O. Roads, ed., *Namias Symposium*, 3–59. La Jolla, CA: Scripps Institute of Oceanography.

NASA Goddard Space Flight Center. (1998). "Assimilation Experiments and Data Sets." Produced by NASA Data Assimilation Office. Available at http://hera.gsfc.nasa.gov/subpages/experiments.html.

National Academy of Sciences. (1966). *Weather and Climate Modification: Problems and Prospects: Vol. 2. Research and Development*. Final Report of the Panel on Weather and Climate Modification to the Committee on Atmospheric Sciences. Publication no. 1350. Washington, DC: National Academy of Sciences.

National Academy of Sciences. (1973). *Weather and Climate Modification: Problems and Progress*. Committee on Atmospheric Sciences of the National Research Council. Washington DC: National Academy of Sciences.

National Advisory Committee on Oceans and the Atmosphere. (1972). *Report*, reprinted in U.S. House of Representatives (1976), *Weather Modification*, Hearings before the Subcommittee on the Environment and the Atmosphere of the House Committee on Science and Technology. Washington, DC: U.S. Government Printing Office.

National Advisory Committee on Oceans and the Atmosphere. (1976, June 30). *Fifth Annual Report*, reprinted in U.S. House of Representatives (1976), *Weather Modification*, Hearings before the Subcommittee on the Environment and the Atmosphere of the House Committee on Science and Technology, 170. Washington, DC: U.S. Government Printing Office.

National Oceanic and Atmospheric Administration. (1999). "NCEP/NCAR CDAS/Reanalysis Project." Available at http://wesley.wwb.noaa.gov/reanalysis.html#intro.

National Research Council. (1983). *Changing Climate: Report of the Carbon Dioxide Assessment Committee*. Washington, DC: National Academy Press.

National Science Foundation. (1965). *Weather and Climate Modification*. Report of the Special Commission on Weather Modification. Washington DC: National Science Foundation.

National Science and Technology Council. (1994). *Our Changing Planet: The FY 1995 U.S. Global Change Research Program*. Report by the Subcommittee on Global Change Research, Committee on Environment and Natural Resources, Supplement to the US President's FY 1995 Budget. Washington, DC: National Science and Technology Council.

Nature editors. (1996). "Climate Debate Must Not Overheat." *Nature* 381 (6583): 539.

Nebeker, F. (1995). *Calculating the Weather: Meteorology in the 20th Century*. New York: Academic Press.

Nelkin, D. (1992). *Controversy: Politics of Technical Decisions*. Newbury Park, CA: Sage.

Nora, P., ed. (1987). *Les lieux de mémoire*. Paris: Gallimard.

Nordhaus, W. D. (1994). "Expert Opinion on Climatic Change." *American Scientist* 82 (1): 45–51.

Norton, G. (1947). "Address of Welcome by Mr. Garrison Norton, Assistant Secretary of State." In WMO, *Conference of Directors, Washington, September 22–October 11, 1947*, 372–376. Geneva, Switzerland: World Meteorological Organization.

Ondaatje, M. (1992). *The English Patient*. London: Bloomsbury.

Oreskes, N., K. Shrader-Frechette, and K. Belitz. (1994). "Verification, Validation, and Confirmation of Numerical Models in the Earth Sciences." *Science* 263: 641–646.

O'Riordan, T., and J. Jaeger. (1996). *Politics of Climate Change: A European Perspective*. New York: Routledge.

Ottino della Chiesa, A. (1985). *The Complete Paintings of Leonardo da Vinci*. Introduction by L. D. Ettlinger. Harmondsworth, Middlesex, England: Penguin Books.

Palmer, T. N. (1993). "A Nonlinear Dynamical Perspective on Climate Change." *Weather* 48 (10): 314–326.

Panofsky, H. A. (1949). "Objective Weather-Map Analysis." *Journal of Meteorology* 6: 386–392.

Parker, D. E., and D. I. Cox. (1995). "Towards a Consistent Global Climatological Rawinsonde Data-Base." *International Journal of Climatology* 15: 473–496.

Parker, R. L. (1994). *Geophysical Inverse Theory*. Princeton: Princeton University Press.

Parry, M., N. Arnell, M. Hulme, R. Nicholls, and M. Livermore. (1998). "Adapting to the Inevitable." *Nature* 395 (22 October): 741.

Peterson, M. J. (1988). *Managing the Frozen South: The Creation and Evolution of the Antarctic Treaty System*. Berkeley: University of California Press.

Phillips, N. A. (1956). "The General Circulation of the Atmosphere: A Numerical Experiment." *Quarterly Journal of the Royal Meteorological Society* 82 (352): 123–164.

Phillips, N. A. (1959). "An Example of Non-Linear Computational Instability." In Bert Bolin, ed., *The Atmosphere and the Sea in Motion*, 501–504. New York: Rockefeller Institute Press with Oxford University Press.

Pickering, A. (1995). *The Mangle of Practice*. Chicago: Chicago University Press.

Pielke, R. A. (1991). "Overlooked Scientific Issues in Assessing Hypothesized Greenhouse Gas Warming." *Environmental Software* 6 (2): 100–107.

Pielke, R. A. (1994, August 24). "Don't Rely on Computer Models to Judge Global Warming." *Christian Science Monitor*, p. 19.

Pielke, R. A. (1998). "Rethinking the Role of Adaptation in Climate Policy." *Global Environmental Change* 8 (2): 159–170.

Plass, G. N. (1956). "The Carbon Dioxide Theory of Climatic Change." *Tellus* 8: 140–154.

Platzman, G. (1967). "A Retrospective View of Richardson's Book on Weather Prediction." *Bulletin of the American Meteorological Society* 48: 514–550.

Platzman, G. W. (1979). "The ENIAC Computations of 1950—Gateway to Numerical Weather Prediction." *Bulletin of the American Meteorological Society* 60: 302–312.

Pocock, J. (1975). *The Machiavellian Moment: Florentine Political Thought and the Atlantic Republican Tradition.* Cambridge, England: Cambridge University Press.

Ponce, P. (1998). "A Stormy Year." Transcript of PBS NewsHour interview, December 29, 1998. Produced by PBS Online NewsHour. Available at www.pbs.org/newshour/bb/weather/july-dec98/weather_12-29.html.

Popper, K. R. ([1934] 1959). *The Logic of Scientific Discovery.* New York: Basic Books.

Porter, T. (1995). *Trust in Numbers: The Pursuit of Objectivity in Science and Public Life.* Princeton: Princeton University Press.

President's Air Policy Commission. (1948). *Survival in the Air Age.* Washington, DC: U.S. Government Printing Office.

President's Council of Economic Advisers. (1998, July). *The Kyoto Protocol and the President's Policies to Address Climate Change.* Available at http://www.epa.gov/globalwarming/publications/actions/wh_kyoto/wh_full_rpt.pdf.

President's Council on Environmental Quality. (1970, August). "Man's Inadvertent Modification of Weather and Climate." *Bulletin of the American Meteorological Society* 51: 1043–1047.

Price, D. (1965). *The Scientific Estate.* Cambridge, MA: Harvard University Press.

Rachels, J. (1990). *Created from Animals: The Moral Implications of Darwinism.* New York: Oxford University Press.

Rahmstorf, S. (1997, February). "Ice-Cold in Paris." *New Scientist* 8: 26–30.

RAND Corporation. (1969). "Weather Modification Progress and the Need for Interactive Research." Weather Modification Research Project, Santa Monica, California. *Bulletin of the American Meteorological Society* 50: 216–246.

Rasch, P. J., and D. L. Williamson. (1990). "Computational Aspects of Moisture Transport in Global Models of the Atmosphere." *Quarterly Journal of the Royal Meteorological Society* 116: 1071–1090.

Rayner, S. (1993, March). "Special Issue: National Case Studies of Institutional Capabilities to Implement Greenhouse Gas Reductions." *Global Environmental Change*, 3: 7–11.

Rayner, S., and E. Malone. (1997, November 27). "Zen and the Art of Climate Maintenance." *Nature* 390: 332–334.

Rayner, S., and E. Malone. (1998). "Why Study Human Choice and Climate Change?" In S. Rayner and E. Malone, eds., *Human Choice and Climate Change.* Columbus, OH: Battelle Press.

Rayner, S., E. L. Malone, and M. Thompson. (1999). "Equity Issues and Integrated Assessment." In F. L. Toth, ed., *Fair Weather? Equity Concerns in Climate Change*, 11–43. London: Earthscan.

Reed, J. W. (1973). Letter. *Bulletin of the American Meteorological Society* 54: 676–677.

Reed, J. W. (1974). "Another Round over Rapid City." *Bulletin of the American Meteorological Society* 55: 786–787.

Rees, W. G. (1990). *Physical Principles of Remote Sensing.* Cambridge, England: Cambridge University Press.

Revelle, R., and H. E. Suess. (1957). "Carbon Dioxide Exchange Between the Atmosphere and Ocean and the Question of an Increase of Atmospheric CO_2 during the Past Decades." *Tellus* 9: 18–27.

Rheingold, H. (1994). "Introduction." In H. Rheingold and S. Brand, eds., *The Millennium Whole Earth Catalog: Access to Tools and Ideas for the Twenty-First Century*, i. San Francisco: Harper San Francisco: Available at http://www.rheingold.com/texts/mwecintro.html.

Richardson, L. F. (1922). *Weather Prediction by Numerical Process.* London: Cambridge University Press.

Robinson, G. D. (1967). "Some Current Projects for Global Meteorological Observation and Experiment." *Quarterly Journal of the Royal Meteorological Society* 93 (398): 409–418.

Rodgers, C. D. (1977). "Statistical Principles of Inversion Theory." In A. Deepak, ed., *Inversion Methods in Atmospheric Remote Sounding*, 117–134. New York: Academic Press.

Rosenberg, A. (1975). "The Virtues of Vagueness in the Languages of Science." *Dialogue: Canadian Philosophical Review* 14: 281–305.

Rosenthal, S. L. (1974). "Computer Simulation of Hurricane Development and Structure." In W. N. Hess, ed., *Weather and Climate Modification*, 523–551. New York: Wiley.

Ross, P. F. (1980). *The Sciences' Self-Management: Manuscript Refereeing, Peer Review, and Goals in Science.* Lincoln, MA: Ross.

Rotmans, J. (1990). *IMAGE: An Integrated Model to Assess the Greenhouse Effect.* Boston: Kluwer.

Rotmans, J. (1992). "ESCAPE: An Integrated Climate Model for the EC." *Change* II: 1–4.

Rowlands, I. H. (1995). *The Politics of Global Atmospheric Change.* New York: Manchester University Press.

Roy, R. (1985). "Funding Science: The *Real* Defects of Peer Review and an Alternative to It." *Science, Technology, & Human Values* 10 (3): 73–81.

Rudwick, M. (1982). "Cognitive Styles in Geology." In M. Douglas, ed., *Essays in the Sociology of Perception*. London: Routledge Kegan Paul.

Ruggie, J. G. (1986). "Social Time and International Policy: Conceptualizing Global Population and Resource Issues." In M. P. Karns, ed., *Persistent Patterns and Emergent Structures in a Waning Century*. New York: Praeger.

Ruggie, J. G., ed. (1993). *Multilateralism: The Anatomy of an Institution*. New York: Columbia University Press.

Sachs, Wolfgang. (1994). "Satellitenblick: Die Ikone vom blauen Planeten und ihre Folgen für die Wissenschaft." In Ingo Braun and Bernward Joerges, eds., *Technik ohne Grenzen*, 305–346. Frankfurt: Suhrkamp.

Sagan, C. (1994). *The Pale Blue Dot*. New York: Random House.

Santer, B. D., T. M. L. Wigley, T. P. Barnett, and E. Anyamba. (1995). "Detection of Climate Change and Attribution of Causes." In J. T. Houghton, L. G. Meira Filho, B. A. Callander, N. Harris, A. Katzenberg, and K. Marshall, eds., *Climate Change 1995: The Science of Climate Change*, 407–443. Cambridge, England: Cambridge University Press.

Santer, B. D. (1996a, June 25). "Letter to the Editor: No Deception in Global Warming Report." *Wall Street Journal*, p. A15.

Santer, B. D., K. E. Tayler, T. M. L. Wigley, T. C. Johns, P. D. Jones, D. J. Karoly, J. F. B. Mitchell, A. H. Oert, J. E. Penner, V. Ramaswamy, M. D. Schwarzkopf, R. J. Stouffer, and S. Tett. (1996b). "A Search for Human Influences on the Thermal Structure of the Atmosphere." *Nature* 382 (6586): 39–46.

Sargent, F. II. (1968). "Weather Modification and the Biosphere." In American Meteorological Society, ed., *Proceedings of the First National Conference on Weather Modification*, 173–180. Albany, NY: State University of New York Press.

Schaefer, V. J. (1968a). "The Early History of Weather Modification." *Bulletin of the American Meteorological Society* 49: 337–342.

Schaefer, V. J. (1968b). "New Field Evidence of Inadvertent Modification of the Atmosphere." In American Meteorological Society, ed., *Proceedings of the First National Conference on Weather Modification*, 163–172. Albany, NY.

Schlesinger, M. E., and J. F. B. Mitchell. (1985). "Model Projections of the Equilibrium Climatic Response to Increased Carbon Dioxide." In M. C. MacCracken and F. M. Luther, eds., *Projecting the Climatic Effects of Increasing Carbon Dioxide*, 81–148. DOE/ER 0237. Washington, DC: U.S. Department of Energy.

Schneider, S. H. (1976). *The Genesis Strategy: Climate and Global Survival*. New York: Plenum Press.

Schneider, S. H. (1979). "Verification of Parameterizations in Climate Modeling." In W. Lawrence, ed., *Report of the JOC Conference on Climate Models: Performance, Intercomparison, and Sensitivity Studies*. Washington, D.C.: WMO/ICSU Joint Organizing Committee, Global Atmospheric Research Programme.

Schneider, S. H. (1989). *Global Warming: Are We Entering the Greenhouse Century?* New York: Vintage Books.

Schneider, S. H. (1992). "Introduction to Climate Modeling." In K. E. Trenberth, ed., *Climate System Modeling*, 3–26. Cambridge, England: Cambridge University Press.

Schneider, S. H. (1994, January 21). "Detecting Climatic Change Signals: Are There Any 'Fingerprints'?" *Science*: 341–347.

Schneider, S. H. (1997). *Laboratory Earth: The Planetary Gamble We Can't Afford to Lose.* New York: Basic Books.

Schneider, S. H., ed. (1991). *Scientists on Gaia.* Cambridge, MA: MIT Press.

Scott, J. C. (1998). *Seeing Like a State.* New Haven: Yale University Press.

Sebenius, J. K. (1983). "Negotiation Arithmetic: Adding and Subtracting Issues and Parties." *International Organization* 37: 281–316.

Sebenius, J. K. (1984). *Negotiating the Law of the Sea.* Cambridge, MA: Harvard University Press.

Sebenius, J. K. (1991). "Designing Negotiations Toward a New Regime: The Case of Global Warming." *International Security*, 15 (4): 110–148.

Seitz, F. (1996, June 12). "A Major Deception on Global Warming." *Wall Street Journal*, p. A16.

Shackley, S., and B. Wynne. (1994). "Climatic Reductionism: The British Character and the Greenhouse Effect." *Weather* 49 (3): 110–111.

Shackley, S., and B. Wynne. (1995). "Global Climate Change: The Mutual Construction of an Emergent Science-Policy Domain." *Science and Public Policy* 22 (4): 218–230.

Shackley, S., and B. Wynne. (1996). "Representing Uncertainty in Global Climate Change Science: Boundary-Ordering Devices and Authority." *Science, Technology, and Human Values* 21 (3): 275–302.

Shackley, S., P. Young, S. Parkinson, and B. Wynne. (1998). "Uncertainty, Complexity and Concepts of Good Science in Climate Change Modelling: Are GCMs the Best Tools?" *Climatic Change* 38: 159–205.

Shackley, S., J. Risbey, P. Stone, and B. Wynne. (1999). "Adjusting to Policy Expectations in Climate Change Modelling: An Interdisciplinary Study of the Use of Flux Adjustments in Coupled A/O GCMs." *Climatic Change* 43: 413–454.

Shankland, R. S. (1955). "New Analysis of the Interferometer Observations of Dayton C. Miller." *Reviews of Modern Physics* 27: 167–178.

Shankland, R. S. (1964). "Michelson-Morley Experiment." *American Journal of Physics* 32: 23.

Shapin, S. (1994). *A Social History of Truth: Civility and Science in Seventeenth Century England.* Chicago: University of Chicago Press.

Shapin, S. (1996). "Cordelia's Love: Credibility and the Social Studies of Science." *Perspectives on Science: Historical, Philosophical, Social* 3 (3): 255–275.

Shapin, S., and S. Schaffer. (1985). *Leviathan and the Air Pump: Hobbes, Boyle, and the Experimental Life.* Princeton: Princeton University Press.

Shapley, D. (1974). "Weather Warfare: Pentagon Concedes 7-year Vietnam Effort." *Science* 184: 1059–1061.

Shiva, V. (1991). *The Violence of the Green Revolution: Third World Agriculture, Ecology, and Politics.* London: Zed Books.

Shiva, V. (1997). *Biopiracy: The Plunder of Nature and Knowledge.* Boston: South End Press.

Shue, H. (1995). "Avoidable Necessity: Global Warming, International Fairness, and Alternative Energy." In I. Shapiro and J. DeCew, eds., *Theory and Practice. NOMOS XXXVII,* 239–264. New York: New York University Press.

Simpson, J. (1975). "Concerning Weather Modification." Testimony to the House Subcommittee on International Organizations and Movements of the Committee on Foreign Affairs. Reprinted in *Bulletin of the American Meteorological Society* 56: 47–49.

Simpson, J., V. Wiggert, and T. R. Mee. (1968). "Models of Seeding Experiments on Supercooled and Warm Cumulus Clouds." In American Meteorological Society, ed., *Proceedings of the First National Conference on Weather Modification,* 251–265. Albany, NY.

Simpson, J., W. L. Woodley, and Robert M. White. (1972). "Joint Federal-State Cumulus Seeding Program for Mitigation of 1971 South Florida Drought." *Bulletin of the American Meteorological Society* 53: 334–344.

Simpson, R. H., and J. Malkus. (1964, December). "Experiments in Hurricane Formation." *Scientific American* 211: 27–37.

Singer, S. F. (1996, July 11). "Letter to the Editor: Coverup in the Greenhouse?" *Wall Street Journal,* p. A15.

Skolnikoff, E. B. (1967). *Science, Technology, and American Foreign Policy.* Cambridge, MA: MIT Press.

Slaughter, A.-M. (1997). "The Real New World Order." *Foreign Affairs* 76 (5): 183–197.

Smagorinsky, J. (1963). "General Circulation Experiments with the Primitive Equations." *Monthly Weather Review* 91 (3): 99–164.

Smagorinsky, J. (1983). "The Beginnings of Numerical Weather Prediction and General Circulation Modeling: Early Recollections." *Advances in Geophysics* 25: 3–37.

Smith, T. (1994). *America's Mission: The United States and the Worldwide Struggle for Democracy in the Twentieth Century.* Princeton: Princeton University Press.

Spence, C. C. (1980). *The Rainmakers.* Lincoln: University of Nebraska Press.

St.-Amand, P., R. J. Davis, and R. D. Elliott. (1973). "Comments on Mr. Reed's 'Dissenting View' on the Black Hills Flood." *Bulletin of the American Meteorological Society* 54: 678–680.

Star, S. L., and J. Griesemer. (1989). "Institutional Ecology, 'Translations,' and Boundary Objects: Amateurs and Professionals in Berkeley's Museum of Vertebrate Zoology, 1907–1939." *Social Studies of Science* 19: 387–420.

Stehr, N., and H. von Storch, eds. (2000). *Eduard Brückner—The Sources and Consequences of Climate Change and Climate Variability in Historical Times.* Dordrecht: Kluwer.

Stehr, N., H. von Storch, and M. Flügel. (1995). "The 19th Century Discussion of Climate Variability and Climate Change: Analogies for Present Debate?" *World Resources Review* 7: 589–604.

Stevens, S. S. (1946). "On the Theory of Scales of Measurement." *Science* 103: 677–680.

Stevenson, Adlai E. (1979). "Strengthening the International Development Institutions." In Walter Johnson, ed., *The Papers of Adlai E. Stevenson*, Vol. VIII. Boston: Little, Brown.

Stone, P. (1992, February/March). "Forecast Cloudy: The Limits of Global Warming Models." *Technology Review*, 95 (2): 32–40.

Storey, W. K. (1997). *Science and Power in Colonial Mauritius.* Rochester, NY: University of Rochester Press.

Stull, R. B. (1988). *An Introduction to Boundary Layer Meteorology.* Dordrecht, Holland: Kluwer.

Suess, H. E. (1953). "Natural Radiocarbon and the Rate of Exchange of Carbon Dioxide Between the Atmosphere and the Sea." In National Research Council Committee on Nuclear Science, ed., *Nuclear Processes in Geologic Settings*, 52–56. Washington, DC: National Academy of Sciences.

Suppe, F. (1985). "Information Science, Artificial Intelligence, and the Problem of Black Noise." In L. B. Heilprin, ed., *Toward Foundations of Information Science.* American Society for Information Sciences publication series, 63–78. New York: Knowledge Industries Publications.

Suppe, F. (1993). "Credentialing Scientific Claims." *Perspectives on Science*, 1: 153–203.

Suppe, F. (1997a). "Modeling Nature: From Venus Science to Health Science." Inaugural Lecture, School of Nursing Distinguished Lecturer Series, Columbia University, November 18.

Suppe, F. (1997b). "Science without Induction." In J. Earman and J. Norton, eds., *The Cosmos of Science*, 386–429. Pittsburgh: University of Pittsburgh Press.

Suppe, F. (1997c, November 7–8). "Scientific Sense and Philosophical Nonsense about Modeling." Workshop on Strategies for Modeling Biological Systems, Center for Integrated Study of Animal Behavior, Indiana University.

Suppe, F. (1998). "The Structure of a Scientific Paper." *Philosophy of Science* 65: 381–405.

Suppe, F. (1999). "The Changing Nature of Flight and Ground Test Instrumentation and Data: 1940–1969." In A. Rowland and P. Galison, eds., *Origins of Atmospheric Flight*. Dordrecht, Holland: Kluwer.

Suppe, F. (Forthcoming a). "Epistemology of Simulation Modeling."

Suppe, F. (Forthcoming b). *Facts, Theories, and Scientific Observation: Vol. 2. Scientific Knowledge.*

Suppe, F. (Forthcoming c). *Venus Alive! Modeling Scientific Knowledge.*

Suppes, P. (1962). "Models of Data." In E. Nagel, P. Suppes, and A. Tarski, eds., *Logic, Methodology, and the Philosophy of Science: Proceedings of the 1960 Congress*, 252–261. Stanford: Stanford University Press.

Takacs, D. (1996). *The Idea of Biodiversity: Philosophies of Paradise*. Baltimore: Johns Hopkins University Press.

Tarantola, A. (1987). *Inverse Problem Theory: Methods for Data Fitting and Model Parameter Estimation*. Amsterdam: Elsevier.

Taylor, P., and F. Buttel. (1992). "How Do We Know We Have Global Environmental Problems? Science and the Globalization of Environmental Discourse." *Geoforum* 23 (3): 405–416.

Thacher, P. (1976). "The Mediterranean Action Plan." *Ambio* 6 (6): 311–315.

Thompson, M., S. Rayner, and S. Ney. (1998a). "Risk and Governance: Part I. The Discourses of Climate Change." *Government and Opposition* 33 (2): 139–166.

Thompson, M., S. Rayner, and S. Ney. (1998b). "Risk and Governance: Part II. Policy in a Complex and Plurally Perceived World." *Government and Opposition* 33 (3): 330–354.

Thompson, P. D. (1961). "A Dynamical Method of Analyzing Meteorological Data." *Tellus* 13: 334–349.

Thompson, P. D. (1983). "A History of Numerical Weather Prediction in the United States." *Bulletin of the American Meteorological Society* 64: 755–769.

Tikhonov, A. N. (1963). "The Solution of Ill-Posed Problems." *Doklady Akademii Nauk SSSR* 151: 501–504.

Tolba, M. (1991). "Address by Dr. Mostafa K. Tolba." In J. Jaeger and H. L. Ferguson, eds., *Climate Change: Science, Impacts and Policy. Proceedings of the Second World Climate Conference*. Cambridge, England: Cambridge University Press.

Toth, F. L., ed. (1999). *Fair Weather? Equity Concerns in Climate Change*. London: Earthscan.

Trenberth, K. E., ed. (1992). *Climate System Modeling*. Cambridge, England: Cambridge University Press.

Treussart, H. (1957). "Technical Assistance and Meteorology." *WMO Bulletin* 6 (1): 18–20.

Tribe, L. (1973). "Technology Assessment and the Fourth Discontinuity: The Limits of Instrumental Rationality." *Southern California Law Review* 46: 617–660.

Tri-State Natural Weather Association, Inc. (n.d. [1975?]). *Cloud Seeding: The Technology of Fraud and Deceit*. Printed as annex in U.S. House of Representatives (1976), *Weather Modification*, Hearings before the Subcommittee on the Environment and the Atmosphere of the House Committee on Science and Technology. Washington, DC: U.S. Government Printing Office.

Tukey, J. W. (1962). "The Future of Data Analysis." *Annals of Mathematical Statistics* 33: 13.

Turner, B. L. II, R. H. Moss, and D. L. Skole. (1993, February). *Relating Land Use and Global Land-Cover Change*. International Geosphere-Biosphere Program, IGBP Report no. 24. Stockholm: International Geosphere-Biosphere Program.

Twomey, S. (1977). *Introduction to the Mathematics of Inversion in Remote Sensing and Indirect Measurements*. Amsterdam: Elsevier.

Udall, S. L. (1966). "Water Resources in the Sky." Speech at the 46th Annual Meeting of the AMS, 1966. Reprinted in *Bulletin of the American Meteorological Society* 47: 275–278.

United Nations. (1992). *United Nations Framework Convention on Climate Change*. Published by UN FCCC Secretariat. Available at http://www.unfccc.de/resource/conv/conv.html.

United Nations Framework Convention on Climate Change Subsidiary Body for Scientific and Technological Advice. (1997). *Cooperation with Relevant International Organizations: Conference on the World Climate Research Programme (Geneva, 26–28 August 1997)*. Report on the Conference, Note by the Secretariat. UN Doc. No. FCCC\SBSTA\1997\MISC.6.

University of Stockholm Institute of Meteorology. (1954). "Results of Forecasting with the Barotropic Model on an Electronic Computer (BESK)." *Tellus* 6: 139–149.

U.S. Department of Agriculture. (1941). *Climate and Man*. Washington, DC: U.S. Government Printing Office.

U.S. Department of Energy. (1980). *Workshop on Environmental and Societal Consequences of a Possible CO_2-Induced Climate Change*, held at Annapolis, Maryland, April 2–6, 1979. Carbon Dioxide Effects Research and Assessment Program. Washington, DC: U.S. Government Printing Office.

U.S. Department of State, Bureau of Public Affairs. (1996, March 19). "Fact Sheet: US Oceans Policy and the Law of the Sea Convention." Website produced by Electronic Research Collections. Available at http://dosfan.lib.uic.edu/ERC/environment/fact_sheets/960319.html.

U.S. Global Change Research Information Office. (1996). *Our Changing Planet: The FY 1997 U.S. Global Change Research Program.* Washington, DC: U.S. Global Change Research Information Office.

U.S. Global Change Research Information Office. (1999). *Our Changing Planet.* Washington, DC: U.S. Office of Science and Technology Policy.

U.S. Global Change Research Program. (1994, May 1–4). *Workshop on Earth System Modeling.* Subcommittee on Global Change Research Integrative Modeling and Prediction Working Group. Washington, DC: U.S. Global Change Research Program.

U.S. Global Change Research Program. (1995). *Forum on Global Change Modeling, October 12–13 1994.* Report no. 95-01. Washington, DC: U.S. Global Change Research Program.

U.S. House of Representatives. (1976a). *National Oceanic and Atmospheric Administration Oversight Hearings.* Hearings before the House Subcommittee on the Environment and the Atmosphere, July 1975. Washington, DC: U.S. Government Printing Office.

U.S. House of Representatives. (1976b). *Weather Modification.* Hearings before the Subcommittee on the Environment and the Atmosphere of the House Committee on Science and Technology. Washington, DC: U.S. Government Printing Office.

U.S. National Committee for the International Geophysical Year. (1956). *Proposed United States Program for the International Geophysical Year 1957–58.* Washington, DC: National Academy of Sciences.

U.S. Senate. (1964). *Weather Modification.* Hearing before the Subcommittee on Irrigation and Reclamation, U.S. Senate, 88th Congress, 2nd session, on a Program for Increasing Precipitation in the Colorado River Basin by Artificial Means. Washington, DC: U.S. Government Printing Office.

VanDeveer, S. (1998). "European Politics with a Scientific Face: Transition Countries, Integrated Environmental Assessment, and LRTAP Cooperation." *Environment and Natural Resources Program Discussion Paper E-98-09.* Cambridge, MA: GEA, Harvard University.

Van Mieghem, J. (1953). "International Co-ordination of Meteorological Research." *WMO Bulletin* 2 (4): 96–100.

Van Mieghem, J. (1955a). "International Geophysical Year 1957–58: Part I. Historical Survey." *WMO Bulletin* 4 (1): 6–9.

Van Mieghem, J. (1955b). "International Geophysical Year 1957–58: Part II. The Programme." *WMO Bulletin* 4 (2): 6–9.

Van Mieghem, J., ed. (1964). *Meteorology.* Annals of the International Geophysical Year 32. Oxford: Pergamon Press.

Victor, D. G. (1999). "The Regulation of Greenhouse Gases: Does Fairness Matter?" In F. L. Toth, ed., *Fair Weather? Equity Concerns in Climate Change,* 193–206. London: Earthscan.

Von Neumann, J. (1955, June). "Can We Survive Technology?" *Fortune*, pp. 106–108, 151–152.

Wamsted, D. (1996). "Doctoring the Documents?" *Energy Daily* 24 (98): 1–2.

Ward, B. (1966). *Spaceship Earth*. New York: Columbia University Press.

Ward, B., and R. Dubos. (1972). *Only One Earth: The Care and Maintenance of a Small Planet*. New York: Norton.

Washington, W. (1992). "Climate-Model Responses to Increased CO_2 and Other Greenhouse Gases." In K. Trenberth, ed., *Climate System Modeling*, 643–668. Cambridge, England: Cambridge University Press.

Watson, R. T., M. C. Zinyowera, and R. H. Moss, eds. (1996). *Climate Change 1995: Impacts, Adaptations and Mitigation of Climate Change: Scientific-Technical Analyses*. Vol. 2 of 3. Cambridge, England: Cambridge University Press.

Weart, S. (1997). "The Discovery of the Risk of Global Warming." *Physics Today* 50 (1): 34–41.

Weiss, E. B. (1989). *In Fairness to Future Generations: International Law, Common Patrimony, and Intergenerational Equity*. Dobbs Ferry, NY: Transnational Publishers and United Nations University.

White, R. M. (1971). "A Centennial—And a Beginning." *Bulletin of the American Meteorological Society* 52: 343–346.

White, R. M. (1990). "The Great Climate Debate." *Scientific American* 263 (1): 36–43.

White, R. M., and R. A. Chandler. (1965). "Project Stormfury: Status and Prospects." *Bulletin of the American Meteorological Society* 46: 320–322.

Wigley, T. M. L., G. I. Pearman, and P. M. Kelly. (1992). "Indices and Indicators of Climate Change: Issues of Detection, Validation, and Climate Sensiti-vity." In I. M. Mintzer, ed., *The Science of Climate Change*, 85–96. Cambridge, England: Cambridge University Press.

Wigley, T. M. L., R. L. Smith, and B. D. Santer. (1998). "Anthropogenic influence on the Autocorrelation Function of Hemispheric-Mean Temperatures." *Science* 282: 1676–1679.

Willoughby, H. E., D. P. Jorgenson, R. A. Black, and S. L. Rosenthal. (1985). "Project Stormfury: A Scientific Chronicle 1962–1983." *Bulletin of the American Meteorological Society* 66: 505–514.

Winner, L. (1986). *The Whale and the Reactor: A Search for Limits in an Age of High Technology*. Chicago: University of Chicago Press.

Winter, J. M. (1995). *Sites of Memory, Sites of Mourning: The Great War in European Cultural History*. New York: Cambridge University Press.

Wirth, T. E. (1996, July 17). "Statement on Behalf of the United States of America." Second Conference of Parties to the Framework Convention on Climate Change, Geneva, Switzerland. Provided by USGCRP Office.

Woodley, W. L. (1970, October 9). "Rainfall Enhancement by Dynamic Cloud Seeding." *Science* 170: 127–132.

Woodward, N. B., S. E. Boyer, and J. Suppe. (1989). *Balanced Geological Cross Sections: An Essential Technique in Geological Research and Exploration.* Short Course in Geology, Vol. 6. Washington, DC: American Geophysical Union.

World Commission on Environment and Development. (1987). *Our Common Future.* Oxford: Oxford University Press.

World Meteorological Organization. (1951a). First Congress of the World Meteorological Organization, Paris, March 19–April 28. *Final Report: Vol. I. Reports.* Document no. 1/I. Geneva, Switzerland: World Meteorological Organization.

World Meteorological Organization. (1951b). First Congress of the World Meteorological Organization, Paris, March 19–April 28. *Final Report: Vol. II. Minutes.* Document no. 1/II. Geneva, Switzerland: World Meteorological Organization.

World Meteorological Organization. (1951c). First Congress of the World Meteorological Organization, Paris, March 19–April 28. *Final Report: Vol. III. Documents.* Document no. 1/III. Geneva, Switzerland: World Meteorological Organization.

World Meteorological Organization. (1951d). Second Session of the Executive Committee, Lausanne, 3rd–20th October, *Resolutions.* Document no. 4. Geneva, Switzerland: World Meteorological Organization.

World Meteorological Organization. (1954). "Technical Assistance Programme." *WMO Bulletin* 3 (4): 128–132.

World Meteorological Organization. (1960). "Special Fund Activities." *WMO Bulletin* 9 (2): 79.

World Meteorological Organization. (1962). *Numerical Methods of Weather Analysis and Forecasting.* Technical note no. 44. Geneva, Switzerland: World Meteorological Organization.

World Meteorological Organization. (1990). *The WMO Achievement: 40 Years in the Service of International Meteorology and Hydrology.* Report no. 729. Geneva, Switzerland: World Meteorological Organization.

World Meteorological Organization. (1991). *Systematic Errors in Extended Range Predictions.* CAS/JSC Working Group on Numerical Experimentation, World Climate Research Programme (WMO/TD No. 444). Report no. 16. Geneva, Switzerland: World Meteorological Organization.

World Meteorological Organization. (1996). *Exchanging Meteorological Data: Guidelines on Relationships in Commercial Meteorological Activities, WMO Policy and Practice.* Report no. 837. Geneva, Switzerland: World Meteorological Organization.

World Resources Institute. (1990). *World Resources 1990–91.* New York: Basic Books.

Worster, D. (1985). *Rivers of Empire.* New York: Pantheon Books.

Wynne, B. (1982). *Rationality and Ritual: The Windscale Inquiry and Nuclear Decisions in Britain.* Chalfont St. Giles, England: British Society for the History of Science.

Wynne, B. (1990). "Risk Communication for Chemical Plant Hazards in the European Community Seveso Directive." In M. S. Baram and D. G. Partan, eds., *Corporate Disclosure of Environmental Risks.* Salem, NH: Butterworth Press.

Wynne, B. (1995). "Misunderstood Misunderstandings: Social Identities and the Public Uptake of Science." In A. Irwin and B. Wynne, eds., *Misunderstanding Science? The Public Reconstruction of Science.* Cambridge, England: Cambridge University Press.

Yanai, M., S. Esbensen, and J.-H. Chu. (1973). "Determination of Bulk Properties of Tropical Cloud Clusters from Large-Scale Heat and Moisture Budgets." *Journal of Atmospheric Science* 30: 611–627.

Yearley, S. (1991). *The Green Case.* London: HarperCollins.

Yearley, S. (1996a). "The Environmental Challenge to Science Studies." In S. Jasanoff and T. Pinch, eds., *The Handbook of Science and Technology Studies,* 457–479. Newbury Park, CA: Sage.

Yearley, S. (1996b). *Sociology, Environmentalism, Globalization.* London: Sage.

Yoshikawa, H., and J. Kauffman. (1994). *Science Has No International Borders: Harry C. Kelly and the Reconstruction of Science in Postwar Japan.* Cambridge, MA: MIT Press.

Young, O. (1994). *International Governance: Protecting the Environment in a Stateless Society.* Ithaca, NY: Cornell University Press.

Young, O. (1999). *Governance in World Affairs.* Ithaca, NY: Cornell University Press.

Young, O., ed. (1998). *Global Governance: Drawing Insights from the Environmental Experience.* Cambridge, MA: MIT Press.

Young, O., G. J. Demko, and K. Ramakrishna, eds. (1996). *Global Environmental Change and International Governance.* Hanover, NH: University Press of New England.

Zehr, S. (1994). "Method, Scale and Socio-Technical Networks: Problems of Standardization in Acid Rain, Ozone Depletion and Global Warming Research." *Science Studies* 7 (1): 47–58.

Zeigler, B. P. (1976). *Theory of Modeling and Simulation.* New York: Wiley.

Zoppo, C. E. (1962). *Technical and Political Aspects of Arms Control Negotiation: The 1958 Experts' Conference.* Santa Monica, CA: Rand Corporation.

Index